Local Positioning Systems

LBS Applications and Services

Local Positioning Systems

LBS Applications and Services

Krzysztof W. Kolodziej
Johan Hjelm

Taylor & Francis
Taylor & Francis Group
Boca Raton   London   New York

CRC is an imprint of the Taylor & Francis Group,
an informa business

Published in 2006 by
CRC Press
Taylor & Francis Group
6000 Broken Sound Parkway NW, Suite 300
Boca Raton, FL 33487-2742

Taylor & Francis Group
is the Academic Division of Informa plc.

Visit the Taylor & Francis Web site at
http://www.taylorandfrancis.com

and the CRC Press Web site at
http://www.crcpress.com

Foreword

We've all heard the three rules of real estate — location, location, location. It's true. The value, utility and enjoyment of your home is indeed based on location. Proximity to good schools, mass transit, shopping, medical services and green space are just a few of the location factors affecting the perceived and real value of your living place. Similarly, the success of public agencies and business places depends heavily on location.

Enter Local Positioning Systems with a new paradigm for knowing "where you are when". By leveraging the location rendering ability of GPS, cellular networks, WiFi access points, and a range of other sensors and networks, we can unlock location services to enable better decision making. Emergency response, asset/property tracking, personal safety, retail sales, and routing are just a few of the domains in which decision making can benefit from location-based information services.

But the need and opportunity for Local Positioning Systems do not end at the door step. A multitude of opportunities and challenges exist within the indoor environment.

What are the current and evolving technologies for rendering location services indoors? GPS devices usually don't work indoors. How can we equip our first responders to give them indoor location information for search and rescue in an emergency? How can we best exploit existing and emerging infrastructure to deliver accurate indoor location services to support businesses, government and the consumer?

This textbook leads us from "macro to micro", taking us on a tour of dozens of existing and possible indoor location based services applications. We get a grounding in the underlying common requirements of such applications and we learn about the relevant features and benefits of current and emerging technologies. The author provides enough technical detail to provide a firm foundation for developers, while stimulating our imaginations with descriptions of dozens of applications.

Many technical textbooks neglect the importance of standards. Kris Kolodziej, who has had experience with geospatial standards development, doesn't make that mistake. Today's location services depend on a set of standard open interfaces and encodings that support plug and play integration and that create market opportunities for the diverse businesses who provide the links in the location services value chain.

This book brings together for the first time everything that should be in a single book about location based services for buildings, campuses, malls, manufacturing centers and other built environments that preclude complete dependence on GPS. The information is timely, because many of the technologies have only recently become available as affordable, capable products that are easily integrated into solutions, and there is a growing awareness of what should be possible. Kris Kolodziej is a reliable guide into a near-future world in which people and things will be less likely to get lost.

Mark Reichardt

Preface

Today, there is a vast array of location technologies that are involved in the calculation of a user's or object's position in a space or grid, based on some mathematical model. Positioning here means allowing a mobile device to be aware of its location with different degrees of precision and accuracy. The technology required for provision of automated location information to mobile devices has been in continual development for several decades. While the majority has its roots in the military (e.g., GPS), modern consumer technology is also rising to meet the challenges, specifically in metropolitan areas. Telecommunication initiatives, like the U.S. FCC's E911 and Europe's E112, have generated a lot of interest in the potential for location based services (LBS) — applications and services that are a function of a person's or an object's location.

Unfortunately, LSB fails because it does not work where people are: indoors and in cities. GPS is great, but not for many of the end-user (consumer facing) and "local" applications that will prove to be the backbone of the LBS market. That is, millions of square meters of indoor space and urban areas are out of reach of GPS systems. Conventional GPS receivers do not work inside buildings due to the absence of line of sight to satellites, while cellular positioning methods generally fail to provide a satisfactory degree of accuracy, resulting in a greater part of the world's commerce and social interaction that is being conducted indoors not being able to take advantage of outdoor positioning systems like GPS. The delivered position fixes cannot even be used for determining whether a target person stays inside or outside a certain building, not to mention that it is by no means possible to locate it with the granularity of rooms or floors.

A multitude of applications and services can benefit from indoor (in-building) positioning and navigation such as logistics, routing, sales, asset tracking, personal safety, and emergency response (e.g., Department of Homeland Security's advanced 3D locator system), as well as consumer handset LBS applications. With the last, location-based advertising is a good example, where vendors care about building a closer relationship to the potential consumer. Google, with billions of dollars in annual revenue generated through targeted ads associated with online searches, might be able to improve the economics of such plans via location-based advertising.

Fortunately, over the past decade, advances in location positioning technology have made it possible to locate users and objects indoors (locally, i.e., in urban centers and inside buildings). These alternative technologies are now being introduced to the market, enabling many kinds of indoor location-aware applications. Different technologies will demand different capabilities from devices, while they bring various constraints. Outside the remit of 2G, 2.5G, 3G, and 4G cellular networks exist other families of positioning technologies that are often referred to as "local positioning", which make use of short-range networks such as 802.11, Bluetooth, RFID, ultrasound, UWB, IrDA, or TV radio signals.

Indoor positioning and tracking applications are not just a vision or found only in the lab. The potentials of location-aware indoor applications were realized in the early 1990s. They were explored in conjunction with research on ubiquitous/sentient computing. Indoor environments present opportunities for a rich set of location-aware applications such as navigation tools for humans and robots, interactive virtual games, resource discovery, asset tracking, location-aware sensor networking, and others. Further, typical indoor applications require different types of location information such as physical space, position, and orientation.

Indoor location-aware applications require micro-detailed geo-referencing to satisfy users' growing needs. It is not enough to geo-reference a building if the position of users and other objects inside the building is also relevant. Objects are used as landmarks, and relationships among the objects are crucial for symbolic representation of the whole system.

This book explores the different types of indoor, urban, and seamless indoor–outdoor location-aware applications, their requirements in terms of the infrastructure needed to support them, and the current limitations. The book gives detailed coverage on the most promising technologies, which are WLAN fingerprinting, RFID positioning, and indoor positioning with non-radiolocation positioning with infrared and ultrasound. The book also addresses the problems created by the lack of a common integrated approach to universal positioning technologies, a lack which drives the current demand for stand-alone, vertically integrated (hardware/software) solutions. LBS has been trying to become the "killer app" but privacy, indoor coverage, and market awareness are still pending issues. This book addresses all of these issues. The TV-GPS positioning technology that is featured in the book has the promise for enabling seamless indoor–outdoor positioning.

Finally, the chapters herein describe the design and implementation of several positioning systems and real-world applications and show how these tools are being used to solve problems that can be related to the reader's own applications.

The motivation for this book came from my MIT graduate thesis and from additional research and projects in this area over the years.

The book is accompanied by a dedicated web page that provides the reader with study guides, development examples, lecture PowerPoint presentations, links, comments, and the latest news in the area of local positioning systems (LPS) and indoor LBS. It can be accessed at http://indoorlbs.com/.

Finally, please email me with any comments or questions at kkolodziej@indoor-lbs.com.

Krzysztof (Kris) Kołodziej,
San Francisco, CA

The Authors

Krzysztof W. Kolodziej received graduate degrees from MIT in information technology engineering and geographic information systems, and an undergraduate degree from Rutgers University in computer applications in geography and urban planning. He has spent the past several years in the industry on consulting projects dealing with the design and development of web mapping and wireless location-aware applications. Specific to indoor location, Dr. Kolodziej has been involved in projects pertaining to homeland security, virtual/augmented reality, city WiFi initiatives, and point-of-interest databases. In his spare time, Dr. Kolodziej is a contributor to industry trade magazines and conferences.

Johan Hjelm, an army captain with 12 years of experience in Swedish journalism, currently works at Ericsson Research. He was one of the first webmasters in Sweden and created some of the first mobile Internet services in the world. He was a visiting researcher at MIT, working at the World Wide Web Consortium (W3C). He was appointed W3C Fellow in 1999. For two years he lived in Japan and managed a research group looking into the next generation of iMode and related technologies. He has written more than 12 books. He is a member of IEEE and ACM.

Acknowledgments

I would like to thank the following organizations that collaborated in this effort: AeroScout, BlipSystems, Ekahau, EPFL, Illinois Institute of Technology's (IIT) HawkTour, Intel's PlaceLab, Massachusetts Institute of Technology's (MIT) Cricket, Nearspace, Rosum, and YDreams.

Special thanks go to the following individuals (in alphabetical order) for their interest and support: James Beal, Cyril Brignone, Antonio Camara, Jose Carlos, John Dmohowski, Tibor Dumitriu, Jay Fortunato, Jeffrey Hightower, Brian Hilton, Creighton Hoke, Jason Hong, James Howard, Jeff Hughes, Cliff Koffman, Nehal Mehta, Santhosh Meleppuram, Harry Niedzwiadek, Mark Reichardt, Joshua Slobin, Steve Swartz, Martin Zalewski, and Todd Young.

Krzysztof W. Kolodziej

I would like to thank my colleagues in Ericsson Research for their understanding and interest.

Johan Hjelm

To my parents

Table of Contents

1 The Three L's: Location, Location, Location

If you are a robot, you are blind, deaf, and do not know where you are. This is equally true if you are a robot moving around on Mars, as if you are a robot moving around in a building at night. In the field of mobile robotics, localization has been referred to as "the most fundamental problem to providing a mobile robot with autonomous capabilities" (Cox, 1991). Autonomous robots are built on mobile platforms, where a new degree of control flexibility is needed. As opposed to industrial robots, they move around in their environment, which is often highly unstructured and unpredictable. Slowly, various markets are emerging for this type of robotic system. Entertainment applications and different types of household or office assistances are the primary targets in this area of development. The Aibo dog from Sony and the Trilobite vacuum cleaner from Electrolux are early examples of this future industry.

Outdoor positioning, using global positioning systems (GPS) or techniques that measure the position of the user in a cellular network, have been well explored and standardized. But GPS does not work indoors and may not give you the accuracy you need to move around. You need some way to measure your position in the building, to make sure where you are and that you are going in the right direction.

This applies to robots as well as people. It may be redundant for the mail delivery person to measure his position, but it may be important to the water mains repairman, who wants to make sure that he is not breaking down walls just to find out where the pipes should go. It may also be useful to the night watchman, who becomes able to create automatic reports of his position and any event that may have occurred there (none, if all goes well).

Localization, or the problem of estimating spatial relationships among objects, has been a classical problem in many disciplines, including mobile robotics (Thrun et al., 2001), virtual reality systems (Welch et al., 1999), navigation systems (VOR), and cellular networks (RAD).

Buildings can have a small area but several floors. These may have the same physical characteristics, but different information associated with them. For the security manager, it makes sense to schedule more rounds on the floor where the engineers sit than where accounts receivable are handled, simply because it is more likely that the engineers handle more sensitive company material — maybe even prototypes. Connecting the indoor location system to an inventory tracking system, marking all sensitive objects with radio frequency identification (RFID) tags, makes sense.

Buildings can also be large structures that cover a big area, but do not have a large extent in height. America is home to the biggest malls in the world, and it makes sense to know where in the mall you are — and where you are going. This is a typical indoor positioning application, which may be equally useful for the delivery person as for the "mall rat" who wants to find his friends.

Positioning systems range from an object is closest to (for example, which shelf in a warehouse a pallet is stacked on) or record when an object crosses a boundary (such as leaving a lot) to sophisticated systems that can determine precise location using multiple readers with overlapping fields of view. You can build indoor positioning systems in many ways, for instance, embedding bar codes where they can be conveniently read, but even if the bar code tags cost next to nothing, you are dependent on finding them to measure your position. This means that the position is only available in certain places, which may be a problem if you are moving along a path and get momentarily distracted (for instance, by an object placed in the path). In this regard, it is better to have a system that enables you to measure your position independently of the location in the building (i.e., you do not have to be close to the positioning device).

This means some kind of field or radiation, and while there are several types available, this book will look more deeply into those that are based on standards and that are practical to deploy, which means radio-based systems. For comparison, we will look into the MIT Cricket system, which uses a combination of radio and ultrasound.

For local-area requirements, such as tracking pallets in a warehouse, people within a building, or vehicles within a limited range of movement, indoor positioning technologies are better suited when compared to wide or global systems like GPS. Examples include triangulation within a wireless local-area network (WLAN) or proximity sensors, including RFID. RFID systems include relatively simple approaches that basically determine which reader is used or accessed. The most economical way to enable indoor positioning is to piggyback it on top of other systems that are used for the communication in the building. It is possible to build dedicated systems, using, for instance, pseudolites, which send out a simulated GPS signal, but this becomes expensive if a large area is to be covered, and the risk of interference with the true GPS signal has to be managed carefully. It makes more sense to use the wireless LAN, which probably is set up in the building anyway.

But placing positioning beacons is only the first step. Once you have determined the positions and made them available to whomever is to do the positioning, you have to relate that position to something. Knowing you are on the sixth floor in Room 135 does not help you much if you do not know how to get where you are going from there. It is even worse if all you know is that you are at a set of coordinates.

This means creating an application on top of the positioning system that can easily be filled with information, and that this information can as easily be retrieved. Actually, it means several applications (for instance, one for the acquisition of information, another for the authorization and authentication, and so on). In addition, you probably want to import information from existing databases and map systems into the application — so you do not have to recreate them. This may mean importing building drawings, which is not just a matter of mapping coordinate points to images, but if it is to work well, it has to involve some more thinking/work.

There is a tendency in the (nascent) indoor location industry to think that you can deliver monolithic applications — systems that in themselves contain everything. This may work for deployments where the vendor is in charge of everything, but it is a model that the wide-area location industry has long left behind. Nobody today even imagines the GPS device provider to be the same as the provider of maps, or the route calculation application. By introducing interfaces between the different components, the industry has taken off since the horizontal specialization lets each part of the chain do what it is best at.

Corresponding interfaces do exist for local-area systems, and we point them out in this book. But much standardization remains for them to be a viable alternative to the totally integrated — or "stovepipe" — solutions.

APPLICATION EXAMPLES AND USE CASES

With the advent of the GPS and the availability of chip-size GPS receivers, all future mobile wireless nodes can be equipped with the knowledge of their location. A user's location will become information that is as common as the date is today, getting input from GPS when outdoors and other location-providing devices when indoors. Availability of location information will have a broad impact on the application level as well as on network-level software. We already have many devices with some location/position awareness or capability — cell phones, personal digital assistants (PDAs), laptops, RFID tags. And we will see many devices that could have location awareness but currently do not — wireless phones (though the pager feature allows one to find it), printers, computers, iPods, bikes, cars, other types of vehicles, etc., as well as whole classes of hybrid products — Wi-Fi cellular phones and iPod handsets, for instance. Plus, we will get further degrees of integration with the devices we find common — a compass or gyroscope in a cell phone, accelerometers in laptops (IBM already has one, I think), etc. Things that sense motion or change in azimuth in conjunction with other location-aware devices. Accurate location-tracking technology helps support a range of wireless applications, such as E911 for tracking voice over wireless phones and wireless asset management. An increasingly popular use of mobile devices is to allow a user the ability to interact with his physical surroundings. For example, handheld computers can provide the user with information on topics of interest near the user, such as restaurants, copy centers, or automated teller machines. Similarly, mobile computers seek to allow a user to control devices located in close proximity to that computer, such as a printer located in the same room. Additionally, specialized computing devices, such as navigational systems found in many automobiles, seek to provide the user with directions based on the user's current location. Also, pushing local information (i.e., in location-based advertising applications) increases business in that companies often have local information that is beneficial for users to have, especially when users are residing within specific areas. If a user's position is known, a central server can then send valuable information to the user based on that location. As the user moves into different areas, the information sent to the user can then change accordingly.

Location-based service (LBS) applications include intelligent information management in Wi-Fi hot spots. A traveler gets off a plane at Boston's Logan Airport and turns on his Wi-Fi-enabled PDA, for example, and immediately the local network pushes out floor plans of the airport, flight and ground transport information, and directions to — or, more ominously, ads from — on-site restaurants and shopping. Airports, for example, can offer to public wireless LAN subscribers client software that shows passengers how to find specific gates, baggage claims, restrooms, and customer service counters. In addition, location-based advertising that identifies where to find the nearest coffee shop or restaurant is possible. Microsoft has launched a web service — MS Location WiFi Finder — that allows Wi-Fi users to obtain details of their location and access to local information. The locations of Wi-Fi access points are pre-recorded (see war driving in Chapter 3) and stored in a database as part of the MapPoint System, which in turn is part of MS Local Live (previously known as Virtual Earth).

LBS products can, for example, provide convention center guests with specific directions from one wing of the facility to another, as shown in Figure 1.1. Figure 1.2 shows how attendees can locate sessions, exhibitors, and event services on

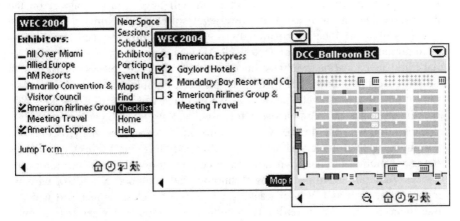

FIGURE 1.1 NearSpace's Meeting Manager.

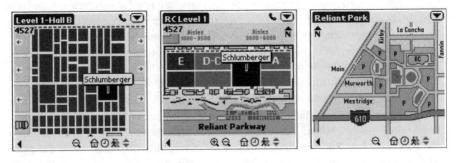

FIGURE 1.2 NearSpace's Meeting Manager.

dynamic maps and floor plans. The exhibition and conference industry constitutes another information-rich environment where indoor mobile location services can be deployed. The growing size of exhibitions (for example, CeBIT 2001, which is the leading event in the IT industry, occupied 432,000 m² of space and attracted more than 830,000 visitors in eight days) naturally creates the need to develop new applications that will benefit visitors, exhibitors, and exhibition organizers alike. The development of applications based on indoor mobile communication technologies will probably transform the *modus operandi* of the exhibition industry in fundamental ways. Visitors will be able to register their personal preferences before visiting the exhibition (for example, through home or work access to an exhibition server). Upon entering the exhibition premises, they will receive a personal hand-held wireless device through which they will be able to obtain information about a particular stand or product, receive personalized notifications by the exhibition organizers (for example, notifications about events that are about to start), navigate along a predefined route generated by the system (based on expressed interests), schedule appointments with exhibitors, and so on. Exhibitors will also benefit by arranging and distributing marketing material more effectively, and by achieving personalized service, thus paving the way for more efficient customer relationship management (CRM). Finally, the exhibition organizers will be able to organize their work force according to events that might occur, provide announcements for specific target groups in a "silent" mode (through the hand-held device), and provide *ex post* statistics to their clients (i.e., the exhibitors).

An LBS makes it possible for a centralized server application to track users and assets. For example, a user's PDA can continuously provide position updates. Universities are very interested in this form of LBS. Areas such as classrooms can be places where a school can block access to the Internet and the ability for students to use chat to share test answers. Of course, the more intelligent students could still use an *ad hoc* form of wireless LAN connectivity for communicating with their neighbors.

In hospitals, staff attach a small radio frequency identification (RFID) tracking device to all the babies that are born. This device is very difficult to remove, and sensors within that area of the hospital track the location of the baby. If anyone moves the baby outside an invisible perimeter (which is just before entering the exit areas of the labor-and-delivery unit), an alarm goes off and the system automatically secures all relevant exits. This makes it very difficult for anyone to steal babies, which happens more often than you would think in some hospitals.

Even without the use of a central server, an LBS makes sense for tracking applications. In this somewhat peer-to-peer form of LBS, users exploit the location of other users to make better decisions. For example, emergency crews responding to natural disasters and terrorist attacks can keep abreast of each other's positions to effectively handle a situation. Each person can carry a PDA that shows the relative location of others.

Manufacturing and logistical operations throughout the supply chain need to efficiently manage numerous logistical optimizations to cut cost and time to market. Manual tracking of assets is labor-intensive and error-prone. Ekahau RTLS solutions let manufacturing plants, warehouses, and distribution centers monitor and manage

	Healthcare	Manufacturing & supply chain	Industrial processes
Finder	Find assets, caregivers and patients. Less time needed to look for people and medical equipment. Reduces inventory and labor. Increased patient satisfaction.	Work-in-process management. Reduction of assembly slowdowns and bottlenecks.	Equipment management. Saving time and money in re-locating assets, tools, and equipment.
	Emergency calls. Fast and automatic locating of caller. Increased patient satisfaction and safety.	Inventory management. Improved asset visibility.	Workflow management. Better utilization of resources.
Tracker	Location-based work dispatching. Faster dispatching of work orders.	Location-based work dispatching. Faster dispatching of work orders.	Staff monitoring, safety & guidance. Improved safety.
	Information delivery. Faster access to patient data.	Information delivery. Faster and more accurate picking of orders.	Information delivery. Location-based alarms.
Logger	Procedure recording. Collecting evidence for legal purposes.	Procedure recording. Locate quality problems.	Procedure recording. Data collection for recovering from emergencies or system failures.
	Analyzing workflows. Improvements in work processes.	Analyzing workflows. Improvements in work processes.	Analyzing workflows. Improvements in work processes.

FIGURE 1.3   Use cases and benefits. (*Source*: Ekahau.)

the movement of work in process, inventory, tools, equipment, and key personnel. Using Ekahau's solutions in manufacturing and supply chain management increases throughput, reduces turnaround times, optimizes equipment utilization, and increases staff productivity. Figure 1.3 portrays the overall possibilities.

PEOPLE/ASSET MANAGEMENT AND TRACKING

Asset management is often suggested as an application of location tagging, the idea being that it is possible to keep an inventory of assets and their locations to make it easier to find assets when needed. Wi-Fi or RFID tags enable the wireless network infrastructure to locate people and assets otherwise not connected to a wireless network. The tag can be used to track people in many valuable applications — child tracking in amusement parks, security personnel in enterprises, hospital patients, tracking the location and motion of consumers carrying a mobile communications device in a commercial retail establishment, employees in office buildings, equipment and parcels in a manufacturing or shipping facility, or attendees at a conference in a convention center. Various types of equipment can be tagged, including vehicles in parking lots; inventory in a manufacturing line; containers, forklifts, and other assets for efficient supply chain management; shopping carts in supermarkets; and medical equipment in hospitals. Some examples include:

Find a person: Ekahau Finder is an end-user web application for instantly finding the locations of people and assets using any standard web browser (see Figure 1.5). It is designed to help workers and managers locate their colleagues, equipment, or patients anywhere in their premises. With Ekahau Finder organizations can streamline and automate processes that would

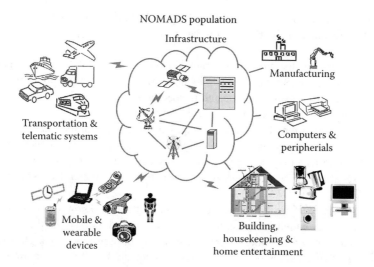

FIGURE 1.4 Application domains. (From Malek, M., The NOMADS Republic: International Conference on Advances in Infrastructure for Electronic Business, Education, Science, Medicine and Mobile Technologies on the Internet, Scuola Superiore G. Reiss Romoli (SSGRR), Telecom Italia, L'Aquila, Italy, 2003.)

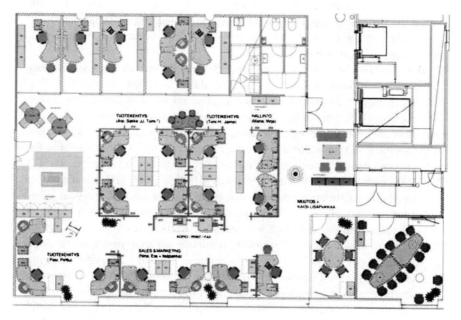

FIGURE 1.5 Ekahau Finder.

otherwise involve time-consuming manual locating of assets or people. Main features of the Ekahau Finder include:

Querying the location of any tracked device based on the specified name, such as Dr. Smith, a vehicle identification number, or shipping ID
Querying assets that are located in specified areas
Query results display the area of the device
The location may be displayed graphically on a map of the tracking area
User interface allows doing batch queries of multiple devices concurrently
The interface may be configured to provide refreshes at regular intervals
Executing queries from any device that supports a standard web browser

Find the asset: If we know where an asset is, then we know where to go to look for it. A member of staff who needed to find an asset could look for it on a map before going to get it. This does not require a very high degree of location accuracy — a few meters is adequate. But it does require good tag battery lifetime of about 1 year. Ekahau Tracker, for example, is an end-user application for monitoring the locations and movements of people and assets. It provides a real-time view of the tracking area with selected objects displayed as icons on a map of the tracking area. With Ekahau Tracker organizations can tell in a single view the locations and directions of valuable personnel and equipment. Plus, Ekahau Logger is used for storing the historical location information to a log file or database. The data can be later analyzed and played back using the Ekahau Tracker or third-party analytical applications. The Ekahau Logger allows organizations to record procedures that can later be analyzed for operations management, safety, or legal purposes.

Detect when the asset is being used: Simply finding an asset is not very valuable if we find it only to discover that it is already in use. By using much more accurate and real-time location information, especially if staff members wear tags, we can detect when an asset is being used. If we have two-way tag communications, we can add protocols to support use assertion and so make use detection more reliable. This is valuable for two reasons: (1) use detection makes it easier to find a free asset, and (2) it becomes possible to measure asset utilization levels and optimize asset provision accordingly, thereby saving costs. Knowing where your company's assets are at all times and limiting the removal of such assets to preauthorized personnel only reduces incidents of theft and loss. Preauthorization is automated by linking particular assets to personnel tags and creating an electronic relationship between the two tags. Also, knowing exactly where all employees are within the building helps to improve customer service, employee safety, and productivity. Large hospitals lose hundreds of thousands of dollars worth of equipment each year and spend countless hours searching for patient care assets. This includes medical devices (such as infusion pumps, portable x-ray machines, and patient monitoring devices), as well as other mobile assets, such as wheelchairs, stretchers, and gurneys. Nurses sacrifice time with patients to seek equipment they need, and maintenance staff lose productive hours searching for specific items that need maintenance. The ability to find assets instantaneously saves staff time and lowers capital or lease costs spent replacing lost equipment.

Page staff when the asset is free: In a system that detects asset use, we can use a paging feature on the tag to alert a member of staff when an asset has become free. If a member of staff is looking for an asset that is in use, we could enable him to select an option that would notify him when the asset becomes free. This would add extra value by decreasing the duration of periods of nonuse, and thus optimize asset provision. This can be especially powerful if we consider applying the same general approach to a wider consideration of assets, such as consulting rooms, or even some people. Wavetrend's ULR RFID technology can be deployed to provide a complete smart building management system for the corporate environment. By enabling the tracking, monitoring, and protection of all assets, including people, a more controlled and productive environment is created in which day-to-day business activities are handled more effectively and efficiently.

Tracking applies to airport passengers and luggage tracking (Figure 1.6). Airline travelers are becoming increasingly angry over baggage-handling status.

RFID, for example, has been slowly introduced into the airport environment. Real-time location systems using active tags have been introduced to track ground service equipment and people around airports. Baggage handling and tracking using passive RFID tags is now being implemented at several airports.

Tracking also applies to tracking players and the ball in sporting events — another area with various interesting possibilities. These include generation of sta-

FIGURE 1.6 Passenger locations.

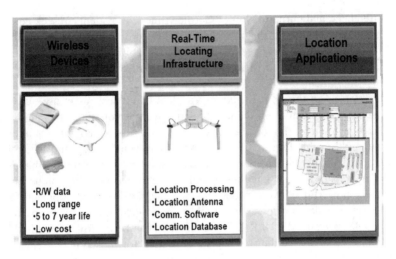

FIGURE 1.7 Location applications.

tistics such as how far a player has run, how fast she is, and so on. Automatic camera tracking and switching of views is another good application, along with recording of games for viewing in a simulated environment, or as a basis for computer games. There is a nice application called Virtual Spectator that lets you watch the America's Cup yacht race as a three-dimensional graphical simulation, either live or recorded. Since most sporting events are played in a fairly open area, some of the technical challenges are not as great as with an indoor LPS in a confined space with obstructions, so a greater range of technology approaches is applicable in this area. For sports over a wider area, such as yacht or motor racing, GPS technology may be applicable.

Also, locating Wi-Fi devices can speed IT maintenance tasks while streamlining simple tasks such as finding the nearest overhead projector for a meeting. The physical security of valuable assets can also be enhanced by affixing positioning tags and monitoring the asset's location.

USE CASE: LOCATION-BASED TRIGGERS (BUDDY FINDER, CONFERENCE ASSISTANT, AND CHILD ALERT)

"While in the shopping mall, I want to use my Instant Messaging Buddy Alert Service to notify me when my friends or family are in the mall."

Indoor positioning technologies open the possibility of relating people, objects, and events in space at, for example, a shopping mall or convention centers. Set meetings in points of interest and facilitate access to those points (i.e., schedule a meeting at a given time in a given spot and provide optimal routing for all involved). Another use is finding people with certain features in the proximity (i.e., date-matching service). Alert-based LBS, also called spatial trigger (proactive) services, are among the most fundamental in importance. The underlying

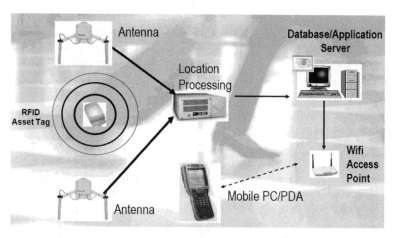

FIGURE 1.8 Infrastructure for location applications.

premise is to notify a user if something (an event or activity) important happens, which in this case is when a friend (buddy) enters the specified perimeter area (or zone). Assume that User1 is traveling to a shopping mall. User1 wishes to consult with User2 (a friend), who will also be going shopping at the same mall. User1 consults his cell phone LBS Buddy Finder service and initiates the alert notification to be triggered when User2 is in the vicinity. (This service would consult User1's user profile from a user profile provider to generate a buddy list that includes User2.) The Buddy Finder service then uses the location information of the communication network's Wi-Fi location capability in both users' cell phones to determine the proximity of their owners.

This spatial messaging concept is similar to the instant messenger service of ICQ, MSN IM, skype, or AOL IM, which can tell the user whether his contacts are currently online. For example, a location-aware version of Microsoft's Instant Messenger software is possible. This will allow users to transmit their location data so that buddy lists will reveal not only when friends are online, but also where they are. This is a virtual meeting place. The same can be true for a physical interaction, whether personal or business. The user is alerted when her contacts are in the vicinity. If prearranged, people can find each other quicker. Teens, for example, will want to receive an SMS that friends or a potential match for a dream date is close by. Moreover, the proactive nature enables impromptu gatherings (*ad hoc* meetings). A real-world example of this kind of service is NTT DoCoMo's Friends-Finder service in Japan. This service was developed for outdoor interaction, which reports people's location when they come within half a mile of their contacts (Boswell, 2000). An indoor LBS version of this service would require the perimeter range (vicinity) to be smaller to a building scale (i.e., shopping mall). Another extension to this application would be a dating service, where dates are arranged by vicinity in addition to user (date) profiling (Hendrey, 2001). This kind of a location-aware service can provide opportunities to informally meet people through a specification language and interface through which the user can specify one or more contexts, in this case location, in which the user thinks such opportunities might exist. For example, if the user needs to ask another person a question, the user could ask the service to notify him when that person returns to some location (i.e., his office room). Microsoft's MSN IM Buddy List service is another real-world example of location-aware services (i.e., "Alert me when you find Harry" or "Alert me when Harry happens to be nearby"). Moreover, if two users are seen together in the same location, the service could trigger further action and play them a video message on their mobile device or the nearest liquid crystal display (LCD) screen display.

Another example of a triggered indoor LBS is the Conference Assistant (Dey et al., 1999), which falls under the community-building application category. The assistant uses a variety of context information to help conference attendees. It also examines the conference schedule, topics of presentations, user's location, and user's research interests to suggest the presentations to attend. Whenever the user enters a presentation room, the assistant automatically (by a location-based trigger) displays the name of the presenter, the title of the presentation, and other related information. Available audio and video equipment automatically record the slides of current presentation, comments, and questions for later retrieval.

Yet another example of a triggered indoor LBS is a Child Trigger, which falls under the safety and security application category. Adult users can gain comfort in knowing that their child is safe, as they will be notified if their child or elderly family member has ventured beyond the security of a prespecified safe region (i.e., child leaves school during teaching hours). These family-oriented security services are priceless to parents and relatives, as all types of time- and location-sensitive information have incomparable value to users. With respect to profit making, the major consumer market driver will be information services, with rapid growth predicted for personal and child safety services (3G).

An example of children tracking comes from Tivoli Gardens in Copenhagen, which deployed a Bluetooth-based children tracking system using the BLIP Systems, enabling the location of a child to be pinpointed down to 20 m. A similar example is the Aalborg Zoo in Denmark which also deployed a Bluetooth-based children tracking system. The system allows parents to keep tabs on their children's where-abouts using cell phones. A small Bluetooth tag is clipped to the child's clothing. At the entrance, parents rent BlueTags' BodyTag BTBT002 for their children and register the required information along with contact details. They receive an SMS to their mobile phone with a code to be used when making inquiries about the location of the individual child. The parent simply sends an SMS with the specific SMS code and, through the tracking software and the SMS gateway, a response comes back with information about the child's location. The BodyTag uses a CSR BlueCore chip. The Bluetooth access points scattered around the zoo detect nearby tags and relay their location to a central database via a wireless local area network. Parents register their cell phones with the system. Parents can request updates on their children's location via text messaging on their cell phones.

HEALTH CARE AND MOBILE PATIENT MONITORING

Clinicians and administrative staff in health care institutions have long faced the dual difficulties of organizing the unpredictable patient care process and optimizing the use of expensive medical equipment. At the same time, the use of mobile wireless devices in health care has exploded as the industry equips mobile professionals with tools to be more productive in their work.

There are many uses for LPS in health care. Staff, patients, and assets in a hospital can all be tracked in real time using LPS. Several systems exist for monitoring the vital signs of mobile patients. Monitored data, especially ECG, is sent wirelessly to a central control room, where it is observed by an expert. The need for location awareness arises when the data exhibit some new abnormality: then a visual assessment of the patient's condition has to be made. The location of the patient is a key factor in how this assessment will be made. Depending on the capability of the location system, it is possible to implement various applications. There are many uses for LPS in health care. Staff, patients, and assets in a hospital can all be tracked in real time using LPS. Below are a few of the many examples that could be implemented:

Find staff and patients: With both patients and staff on the move, hospitals face significant challenges in managing the fluid process of patient care. With patients often scheduled for multiple, consecutive procedures, knowledge of their location helps optimize the patient care process and helps manage schedules in real time. Patients in critical care situations, such as those recovering from cardiac surgery, particularly benefit from constant location tracking, as it enables rapid medical assistance if the patient's condition abruptly deteriorates. Location of medical staff is also important, both in emergencies in which a physician must be summoned and in cases in which staff themselves need help.

If we can find where the patient is, the expert can telephone a nearby ward or department to request that a suitable member of staff finds the patient and performs an appropriate physical assessment. This feature only requires location of the patient to a few meters' accuracy. In practice, this can be a poor solution because it can take a long time to get hold of a suitable member of staff who is actually free.

In the hectic environment of health care, patients, coworkers, and medical devices easily get out of synch and require manual coordination. Automatic asset and people tracking allows hospital staff to locate patients, caregivers, and medical equipment. Ekahau RTLS solutions let hospitals synchronize workflows, reduce equipment inventories, and increase patient throughput while at the same time improving quality of care and staff satisfaction and safety. This ensures significant cost savings through improved operations efficiency and resource management in demanding and constrained health care facilities.

Page a nearby staff member: If we can find where the patient is, and we can find the locations of members of staff who are qualified to perform visual assessments, then we can locate the nearest member of staff qualified to perform the assessment and page him. This member of staff then finds the patient, performs the assessment, and telephones the results to the expert. This feature still only requires location accuracy of a few meters, but also requires paging. It works better than the base application because a qualified member of staff is automatically selected, but it still requires that the staff member is not busy, is able to find the patient, and is able to get through to the expert with the visual assessment.

Film the patient automatically: At the other extreme, we could use digital pan–tilt–zoom cameras to film the patient and put video or still footage of the patient on the screen in front of the expert, allowing the expert to make a visual assessment of the patient directly. This feature requires very high accuracy, real-time performance, and integration support. It could potentially be much more effective because it would remove the need for communication between members of staff.

PDA-toting physicians could have patient charts automatically sent to their PDAs as they walk into a hospital room. Health care institutions are also a key market for asset-tracking technology. In fact, they will likely start with asset tracking because the return on investment is clearer than it is for intelligent information management. Hospitals expend hundreds of hours of staff time a year looking for wheelchairs, pumps, diagnostic and monitoring equipment, and other big-ticket assets on wheels. They also typically maintain a 30% excess in inventory on these items to allow for loss and theft. Wireless asset tracking saves worker time, helps locate sometimes urgently needed equipment more quickly, and allows hospitals to reduce inventory. Many have already installed proprietary wireless asset-tracking systems or are seriously considering it. These systems involve attaching a wireless "tag" to the asset. The system can then triangulate the tag's location on demand. Proprietary systems can cost up to $1 million. For hospitals that are already deploying Wi-Fi networks for other applications — as a growing number are — adding a location-based asset-tracking capability would be orders of magnitude less expensive. Plus, the hospital would not have two wireless networks to manage and maintain.

SECURITY

There are a wide variety of applications of LPS relating to security. In an office environment, employees and visitors can wear tags that allow them to be tracked by the LPS. Some offices have security policies with regard to visitors; for example, visitors may not be allowed in certain areas, or certain visitors must be accompanied by an employee at all times. These policies can be enforced much more reliably by using an LPS than by existing mechanisms. Currently, access to secure areas is commonly controlled via badge-locked doors, but this can easily be defeated through tailgating. Other security applications include:

Combating handheld computer and shopping cart theft: Retailers typically experience annual loss rates of 15% for shopping carts, which can add up to millions of dollars for large chains. At the same time, the widespread introduction of handheld computers for employees to perform mobile point-of-sale transactions and inventory checks is being slowed by fear of theft.

By tracking this equipment both in the store and in the parking lot, retailers can sharply reduce theft. Corporate enterprises benefit greatly from location-enabled wireless LANs. Not only can they gain from productivity enhancements that come with mobility, but location visibility also enhances wireless security.

Rogue access point location: Rogue or unsecured access points connected to a corporate network can create gaping holes in a network security system. Detection of rogues is important, but accurate location of rogue access points also allows their speedy removal to eliminate security breaches.

Active tracking systems can be combined with other traditional security devices, such as movement sensors. If a passive security system detects the presence of a person in an area, that information can be checked against the LPS. If there is a person in the relevant area with a valid tracking tag, the alarm will not be triggered; otherwise, it will. Other security systems involve cameras that allow an area to be remotely monitored. Such systems often have software that detects motion in the video feed, and that feed can then be brought to the attention of the person doing the monitoring. This approach allows a single person to monitor input from many more cameras than would be possible without motion detection. Again, an LPS could further improve this approach by only triggering an alarm in a situation where the detection of movement is inconsistent with the current locations of authorized personnel. LPS enables the possibility of an always-on burglar alarm system. Breaches of security policy can be handled in various ways depending on their severity; either an immediate alarm could be triggered or more minor violations could just be logged and reported later. A related area is critical asset tracking and audit. Many organizations, particularly defense contractors, have parts and equipment of a sensitive, secure, or hazardous nature. These parts need to be monitored and audited to record their movements and who had access to them, as proof that they have not been tampered with or viewed by unauthorized personnel. When an asset is moved, the system can check that a member of staff with appropriate clearance accompanies it. If the asset moves without a member of staff present, or was moved by an unauthorized member of staff, this represents a potential security breach and must trigger an alarm.

Visitors could also be tagged for security so the system could register if a visitor was in an area other than a general public area or the vicinity of the person they were visiting. If a baby was carried out of a predetermined area by someone other than one of its parents or a member of staff, then an alarm could be raised.

Asset tracking can also be important in hospitals. One study said that most hospitals have twice as many IV pumps as they use at any given time; they need this many to make sure that they can easily find one when they need it. This study suggested that a typical hospital could save $2 million a year if it could accurately locate all IV pumps at any time, and thereby significantly reduce the number of pumps needed.

EMERGENCIES

In emergencies and every day, you need to know the status and exact location of patients, caregivers, and essential equipment. Existing infrastructure for responding to an emergency situation, such as a building fire, is inadequate. When the first responders come, they get very limited information about the building or the location and nature of the fire on-site. Also, during the second phase of the operation, when other firefighters join the first responders, there is no way to effectively track the personnel and harmonize the operation. Due to the lack of location information, many trapped firefighters have lost their lives in the past. The first responders would download this prestored information to fight the fire efficiently. Incident officers would be able to see a three-dimensional visualization of the site and effectively organize the operation. Rosum TV-GPS is an invaluable tool for tracking and locating first responders in fire ground or other emergency scenarios. High-power, low-frequency TV signals allow incident commanders to see into buildings and other structures in order to ensure the safety of their teams. Rosum also provides a high-accuracy three-dimensional positioning system for emergency responders. This system is deployed locally at the emergency site and provides high-accuracy (room-level) location and tracking in the local area.

Training exercises for emergency workers can be significantly enhanced by using LPS. When an exercise is reviewed, the system can show exactly who was where at all times, and present this in a variety of two- or three-dimensional graphical formats. LPS can be used to automatically associate video with individuals. A number of fixed digital video cameras can be located around the training facility, and the field of view of each of those is known. Since the system knows the location of each person at all times, it can automatically generate all relevant video data for each individual, optionally filtered by time or location, which allows much more efficient review of the exercise. Where appropriate, LPS can also be used to automatically pan movable cameras to follow certain individuals.

The LPS can also be used during the training exercise in various ways. The location of participants could be used to trigger certain actions, such as a simulated

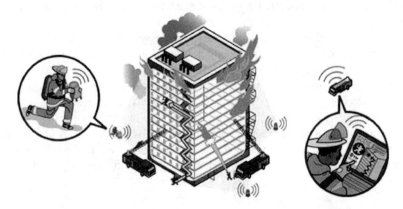

FIGURE 1.9 Rosum three-dimensional positioning for emergency responders.

explosion. Such an event would need to be triggered when people were in a position to be appropriately affected by it, but there may also be certain safety considerations —for example, it should not be triggered if people are too close to its location. In military exercises the LPS may also be used to determine if people have been "killed." There is existing technology that will do this when there is a direct line of sight, such as laser rifles, but these do not work if the shot goes through a wall, floor, or ceiling, so they are not as useful for indoor training.

LOCATION-BASED NETWORK ACCESS/SECURITY

While mobility brings freedom to the end user, it can make the life of an IT administrator a nightmare. How do you track down rogue devices or malicious users attacking a corporate network? How do you keep track of wireless assets (e.g., laptops) that are mobile? Will your wireless network support emergency 911 requirements when delivering voice services?

The problem: Students sit in classrooms surfing the web over campus Wi-Fi networks instead of listening to the professor. If IT staff can identify and track the location of wireless clients, they can improve the accuracy of wireless LAN planning and deployment, optimize ongoing network performance, enhance wireless security, and improve both the visibility and control of the air space. As a result, location tracking has become an essential feature for enterprises that are building business-critical wireless networks. Airespace's and PanGo's location-based access system allows network administrators to change access privileges based on time and location. They could even filter content so that only web sites associated with the course being taught are accessible. This way, location tracking is tied in with wireless service delivery, intrusion protection, radio frequency (RF) management, capacity management, and RF prediction for better WLAN planning, performance, management, and security. Airespace uses advanced RF fingerprinting technology to identify and track users to within 10 m of their exact location — anywhere they roam throughout an enterprise environment. This enables IT staff to establish access control policies that are based on geographic location, immediately identify the source of unauthorized WLAN activity such as rogue access points, and adapt to network conditions in real time for dynamic capacity management.

Corporations might find it helpful to chart internal locations in order to manage their wireless devices. If you know where these devices are inside your infrastructure, you can assign user rights to the network based on location. For instance, once you leave the building, your Wi-Fi connection could be automatically terminated, or you might have different rights in the lobby or in a meeting room or an office.

GAMES

For example, Ydreams' *Undercover* took wireless gaming to new grounds in 2003 with location-based game play, real-time multiplayer action, and pioneering online features. One of the main features is a three-dimensional map of player locations. Also, the Technical University of Graz (TU Graz) recently installed Ubisense and is investigating the range of rich and meaningful augmented reality experiences enabled by a combination of Ubisense's UWB-based technology with those more traditionally associated with virtual reality applications.

At CMU, Ekahau is used as the positioning technology as part of a game that combines real-world locations with input from a portable computer. You would be able to take a mobile computer and walk through different rooms, and depending on where you are in the physical space, you would be moving your avatar in the virtual space. In the Lighthouse location-aware museum guides by Galani and Chalmers (2003), mobility is a resource for interaction as participants use motion to communicate between each other, to signal presence, and to support contextual awareness. In location-based games *Can You See Me Now?* (Flintham et al., 2003) and *George Square* (Chalmers, 2004), mobility also represents a tool for interaction. These systems use both GPS and WLAN beaconing to enrich the gaming experience, and the imprecise nature of this kind of location sensing is exploited by the designers to enrich the game by creating uncertainty. In these location-based games mobility also represents a tool for interaction. These systems use both GPS and WLAN beaconing to enrich the gaming experience, and the imprecise nature of this kind of location sensing is exploited by the designers to enrich the game by creating uncertainty. Imprecision and ambiguity are very important in Reno, as they afford an essential space for privacy. Thus, the imprecise nature of cell phone tower-based localization can be viewed as both a problem and an advantage.

There are a number of games that could be enhanced by LPS, such as paintball and similar simulated fighting games. Applications here are similar to military training, including after-the-fact review of a game by participants, both looking at movements on a map and accessing video that is referenced to each player. In general, there could be interesting possibilities in games that combine real-world players and movement with virtual worlds in a computer.

RETAIL AND SHOPPING

Grocery retailing constitutes a rapidly growing and highly competitive industry in all western economics (Helms et al., 2000). In order to attract consumers and increase their profit margins, retailers are constantly seeking ways to increase supply chain efficiency through the adoption of innovative applications (Chandra and Kumar, 2000). An example of such an application is MyGrocer (Kourouthanassis et al., 2001), which addresses some of the inefficiencies of the retail grocery supply chain through mobile shopping of electronically referenced grocery products. The system combines RF-ID, Bluetooth, and WLAN technologies to create an enhanced shopping experience for the consumer. Among the innovations of the system are: the elimination of checkout lines, the personalized shopping cart with built-in data display, the ability to launch personalized product promotions based on consumer profile and current location, and others.

Providing product information to retail customers, such as directions to the selected retail item via a Product Finder application, is very useful. Some of the services that are possible include navigation to points of interest (take me to the nearest Chinese restaurant), locating friends or family who are nearby, and the ability for vendors such as restaurants or stores to send special offers to potential customers who are nearby (by text messages or other means). Retail customers in a large retail establishment, e.g., a department store, often do not know precisely where within the store to find a desired item. In order to locate the item, the customer begins by attempting to guess in which department of the store the item is located. The customer then typically attempts to locate store personnel to ask directions or, alternatively, to locate a listing of departments, which is typically found proximate to an escalator bank in the middle of the store. Such a listing often takes the form of an alphabetical display board that simply lists the floor for each department. Even if a store map is available, the customer must find the map and then attempt to determine a path from his or her present location to the desired department. These systems for providing product location information to a retail customer suffer from numerous disadvantages. Directions are not easy to obtain if the department in which the item is located is not known, if store personnel cannot be located, or if the person found does not know a particular product's location. Further, a customer who is blind or has limited English-language ability may have difficulties in reading available maps or signage. Therefore, there is a need for an easy-to-use solution. Using a computerized Product Finder application for conveying product location information, a retail customer enters a request for the location of a selected retail item into his mobile device (e.g., a PDA) or a retail self-service terminal, which is connected to an in-store computer system having product location information stored thereon. Interfaces are provided between the computer system and the mobile device or retail self-service terminal. The interfaces have access to database tables having stored therein additional product location information, including department and aisle or shelf information. The Product Finder application further includes interfaces between the in-store computer system and retail self-service information terminals located throughout the store. Each retail self-service terminal may suitably be a multipurpose unit that provides information to retail customers as to product location and also as to other topics of interest to shoppers, including pricing information and sales promotions. In addition, each self-service terminal could also include various output devices, including a speaker and printer.

The retail customer enters an information request on his mobile device or into the terminal. The interfaces further incorporate path-finding methods (Chapter 4), performed by way-finding algorithms, that allow a path from each mobile device/ retail terminal to the ultimate destination to be defined through a series of intermediate points. The system uses (x, y) coordinate values on an underlying map, which is designed to be compatible with GIS. In other words, the system's database tables are stored in a format that is compatible with a GIS. Location points, expressed as IDs that point to map coordinates, need to be defined at least for each ultimate destination. To allow for customized directions from a given starting point, the location points of all retail information terminals in the store also need to be defined. Intermediate points can also be defined, if desired. In such a case, careful planning and maintenance are required if the intermediate points are to be associated with text or audio clips that provide directions for a portion of an overall path. Intermediate points may prove most useful to trace nondirect paths on a base map. In any case, the underlying logic determines a path and the points on the path given any two distinct locations.

The Product Finder application is designed to integrate with third-party application packages, including planogramming and GIS products. Planogramming databases often provide shelf position information for specific item codes, with the shelf's location in the store possibly defined as well. GIS databases, although usually applied to allow way finding on geographic maps, could be scaled down to define a store layout as the underlying map on which paths are traced. The directions to the selected retail item are provided as an output at the retail or self-service terminal. This output can take the form of text, audio, and graphical images.

Such applications could be enabled within local areas, such as shopping malls, using LPS, providing that they can persuade customers to carry tags that can be tracked. This could be done by offering people discounts and special offers, as stores currently do with club cards. LPS has some potential advantages over other technology approaches, as well as enabling significant additional applications. Other technologies are generally not very accurate indoors, and may well not be able to accurately identify which floor a person is on in a multistory building, such as a typical mall, which limits the use of such technologies in this environment.

LPS, on the other hand, is sufficiently accurate that it could enable navigation not just to a certain store, but to specific items within a store. It can enable people to find their friends or family within a store. And the concept of tailored special offers from stores nearby seems much more likely to be successful in an environment where people are known to be shopping, rather than in a more general situation where people may be just driving along a street on the way to somewhere else. In addition, the shopping mall, and individual stores, get very valuable information about exactly where each visitor goes, which can help with improving store layout, marketing, etc. They could also communicate with customers using screens located around a store — when someone walks in front of a screen it could display information known to be of interest to that person. For example, YDreams' Virtual Promoter uses proximity sensors to prompt the customized three-dimensional character, which then invites the surprised passer-by to come closer and take a look. After the first contact is made, the three-dimensional character will react to both proximity and movement: when a visitor comes closer or walks away, different pieces of information are triggered — further insights about the company, product or brand, videos or ads.

This would enable stores to combine the advantages of a personalized recommendation system like that used by Amazon.com with the advantages of a physical store, such as the ability to physically inspect merchandise and take it with you immediately. Personalization of location-based information in a wireless indoor environment is one of the current trends in the wired world today. With personalization, information can be tailored to individual user preferences and characteristics. The main objective of personalization is to provide services and information to users more effectively and efficiently, thus making user interaction much faster and simpler. With the emergence of higher-bandwidth wireless technologies, the use of mobile devices will grow more rapidly. One factor that sets wireless technologies apart from wired technologies is user mobility. Mobility provides the ability for users to get connected anywhere, anytime, which is the driving force behind the popularity of wireless technologies. Mobility brings another aspect into the concept of personalization: location-dependent information. With the precise knowledge of the location of users, information can be further tailored to the needs of the users. This will enable service providers to provide real-time services that cater not only to the preferences of the users, but also to their locations.

Current Wi-Fi tags, for example, are about the size of a small stack of business cards and cost less than $40 for inventory control applications in retail. The idea would be to tag every item in the store. Taking inventory could be done automatically in seconds. There are huge challenges, though. The tags would have to be much, much smaller and much, much cheaper. The cost for tags has to be less than a penny, and the problem is, when you get to that level, you lose a lot of the intelligence.

Location technology from Ekahau, for example, was used in a Personal Shopping Assistant (PSA) application developed for the Metro Group Future Store Initiative. PSA provides accurate shopper location information in the Metro Future Store and displays location-specific personalized shopping lists, favorites, and special offers. The system can offer discounts on items related to those put in the cart. It can also trigger in-store signs. So if the shopper puts Pringles in the cart, an ad for Coca-Cola might be displayed. Shoppers who scan all their items can have the information communicated to a cash register wirelessly and check out quickly. (Metro Group is the largest retailer in Germany and is experimenting with RFID tagging and other advanced technologies, including smart scales and wireless computers on shopping carts.) The PSA is a custom-designed tablet PC with integrated Wi-Fi capability and bar code reader. The PSA is a trolley-mounted touch-screen device that the customer takes from the charging rack and clips onto her trolley as she enters the store. The system tracks the shopper's movement and the screen displays location-specific relevant promotional items, coupons, and special offers. The PSA then helps guide the shopper through the store based on her shopping list and favorites. In addition, the PSA can allow delicatessen and other prepared items to be preordered directly from the device. The PSA also offers reminders for infrequently purchased items and suggests complementary items to accompany customer purchases. The PSA can present recipe and menu planners, with lists of ingredients and nutritional advice. Finally, the device can be used to self-scan items for quick checkout and to track current total balance.

Critical to the value of the PSA is accurate positioning information. Knowing an individual shopper's exact position allows the PSA to deliver exactly the right information and target promotions to the shopper that relate to the products she is actually in front of on the shelf. Customers must be Metro Group loyalty card customers to use the service. As soon as they swipe their card over the PSA's bar code reader, the unit retrieves — over the store's 20-access point Cisco Wi-Fi network — data on any past shopping activities or shopping lists they might have created at the store's web site before coming in. As they move around the store, the device feeds them information based on which aisle they are in. It will remind them of items that are on their shopping list or items they frequently purchase. It will also remind them of infrequently purchased items in that aisle that they may have forgotten but need. Of course, it also uses the system to advertise in-store specials.

A search function allows customers to key in the name of an item and then shows them where it is in the store on an on-screen map.

As they pick up items, customers swipe them over the bar code reader. When they check out, they go through a special express check out, where their order is already added up and all they have to do is pay. The benefits for the customer are obviously the convenience of not having to ask where things are, the speed, and, of course, the savings from the promotions. It also reminds customers of things they might otherwise forget, so the store does not lose some of their consumers' spending to convenience stores and gas stations. Finally, the service is an effective way to promote store brands or sponsored specials and increase total basket size. For the store, among the benefits is getting loyal customers, because only by using the service regularly are customers' shopping patterns and preferences learned by the system.

Because the wireless network itself is not particularly stable and the Wi-Fi client devices chosen are not optimal for the location part of the application, the software-only Ekahau positioning technology can only pinpoint the location of a customer to an aisle — most of the time. It cannot tell what product section she is standing in or passing. A real-world implementation scenario is desired where the environment would be a little more static. This would require the positioning to be down to the level of product section rather than just aisle. To deploy a PSA system like the one at the Future Store costs a typical grocery chain about $150,000 per store. That includes hardware and software, including PSAs and the positioning system, plus training and full implementation by a systems integrator. Systems like Ekahau are preferred for developing such applications like the PSA because Ekahau is a software-based Wi-Fi location platform, Ekahau Positioning Engine, which is an ideal solution for application developers. It does not need any additional hardware installation to work — the standard Wi-Fi data network is enough.

The Personalized Shopping Assistant is portrayed in Figure 1.10.

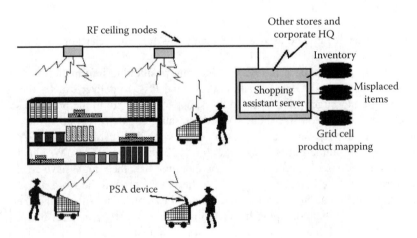

FIGURE 1.10 Personalized Shopping Assistant.

USE CASE: MOBILE COMMERCE (PRODUCT FINDER, PERSONALIZED SHOPPING ASSISTANT)

1. "I want to find a specific product (and its store) that I know of." (This product might or might not be on sale.)

2. "I want to be notified when I am in the vicinity of the store that has the product I am looking for."

3. "I want to be notified about the product that I am looking for *only* when it is on sale. I might allow the service to use my user profile to target me for similar products, special promotions, etc."

1. The user is interested in locating a particular store that has a product of interest. The user can either ask for the *nearest* (proximity) store or *pinpoint* a store. The former case is when the user wants to know what is the location of the nearest store of a known type. Following this, a proximity search for the nearest store is executed against a Yellow Pages server. The latter case is where the user identifies the store by specifying its name or the type of business that it is (the products that it sells). If known, the user could also specify the store by a phone number or some other unique ID. Following this, a search for the specific store location is executed against a Yellow Pages server. For both cases, the fetched location of the store is displayed on a map on the mobile device, relative to the known device location. Optionally, the user can then request the best route to the store, display directions to the store, or call the store.
2. The alert service is used by a store that has the product to notify interested customers who gave access permission rights to the service provider (i.e., stores in the mall or a third party). Users will receive a message about the sale as they enter the vicinity of the store. Therefore, customers will receive the message before they start shopping and not afterwards, giving the retailer a competitive edge.
3. This is a special case of the previous action. Here, the user wants to be notified only when the product is on sale. This could be a live service where the user bargains/bids (similar to Ebay.com or Priceline.com) with the store vendors, letting them know what he is interested in buying and the price he will pay via a user profile. This is a dynamic service not only because the user's position changes as he moves around the mall and is targeted by vendors based on his location, but also because the prices of products change "live," or on the fly. Only the stores that meet the user's price requirement would be allowed to notify him.

A real-world example of such an indoor LBS application is the Personalized Shopping Assistant (commerce application category) (Asthana et al., 1994). The current methods of advertising in stores rely on public address systems or programmable light-emitting diode (LED) displays in the aisles. These methods are not effective due to the noise, distractions, and impersonal nature of the method. Shopping by catalog or from home (such as using Prodigy or TV-based home-shopping networks) has great appeal but is not going to replace physical shopping. Also, mobile text messaging (or SMS) allows companies to distribute advertising messages to mobile phones, targeted by user preference profiles. What customers are frustrated about is where to find things, determining the right price, etc.

Overall, location will soon be an aspect of this targeting, limiting the campaign to people currently in a specific location. LBS are more refined and, in turn, will have a higher value on return. Coupled with this can be the concept of an e-voucher within the message, to enable further enticements of discounts in the store for other products. Advertising messages are another marketing channel to customers and therefore have the potential of very high returns for the communication infrastructure provider (i.e., wireless carrier/operator). Therefore, these providers will eagerly offer this service widely to their clients.

The personalized shopping assistant is based on two products: a very high-volume handheld wireless communications device, the PSA (Personal Shopping Assistant), that the customer owns (or may be provided to a customer by the retailer), and a centralized server located in the shopping center with which the customer communicates using the PSA. The centralized server maintains the customer and store databases and provides audio/visual responses to inquiries from tens to hundreds of customers in real time over a small-area wireless network. By integrating wireless, video, speech, and real-time data access technologies with location positioning, a unique shopping assistant service can be created that personalizes the attention provided to a customer based on individual needs, without limiting his movement or causing distractions for others in the shopping center.

The objective of this service is to enhance the shopping experience of users by exploiting their contexts in a store. Each shopper carries a specialized device. As the shopper wanders around in the store, the device automatically displays the description of the items that the shopper is currently seeing (location context). In addition to helping the user locate an item, the service can also recommend sales items that match users' interests without any explicit user instructions (personal context) and do a comparative price analysis. In fact, MSN Shopping Assistance (MSN Shopping Assistant), while not location based, offers a live chat with a real-person shopping assistant that can help the user find a great gift for anyone on the customer's list. Also, Microsoft's OnSale Mall Buddy service has personalized sales announcements, e.g., "Alert me when electronics are on sale." These applications can make use of Yellow Pages services to request content of interest (COI).

Another use for being able to locate and track customers is for commerce, analyzing customer shopping behavior. Shopping malls and individual stores get very valuable information about exactly where each visitor goes, which can help with improving store layout, marketing, etc. They could also communicate with customers using screens located around a store — when someone walks in front of a screen it could display information known to be of interest to that person (by means of accessing that person's user profile, saved on his mobile device). This would enable stores to combine the advantages of a personalized recommendation system like that used by Amazon.com with the advantages of a physical store, such as the ability to physically inspect merchandise and take it with you immediately. With respect to profit making, m-coupons, cited as a strong growth area in industry reports, will remain niche markets with slower than expected growth (3G).

USE CASE: NAVIGATION

1. "I want to be navigated to the store that has a product on sale."

2. "I want to be notified along the way when I am near someone (or something). "

1. Having two (or more) positions (the user's initial location and the location of the destination), the service can calculate the route between them. The service would compute the route (best/shortest) between the points or "spaces" and display the route (map) or directions (text). Typically, the route will be determined between two locations, one being the starting point or space, which is the current device location, and the other being the end point (the destination).
2. There is also the case where the user has several stops along a route (i.e., a store that might have something on sale that is of interest to the user or a friend that might be in the vicinity), so the user must specify waypoints in addition to an endpoint.

The endpoint and waypoints may be determined by any number of ways. For example, the user might use the pinpoint or nearest (proximity) services described above to establish these points or spaces (nodes). Once the route has been calculated, the user has two options:

• Display route: Show a map of the floor plan layout.
• Display directions: This might be as important as maps and actually preferred by some users, as it may be easier for them to follow instructions than to read a map.

An example of an application is WayFinder from MIT Cricket. This application, running on a handheld computer, can help sighted or blind people navigate toward a destination in an unfamiliar setting. For example, the WayFinder might lead a person from the entry lobby of a building to the office of the person hosting the visitor or to a seminar room. WayFinder gives incremental directions to the user on dynamically active maps using the user's position and orientation with respect to a fixed set of wireless beacons placed throughout the building. In addition, when considering a navigational application, a wheelchair or the Segway (Segway) can be considered the user's mobile device, which can be designed to (automatically) navigate the user to her destination.

In terms of the positioning infrastructure, positioning aids such as spatial bar code (SBC) or emitters are needed around a building. Figure 1.11 shows an example of such SBCs placed in the Colombo Shopping Center in Lisbon, Portugal. Using

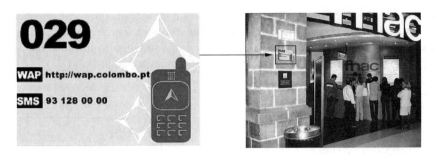

FIGURE 1.11   Spatial code bars. Tags placed in a store in Colombo Shopping Center, Lisbon, Portugal.

Wi-Fi or Bluetooth access points (emitters) as anchors will allow users to automatically pinpoint their location without having to read or write any SBC.

Moreover, locating a user based on Wi-Fi cell ID might be adequate for navigating the user to a store, but it might not be precise enough for a shopping experience inside the store. More precise positioning technologies could enable navigation to specific items within a store's aisle. The mapping infrastructure is used for exploration and navigation of a place, which is usually associated with the use of symbolic representations such as maps (of a shopping mall), which are then matched against the reality they try to symbolize. This matching is facilitated by the use of anchors (points of interest) and paths displayed in maps that are easily recognizable locations. These maps and their features (i.e., anchors, paths) can be modeled by using either symbolic representations or geometric features. If the service is using a geometric map, the user may optionally display distances along segments of the route. If the service is based on a symbolic model (spatial relationships are based on either associations, containment, or proximity), directions will not be given precisely as to how many feet before making a right turn. In either case, geometric or symbolic, the directions should probably be given by means of symbolic reasoning (i.e., "When you reach store A, make a right turn," instead of "After 10 ft, make a right turn"), as people associate with this type of reasoning better. In the case of the Segway or an automatic wheelchair, directions need to be given in terms of geometric reasoning since the machine is not capable of symbolic reasoning.

Overall, indoor navigation is going to be more often the desire than the need until an infrastructure of positioning/navigating aids (i.e., anchors) is in place. In terms of the software (service) infrastructure, the user will most likely invoke a location-based concierge application that would determine his location and he would subsequently receive navigation/routing details about getting to the destination. Optionally, the service is able to specify some waypoints (or these points have to be specified by the user), and the user may specify route determination criteria. This criteria might be fastest, shortest, etc., and can also specify the preferred mode of transport of the user (elevator vs. staircase). The details of the returned information might include directions to the site and other relevant information (according to the user's profile, for instance). The route can also be optionally stored on the terminal or application server. The user may store it for as long as needed, thus requiring the means to also fetch a stored route. Regardless of how the endpoints and waypoints are established, the information is then sent to a route server, which calculates the route. Applying OGC's OpenLS services, a subscriber would seek navigation advice from the service provider via a personal navigation service.

RETAIL, ADVERTISEMENT, AND MARKETING

Today, most people carry their mobile phones and PDAs everywhere they go due to high demands for mobility and flexibility. The ability to deliver information to a specific place at a specific time makes it possible to communicate with people at moments where they normally would not be reachable. Creating an environment where it is possible to find, identify, and send information to and from mobile phones and PDAs opens up completely new ways of communicating with your customers or employees. Location-based advertising can also play a big role in retail and shopping. Many Internet services derive their revenue stream from advertising. Advertisers pay to have their content delivered to users who access web site and web server interfaces.

Barnes (2003) introduced the concept of tempting nearby users into the stores and delivering geographic messaging related, e.g., to security in a particular area of a city. Varshney and Vetter (2002) suggested mobile advertising to be a very important class of mobile commerce. They augmented location information with the personalization of the delivery by obtaining the history of the user's purchases or consulting the user at an earlier stage. In addition, the users might be able to either receive push advertisements or actively pull the messages.

Advertisers desire to target their audience at the most appropriate time. In marketing, mobile advertising has two distinct meanings: advertisements moving from place to place (e.g., displayed on the sides of trucks and buses) and advertisements delivered to mobile devices (e.g., mobile phones and personal digital assistants (PDAs)).

Permission-based advertising is key, ruling out unsolicited advertising (i.e., spamming). Sometimes, wireless advertising is used to refer to mobile advertising. Proximity-triggered mobile advertising is a special case of location-based notification services (Munson and Gupta, 2002). Usually, notification services are user driven, e.g., getting a notification when a set of conditions is met. Advertising, on the other hand, is typically not user driven; i.e., the recipient does not request or pull the advertisements from a server, but they are pushed to her instead (Ranganathan and Campbell, 2002).

Knowing the location of a user as relevant to a particular advertisement is desirable. Automating the delivery of the content is desirable. For example, restaurants could welcome their guests and get a notice that they have arrived, offer information, web access, and entertainment for the children while they wait, offer them a loyalty coupon, and welcome them back.

WideRay's Jack Service Point (WideRay Jack, 2003) is a product for delivering local content, such as advertisements, using Bluetooth or infrared. A number of these devices have already been deployed at mass events, such as sports and conferences, for distributing event-related information. We are not aware of any published research papers that would involve Jacks. The consumer tells the service system that she is hungry, for example, in which case the system provides her only with mobile advertisements matching this status. The system recognizes the consumer when she passes a nearby base station and automatically sends a mobile advertisement as the consumer is approaching the coffee shop in question. Thus, the advertiser defines whom she wants as the receiver of the mobile advertisement. Correspondingly, consumers tell the system what kind of messages they are willing to receive. So from the consumer's perspective, it is not about receiving junk mail, but desired information.

People moving in the city center are offered only mobile services that are purposeful and useful in view of both the customer and the enterprise. There are numerous examples: The system could, for example, send an automatic advertisement about an umbrella offer to the customer when a sudden shower occurs. When a tourist or visitor is looking for a suitable place to eat lunch or for the best clothes stores, he finds the place on a mobile map. The application would then guide the tourist to the location.

Similarly, shopping malls can present information to shoppers about upcoming events and, e.g., traffic information, while gathering information about movement patterns; increase revenue through direct marketing to mobile phones; and reach the customers when and where they choose. Randell and Muller (2000) presented the Shopping Jacket infrastructure, which used GPS and local pingers in stores for positioning. Wearers were alerted when passing an interesting shop. The system could also be used to guide the user around a shopping mall.

Public transportation can provide travel information, news, and entertainment, and fill the idle minutes of busy travelers. The system automatically learns the movement patterns of frequent travelers and sends out messages if they are affected by delays.

Major retailers are introducing computing devices mounted to shopping carts. These devices help customers find products in a large store while at the same time allowing the retailer to target customers with advertising related to products in the cart's immediate vicinity. By linking a cart's location with a frequent-shopper card, retailers can tailor promotions to the individual shopper. Another example is having a shopper inside the store approach a retail store with a mobile RDPS; a configured advertisement of a special deal at the retail store can be proactively delivered (i.e., pushed) to the user automatically on behalf of the store and direct the shopper to specific sales items as the shopper moves about the inside of the store. Stores can also push information about special offers to customers on their mobile phones, as they pass by the shop or enter a shopping center. Additionally, stores can implement customer loyalty programs by pushing personal offers to their best customers. Main benefits for end users include personalized information upon signing up for services fitting personal needs. Main benefits for operators include the offer of additional services, location-based services (via localized WAP Push when using Bluetooth for applications, for instance), and localized information and entertainment via MMS or SMS.

FIGURE 1.12 Phone marks and anchor points: product querying and interaction (left) and storing the location of a car (right). (From YDreams.)

Kaasinen (2003) analyzed user needs for mobile location-aware services. Most users did not mind being pushed information, as long as they really needed the information. Thus, location itself is not enough to trigger pushed advertisements, but it has to be complemented with personalization. Push of personalized content and location-dependent information to mobiles that are equipped with, for example, a Java API, supports J2ME and Symbian applications. In order to accelerate queries, particularly for mobile users, LBS provide personalization mechanisms that consider individual preferences of a user. By this technology, less interaction steps are needed to identify the desired information, as, for example, certain categories are excluded already from the beginning or may be automatically refined. The user may comfortably control and maintain his profile from any Internet PC and then automatically access his profile in each query via all supported devices. This way, all categories a user is currently not interested in can be excluded from, e.g., event queries for concerts. In addition, the user may indicate further refinements within a category: if someone is only interested in a particular product when shopping, the LBS only indicates these kinds of events to him. Personalization can also apply to groups. In an academic setting, for example, information (such as notes, exams, and dynamic audiovisual content) may be delivered to a student or professor as a function of the classroom he occupies (e.g., as a locale) and the time of day. Or in a medical setting, patient information delivered to a doctor or nurse may be delivered as a function of the patient then assigned to the room or bed that the doctor or nurse is visiting.

RETAIL AND POSTANALYSIS OF SHOPPING BEHAVIOR

Retail can also benefit from store layout analysis: by analyzing the path chosen by shoppers in a store, retailers can change store layout to alter purchasing behavior and boost sales. Linking layout information to actual purchase data can lead to further improvements in store and product layout. This is an example of macroscopic or behavioral analysis of the mobile communication terminals or devices where traffic monitoring would take place in a supermarket, shopping mall, or convention center to better market, position, or place products and services in the future.

TOUR GUIDES

Museums are also highly suitable application areas for deploying indoor LBS services. Large museums occupy vast areas filled with numerous exhibits that are almost impossible to assimilate in a single visit. Therefore, museums are always seeking ways to ease the experience of the visitors through, for example, guided tours and personalized audio devices. However, a number of more innovative location-aware applications have also started to emerge. Oppermann and Specht (1998) describe such a service. Visitors may prepare their excursion to a museum at home by logging on to the system through the Internet and selecting the exhibits in which they are most interested. Upon entering the museum, users are provided with a mobile device that is connected to a server through WLAN and Bluetooth technology. By being location-aware, the device can provide the visitor with a personalized guided tour. Additionally, the navigation through the museum becomes a superior experience altogether because description and comments in audio and video format can be delivered to the visitor, based on position and personal interests. For example, a visitor who stands in front of a painting can be provided with information about the work itself, the painter, the artistic style of the era, and so on. Similarly, by monitoring visitors' navigation patterns and by instantly becoming aware of congestion spots within the museum, management will be able to organize exhibits more effectively and even make real-time decisions regarding employee allocation to sections where additional needs emerge.

In a fashion similar to museums, libraries are vast indoor environments where information is abundant but not necessarily easy to identify and locate. A typical scene in a busy day of a library would involve people waiting in line in order to use a computer or ask the librarian to obtain information about a particular book or a certain thematic area. Such problems can be overcome, or at least be significantly reduced, through the introduction of an indoor mobile location tracking system. Visitors will be able to specify requests to a personal hand-held mobile device in order to query the library database and identify the exact position of a given title, as well as a possible route to it. Moreover, visitors will be able to express requests regarding similar books (for example, on the same thematic area or by the same writer), and the information will be delivered to them during their navigation through the library, thereby alleviating them of the burden of walking back and forth to an information kiosk or a library desktop computer.

A project is underway with the Metropolitan Museum of Art in New York wherein Ekahau's technology would be used to identify the locations of specific works of art. Given a device that can recognize these locations, a visitor could then enjoy an interactive guided tour.

Tracking can be used when migrating through an exhibit or museum; content or data describing or relating to a proximate exhibited item may be delivered, changing as the user changes location. Another example is the Smart Library application at the OULA library in Finland. A library customer is able to browse the OULA library catalog wirelessly with mobile devices. OULA search is operable with mobile phones equipped with an XHTML browser, as well as PDA devices and laptops with WLAN capability. Users request map-based guidance to a desired book or collection. NearSpace specializes in producing campus and venues mapping software (see Figure 1.13 and Figure 1.14).

FIGURE 1.13 NearSpace campus map.

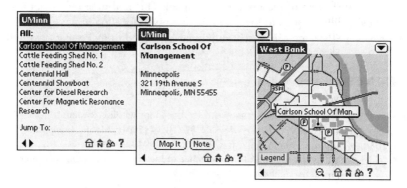

FIGURE 1.14 Navigation GUI (NearSpace).

Another example of tour guides is the Hyogo Prefectural Historical Museum where a Bluetooth positioning system is used along with BlipNet access points. The PDAs provide maps and the location of the user, as well as a list of all the items in the museum. Text, pictures, audio, and graphics are utilized to provide information about the museum. Text is HTML-based, and the push of objects is accomplished via the BlipServer API. The museum is also able to obtain tracking information via the positioning system. The tracking information is useful for optimizing the museum's floor plan. This system makes it easier for the museum to update information for their visitors, instead of updating brochures and posters. The Bluetooth positioning system determines proximity based on the measurement of the power from the access points, not the client devices. The power control of the access points can be modified to "tune" the system according to the needs for appropriate positioning of visitors.

OTHER

In addition to contextually aware functionality presented to the mobile user in the defined space, information can be collected and utilized to analyze where various workflow bottlenecks may exist or other spatially related challenges arise. For example, knowing the motion of the mobile device can be used to allow the communication network itself to anticipate handoff to a set of communication access points and preemptively prepare for a possible network handoff of the mobile devices' communication.

While cities such as Philadelphia have just announced long-range plans to install WiFi nodes on every street lamp, MIT has already activated over 2000 access points, and will soon bring coverage to 100% of the campus. The effects of complete wireless coverage are monumental, as traditional work spaces are being abandoned in favor of more enjoyable environments such as campus lounges and public spaces. The iSPOTS project, produced by MIT's SENSEable City Lab, documents these changes in real time using log information from MIT's wireless network. This is beneficial for college students wondering whether the campus cafe has any free seats or their favorite corner of the library is occupied. This real-time information shows exactly how many people are logged on at any given location at any given time. It even reveals a user's identity if the individual has opted to make that data public.

REFERENCES

Asthana, A., Cravatts, M., and Krzyzanowski, P., An indoor wireless system for personalized shopping assistance, in *Proceedings of IEEE Workshop on Mobile Computing Systems and Applications*, Santa Cruz, CA, December 1994, pp. 69–74.

Barnes, S.J., Known by the network: the emergence of location-based mobile commerce, *Adv. Mobile Commerce Technol.*, 171–189, 2003.

Boswell, R., Location-Based Technology Pushes the Edge, *Telecommunications*, June 2000.

Chalmers, M., *George Square*, demo game at Mobile HCI 2004 Conference, 2004.

Chandra, C. and Kumar S., Supply chain management in theory and practice: a passing fad or a fundmental change? *J. Industrial Management Data Syst.*, 100, 3, 100–113, 2000.

Cox, I.J., Blanche: an experiment in guidance and navigation of an autonomous robot vehicle, *IEEE Trans. Robotics Automation*, 7, 193–204, 1991.

Dey, A., Futakawa, M., Salber, D., and Abowd, G., The Conference Assistant: combining context-awareness with wearable computing, in *Proceedings of the 3rd International Symposium on Wearable Computers (ISWC '99)*, San Francisco, October 1999, pp. 21–28.

Flintham, M., Anastasi, R., Benford, S.D., Hemmings, T., Crabtree, A., Greenhalgh, C.M., Rodden, T.A., Tandavanitj, N., Adams, M., Row-Farr, J., Where on-line meets on-the-streets: experiences with mobile mixed reality games, in *Proceedings of CHI 2003*, 2003.

Galani, A., Chalmers, M., et al., Developing a mixed reality co-visiting experience for local and remote museum companion, in *Proceedings of HCI International (HCII2003)*, 2003.

Helms, M.M., Etkin, L.P., and Chapman, S., Supply chain forecasting: collaborative forecasting supports supply chain management, *J. Bus. Proc. Management*, 6, 5, 392–467, 2000.

Hendrey, G., Managing the Wireless Internet, *RF Design*, March 2001, pp. 50–56.

Kaasinen, E., User needs for location-aware mobile services, *Personal Ubiquitous Comput.*, 7, 70–79, 2003.

Kourouthanassis, P., Koukara, L., Lazaris, C., and Thiveos, K., Last-mile supply chain management: MyGROCER innovative business and technology framework, in *Proceedings of the 17th International Logistics Congress: Strategies and Applications*, Thessaloniki, Greece, pp. 264–273, 2001.

MSN Shopping Assistant, http://shopping.msn.com/softcontent/softcontent.aspx?scpId=3565&scmId=1422.

Munson, J. and Gupta, V., Location-based notification as a general-purpose service, in *International Conference on Mobile Computing and Networking, 2nd International Workshop on Mobile Commerce*, 2002, pp. 40–44.

NTT DoCoMo, iArea: Location Based Services, available at http://www.nttdocomo.com/corebiz/imode/services/iarea.html.

Oppermann, S. and Specht, M., Adaptive support for mobile museum guide, in *Proc. Interactive Appl. Mobile Comput. Conf.*, Rostock, Germany, November 24–25, 1998.

Randell, C. and Muller, H., The shopping jacket: wearable computing for the consumer, *Personal Ubiquitous Comput.*, 4, 241–244, 2000.

Ranganathan, A. and Campbell, R., Advertising in a pervasive computing environment, in *International Conference on Mobile Computing and Networking, 2nd International Workshop on Mobile Commerce*, 2002, pp. 10–14.

Segway (the human transporter), http://www.segway.com/. 3G, http://www.3g.co.uk/PR/Oct 2003/5950.htm.

Thrun, S., Fox, D., Burgard, W., and Dellaert, F., Robust Monte Carlo localization for mobile robots, *Artif. Intell.*, 128, 99–141, 2001.

Varshney, U. and Vetter, R., Mobile commerce: framework, applications and networking support, *Mobile Networks Appl.*, 7, 185–198, 2002.

VOR (Very High Frequency Omnirange), http://www.allstar.fiu.edu/aero/VOR.htm.

Welch, G., Bishop, G., Vicci, L., Brumback, S., Keller, K., and Colucci, D., The HiBall Tracker: high-performance wide-area tracking for virtual and augmented environments, in *Symposium on Virtual Reality Software and Technology*, December 1999, pp. 1–10.

WideRay Jack, http://www.wideray.com/product/hardware.htm, referenced October 22, 2003.

2 Preconditions and Frameworks for LBS Development

PRECONDITIONS (LBS MARKET AND DEVELOPMENT)

Starting with the outdoor LBS market, LBS had originally been labeled a killer application for wireless applications, and widespread impact has been repeatedly predicted by vendors since the wireless Internet bubble of the late 1990s. These predictions were disappointed as the market delivered the harsh lesson that location might be a killer capability for certain mobile applications, but was not a killer application in its own right. In particular, consumer-oriented applications such as location-based pushed advertising have seen disappointing, if not entirely unexpected, results, due to slow adaptation, lack of innovative applications, and privacy-related issues. Japan has witnessed some success in LBS: person-to-person LBS have been marketed by DoCoMo in the form of location-augmented iMode web sites (used for, e.g., dating) (NTT DoCoMo). Ubiquitous cell phone tracking was commercially launched by KDDI with GPS MAP, a person finder service targeted at corporate customers, in 2002 (KDDI, 2002). This system works with GPS-enabled handsets that can be tracked through client software on PCs. Location-based services are proving slow to attract users, with only 10 % of Japanese A-GPS mobile handset owners signed up for the service. Only 10% of KDDI (the Japanese mobile operator and leader in the GPS phone market) GPS-enabled phones are subscribers to location-based services. Overall, the slow take-up of location-based services is largely attributed to the time it takes to get the phone's first fix, about 40 sec.

Operator NTT DoCoMo does not use the CDMA network — and thus cannot use its time synchronization ability — and the first time to fix can take over a minute. Location information is stored on a server system managed by the operator. A more general cell phone-based person finder application has been developed by Kivera, Inc., for AT&T Wireless (now part of Cingular). The system, called Find People Nearby (formerly known as Find Friends) (AT&T Wireless) allows a user to build a buddy list and to locate other service subscribers in any area covered by AT&T. After locating a friend, the user can call the person, send a message, or invite him or her to some meeting point chosen from businesses in the AT&T *Yellow Pages*. The success of this application has been arguably limited by the lack of interoperability with other providers.

The market for LBS has begun to emerge but has been hampered to date by the limitations of traditional positioning systems. These systems call for new hardware in mobile devices, require line-of-site views of orbiting satellites, and suffer from reliability issues in urban areas.

Fortunately, in recent years, strides were made in positioning technologies related to higher accuracy, locator miniaturization, battery life, multipath effects, ability to locate indoors, and the economical use of RF bandwidth. The vision of pervasive ubiquitous computing where users have network access anytime, anywhere, is being enabled by deployments of high-speed wireless networks in common places of congregation, such as airports, malls, hotels, parks, arenas, and so on. Whereas traditional interfaces with computers have been through a keyboard and monitor, pervasive computing seeks to weave computers seamlessly into our daily lives. In a world with pervasive computing you will be able to go shopping and automatically get more information about a product that is tailored to your preferences. You will be able to open your notebook computer and automatically discover network services that are present in your area. You will be able to start complex computing jobs from your handheld device and not have to worry about where the results are going to be stored. These are just a few of the things that will come as a result of living in a truly pervasive world.

Why pervasive computing? As hardware integration increases to amazing new levels with each generation of design, devices increase in their functionality and versatility. They also become cheaper, smaller, and handier. Why now? Several factors have fueled this vision. Recent advances in CMOS IC, wireless communication, and MEMS technology have led to dramatic reductions in size, power consumption, and circuitry cost. Various functions such as sensing and signal processing can now be integrated into a single wireless node. Coordination and communication among such nodes will not only enable seamless computing, but also revolutionize information technology, especially applications related to sensing and controlling physical environments. Small active devices or sensors can coordinate to perform larger sensing tasks (i.e., distributed microsensing tasks), which could not have been achieved with individual node capabilities. Wireless service provisions are becoming global, full coverage, no drop, and low latency, supporting a wide range of services for applications involving voice, data, message, video, web, GPS, etc. Wireless customers are interested in wireless communications supporting sensor networks, which will become increasingly practical with the advent of new power technologies and new receiving and transmitting technologies. Potential applications of such sensor networks span many domains: smart spaces, factory instrumentation, inventory tracking, and location-based services. Coordinated efforts in exploring such applications have already begun, for example, the Center for Embedded Networked Sensing (CEN) at UCLA, MIT's Project Oxygen (OXY), Berkeley's CITRIS effort (CIT), and the Aware Home at Georgia Tech (AWA), to name a few. Fundamental to such seamless coordination in these systems is location awareness. Localization is a mechanism to establish spatial relationships in these devices. Because these systems are coupled to the physical world, location measures and gives a context to that physical coupling. Many of these envisioned systems are embedded to monitor or control the behavior of physical systems (compared with strictly virtual information systems), and therefore nodes often need to determine their action based on their physical location (am I the correct sensor to monitor a particular object?).

Intel's Place Lab and Skyhook's Wi-Fi Positioning System (WPS), for example, leverage the existing 802.11 hardware already resident in over 100 million devices today. Once a user has Wi-Fi, she will not have to buy extra hardware to use Place Lab, and the software can be downloaded for free from the Place Lab web site. Increasingly, laptops, cell phones, and PDAs are being sold with Wi-Fi capability already installed for around an extra $30. WPS is an all-software solution, which means it can be rapidly distributed and deployed. WPS takes advantage of the fact that almost all access points are configured to transmit their unique IDs whenever a Wi-Fi client sends out a scan request. Skyhook exploits this capability (which does not drain the resources or compromise the security of access points) by combining it with the data collected by war drivers, who use GPS and laptops to discover and map out access points in towns and cities. Once installed on a laptop or mobile device, the WPS software measures the time it takes for signals from nearby access points to return a scan request, and then uses triangulation to pinpoint the location.

Firms that use existing mobile equipment like Palm or Pocket PC can slot PC card RFID readers into their mobile devices. The computing power of the hosting mobile devices allows logistics firms like Tibbet & Britten to use this technology in a disconnected mode when delivering items to their customers. They can store the EPC code, look up additional data, and add more data, like the state of the object, to a mobile enterprise application like Aligo on the handheld device.

Adding GPS to a phone requires that the manufacturer add a chip. In a wireless system, the calculations to convert signal relay times into geographic location will get performed on a phone's processor (although for now, the Federal Communications Commission (FCC) has mandated that phones in the future have GPS functionality). Not only does Intel promote Wi-Fi, but it has also recently begun to gain momentum in the market for cell phone chips. GPS chip sets are being integrated into mainstream cell phones and PDAs, lowering cost and decreasing the users' barrier to entry. To help meet the E911/E112 requirements, cell phone manufacturers are now producing handsets that use a mix of location technologies. When GPS is not available, the locations of nearby cell towers are used to produce a coarse location estimate. Based on this coarse location, the phone can download the expected position of the satellites, allowing the handset to lock on to GPS much more quickly (on the order of seconds instead of a minute or more) when GPS does become available.

It is significantly harder to develop an LPS that works indoors rather than outdoors. This is because signals reflected from walls, floors, or ceilings can cause errors in calculations, and often there may be obstructions in between the tag being tracked and the sensors.

For many LPS applications, accuracy requirements are more precise than one might think. For example, for most applications it is essential to be able to tell with certainty which room a person is in, or which side of a partition he is on. For this reason, accuracy to within a foot or so is required for many of the applications mentioned. Granular location capabilities that come with accurate positioning techniques like RF fingerprinting helps support a broad range of wireless applications. A few brief examples include:

E911: Precise location information for emergency personnel tracking for voice-over (Vo) WLAN phones.

Asset management: Using 802.11 RFID (asset) tags, an IT administrator can rapidly find a valuable asset such as a projector or a server.

Productivity: Locating portable EKG machines or defibrillators in a hospital.

Security: Determining the RF perimeter of a building and determining when an attack is occurring within the enterprise, enabling IT staff to take appropriate action.

Some of the applications can be implemented at least to some extent by using alternative technologies rather than a full LPS. One set of alternative technologies could be described as proximity sensors. RFID tags are one example of such a technology. These can detect a tag when it passes within a relatively short distance of a sensor (usually a few feet or so). Other technologies can detect when a tag is within a room, but cannot identify its location within the room.

In most cases where a simpler approach might suffice for some basic applications, there are additional applications of interest that can only be implemented with a full LPS. It is significantly cheaper and simpler to have a single tag and sensor infrastructure instead of multiple ones, so in such a situation it is important to consider whether the simpler approach will suffice for all applications that may be needed. If not, it makes sense to implement an LPS infrastructure that will handle all applications from the simple to the sophisticated. The LPS approach also provides much greater flexibility. With proximity approaches, every location where a tag needs to be detected must be defined in advance. Any time that this changes, new hardware needs to be installed. With an LPS, the sensor infrastructure detects tags everywhere, and so application events can be changed very simply using software tools, rather than having to change the hardware installation.

Reasonably high update rates are also required; in general, one or more location updates a second are required for many of the applications mentioned here. An important aspect of a practical production system is being able to intelligently vary update rates of individual tags in a dynamic fashion — this is discussed further in the section on power consumption.

SCALABILITY

An indoor environment contains walls and other clutter that cause attenuation in transmitted signals, limiting the area over which the LPS can operate. To cover larger buildings, individual LPS cells must be "tiled," in much the same way Global System for Mobile Communications (GSM) tiles radio cells to provide telephony coverage. In addition to the sensor system, applications also need to scale, from supporting a handful of users in a small office up to possibly hundreds of people and thousands of objects in a larger building, such as a hospital.

REAL-TIME OPERATION

Unlike outdoor location-based services, such as those based around the location of mobile telephones, an in-building location system represents a true "human in the loop" control problem. This requires low-latency, real-time operation, in order to maintain a sense of involvement with the user.

POWER CONSUMPTION

For a real-world system, managing power consumption to prolong the battery life of tags is an important consideration. One of the keys to power management is to intelligently vary the update rates of individual tags. A fast-moving tag may need to have its location updated multiple times a second, whereas a stationary one does not. Certain tags may need more frequent updates than others. Intelligent management of update rates is an important task of the system infrastructure.

Integration

The LBS application might be implemented as a stand-alone application or integrated as a module within a larger communication framework. The whole framework can consist of different modules. One of these modules might be a session manager, which will perform all registration and session management operations.

In addition to this, it will perform presence and instant messaging (IM)-related functions. Serial Interface Protocol (SIP) can provide the signaling protocol between modules because of its simple and extendable structure. Presence and IM are two of the extensions supported by SIP (Campbell et al., 2002; Dey et al., 1999). Standard SIP clients are able to register with the session manager and receive and send presence and IM packets. In addition to these services, standard eXtensible Markup Language (XML) syntax can be extended to support location-based presence and messaging.

A streaming server within the system might be used, for example, the Wireless Information Delivery Environment (WIDE) system, which is based on the Infostation concept (NetLab, 2003).

Streaming data can be sent via Real Time Protocol (RTP). A Directory Services Manager (DSM) will be used to provide LBS to users and will work in coordination with a Directory Services Database (DSDB) and a User Profile Database (UPDB) to manage the LBS. Among the services provided will be voice or video streaming, notifications such as flash news, and commercial advertisements. Many applications described here require integration between the LPS and other software systems, so it is essential that the LPS platform is designed with this in mind. For example, a hospital may want applications for security, office productivity, and intelligent building infrastructure, in addition to specific health care applications. It is important that the LPS platform can handle multiple applications developed by different vendors or organizations, running concurrently.

Development

The development of a location-based system such as IIT's Hawk Tour is relatively complex, not something you should jump into without a solid background in wireless networking. The perils of wireless networks (such as spotty coverage) require careful exception handling as users roam.

System Use Cases, Scenarios, Business and Usage Models

Place Lab usage scenario involves users with Wi-Fi-enabled (also Bluetooth, GSM, and GPS) mobile devices (notebook computers and PDAs being most common) and a Wi-Fi positioning database that determines users' positions. A Place Lab user subscribes to databases, potentially from multiple providers, that the client Wi-Fi positioning algorithms use to convert an access point BSSID (plus signal strengths) into a geographic position. Users first download a Wi-Fi positioning database and refresh it periodically. Then, as they travel in and around Wi-Fi access points (APs), they may visit location-enhanced web services and use the Place Bar to push location information into these web services while controlling, to a fine degree, the personal information being revealed. A Place Lab user subscribes to databases, potentially from multiple providers, that the client Wi-Fi Positioning algorithms use to convert an access point BSSID (plus signal strengths) into a geographic position. A downloaded and continually updated distributed contributor database of all the Wi-Fi access points in the world will allow clients to compute their own positions and divulge their location information only when they want to. Over time, this collection of Wi-Fi positioning databases is foreseen to grow to include every AP in the world. Given such a collection of databases, whenever the client receives BSSID beacons, she is able to calculate position without additional network communication. This positioning might be integrated into GIS client mapping software (e.g., ESRI BusinessMap or Microsoft's Streets & Trips), or it may be integrated into a web browser to communicate with a location-enhanced web service like Google Earth or MSN Virtual Earth.

Place Lab's business model is a positioning infrastructure designed with the following objectives:

1. Maximizing coverage across entire metropolitan areas.
2. Providing a low barrier to entry by utilizing predeployed hardware. This includes low-cost, convenient location-finding technologies.
3. Making users comfortable with respect to their location privacy.
4. Having existing web content easily customized to geographic locations.

The argument of Place Lab is that in order for LBS applications (e.g., location-based searches) to reach critical mass levels, coverage is more crucial than accuracy and cost. Despite the dozens of research and commercial positioning systems that have been built to date, and the efforts within to improve accuracy, building and deploying LBS applications that are usable in everyday situations is arguably no easier now than it was 10 years ago. The objective of the Place Lab initiative is to bootstrap such an activity through low-cost positioning technology in conjunction with a broad community-building effort that will create the large collection of location-enhanced web services needed to catalyze business models, specifically in terms of the following:

1. *Coverage*: In terms of coverage, traditional systems primarily focus on optimizing accuracy rather than wide-scale deployment and have accuracies in the 5- to 50-cm range. They have coverage constrained to a room, building, or campus environment. While this availability is sufficient for many home or office scenarios, limited coverage rules out many personal and social applications targeted at people's daily lives. While this precludes using Place Lab with some applications, there is a large class of applications that can utilize high-coverage, coarse-grained location estimates. For example, Dodgeball.com hosts a cell phone-based social meet-up application, which relies on zip codes to represent users' locations. Such applications have thousands of daily users and have accuracy requirements that can be met by Place Lab even using limited calibration data.

2. *Low cost/low effort (calibration)*: Since Wi-Fi is widely used across a broad variety of mobile devices, this positioning technology is extremely low cost. Note that Place Lab still believes it is important to model location uncertainty and minimize it to the extent possible. The bottom line, however, is to achieve this without requiring custom hardware or limiting the operation to controlled environments. Moreover, the primary challenge in expanding the deployment of a location positioning system is the installation and calibration cost. Certainly, with limited calibration, Place Lab will estimate a user's location with lower accuracy. The underlying trade-off is the accuracy of the positioning infrastructure vs. the calibration effort involved. Place Lab makes the trade-off of providing positioning on the scale of a city block, but manages to cover entire cities with significantly less effort than traditional indoor location systems. Products like Ekahau, for example, do not ship with any radio maps and require time-consuming site and system calibrations. Also, other location systems like MIT Cricket and MSR RADAR only work in controlled (limited) indoor environments. This is due to the large amount of calibration data that needs to be collected and maintained, which requires considerable effort to deploy on a significantly larger scale. These other systems provide higher positioning accuracies, but at the cost of many hours of installation and calibration (e.g., over 10 h for a 12,000 m^2 building) (Haeberlen et al., 2004), and consequently have resulted in limited deployment. In contrast, Place Lab uses sparser calibration data that can be collected while walking (or driving) and is contributed by a community of users.

MIT Cricket

Applications that can be developed using Cricket include resource discovery, human/robot navigation, physical/virtual computer games, location-aware sensing, hospital/medical applications (e.g., equipment and patient tracking/monitoring), stream migration, pose-aware applications like the software flashlight/marker, etc. It can provide distance ranging and positioning precision of between 1 and 3 cm, so applications that benefit from better accuracy, such as cellular E911 services and GPS, will also find Cricket useful. Cricket is designed for low-power operation and can be used as a location-aware sensor computing node (running TinyOS), to which a variety of sensors can be attached. At the beginning MIT researchers were hopeful that a purely RF-based system could be engineered and made to work well, providing location information at the granularity of a room, and ideally portions of rooms. This approach attempted to limit the coverage of an RF transmitter to define the granularity of a geographic space, using received signal strength to infer the best location. Despite many weeks of experimentation and significant tuning, this did not yield satisfactory results. This was mainly because RF propagation within buildings deviates heavily from empirical mathematical models. In its MIT Cricket environment, the corresponding signal behavior with the inexpensive, off-the-shelf radios was not reproducible across time. It was therefore decided to use a combination of RF and ultrasound hardware to enable a listener to determine the distance to beacons, from which the closest beacon can be more unambiguously inferred. This is achieved by measuring the one-way propagation time of the ultrasonic signals emitted by a beacon, taking advantage of the fact that the speed of sound in air (about 1.13 ft/msec at room temperature) is much smaller than the speed of light (RF) in air.

MSR RADAR

The Microsoft Research RADAR system is a wireless LAN-based indoor location determination system. The key highlights of RADAR are (1) it runs over existing off-the-shelf wireless LAN hardware; (2) it can estimate user locations to within a 2.3-m accuracy; and (3) it can adapt its performance to the dynamically fluctuating nature of the wireless network. Furthermore, applications like RADAR have influenced network and operating system designers to include wireless LAN-specific hooks in the implementation of their networking stack. For example, the NDIS 5.1 networking stack in Windows XP includes wireless LAN-specific management objects that can be accessed by applications that need to constantly monitor and control parameters about the wireless network.

A FRAMEWORK FOR DEVELOPING INDOOR LOCATION-BASED SERVICES

Technology is not the only thing used in developing a service. The development process also consists of creating relations to other organizations, and creating a framework for this relation. This is probably more work, and harder work, than creating the technical implementation of the service. If the service involves economical relationships, business agreements have to be set up.

The Location Interoperability Forum (LIF) Mobile Location Protocol (MLP) API and OpenLS APIs have solved a majority of the integration challenges that slowed down the outdoor LBS market in its shaping years. Today, developers usually have to create stand-alone applications, even if their services leverage data from sources outside the application. The applications are limited only by the external data that are leveraged in the development of services.

There are two possible business models that the location-based services market can follow: the closed-services model of the telecommunications market or the open-services market model of the Internet. The architectural implications of these services are very different.

In the outdoor world, the change to the open-architecture model was predominantly driven by past failures, successes, and trial and error, rather than by proactive innovation and creativity.

If the same services could be used indoors as well as outside, one of the major reasons for indoor location services would disappear. The main advantage to the technologies used for indoor location services is that in addition to cover where there is no coverage from outdoor systems, they can offer higher precision in the location. Which precision you need depends on the application and what type of information you are looking for. If you want to locate a doorway, there may not be more than two or three per city block, but there will be one per room in a corridor, for instance.

As an example of services, we will use an application at MIT, which works in both the indoor and outdoor domains. The scenario has two communication providers, one for each zone. For the indoor world zone, MIT provides the Wi-Fi network. For the outdoor world zone, T-mobile (as the mobile operator) provides location services using its GSM network. Either of the services could extend into the domain of the other. For example, the Wi-Fi network can be extended to the outdoor world. In that case, Wi-Fi hot spots would emerge that would enable communication access from the Wi-Fi network. Different business models are possible. The university could provide services for free to students and faculty, or a provider could charge for access to the services (as, for instance, Wi-Fi access at Starbucks, which, incidentally, is provided by T-Mobile). The access range from a Wi-Fi access point is about 300 m in open areas.

USING OPEN STANDARDS

Standardization has been important in creating and growing global markets for computing and communications systems, as well as for mobile telephony and the associated mobile location systems. However, a market force works against standardization in the fact that lack of standards creates opportunity for companies to earn profit and market share by branding a specific (proprietary) solution to a problem. Microsoft is the classic example, capturing the market with its Windows operating system, in essence becoming the standard while the industry was still too young to have developed a more democratic (industry consensus-based) solution. With its popular product and first-mover advantages, the company captured a large user base. By becoming a closed industry standard, the company prevented future competitive threats as competing similar applications are not compatible with the popular application developed by the company. If users of indoor location services want to avoid being caught in the same trap as the users of office software, they have to actively support open systems as part of setting up the systems.

Another market force on a smaller magnitude that is for standardization is a leading-edge user community where developers and content producers have a forum to innovate and introduce higher-value applications based on open interoperability standards. This user group can break down the cycle of debating whether the infrastructure is needed first or the "killer application."

In the U.S., Wi-Fi technologies are being deployed to provide a multitude of high-value services. PDAs have been successful as the handheld platform of choice, typically using Wi-Fi as the connectivity method. In the rest of the world, mobile telephones have become the leading handheld terminals. Currently, mobile telephone technology is being combined with Wi-Fi, e.g., by Motorola. The E911 regulation for emergency services (which mandates that a mobile telephone has to be automatically positioned within 50 m when calling the emergency number 911) dictates that mobile operators will have to be in the forefront on positioning technologies, a fact that is a driving force for location-based services. It is still not clear how this will impact Wi-Fi providers, both telecom operators and other companies, such as NYCWireless.

Another federal government initiative that could become a major driving force is Homeland Security, which can have an impact on the communication and positioning infrastructure, as sensors are installed in subway tunnels and other indoor spaces. It is possible that the impact can be as large as that on outdoor location systems by GPS.

Nevertheless, until Wi-Fi hot spots are widespread, and roaming agreements are in place between them, public Wi-Fi will continue to be more like the Internet cafe experience, where the users have to seek out access, than the cell phone experience, where it finds the user.

In order to address these issues and uncertainties connected with each infrastructure type and driving forces that may or may not have an impact, this chapter proposes a framework that offers a set of viewpoints and steps for generating information to help indoor location-based service designers choose between business strategy and service portfolio options. These steps do not necessarily follow a linear sequence of activities, but rather a set of information gathering and processing tasks that can be used flexibly to understand indoor LBS and mitigate against risks that may accrue from selecting an incorrect design option.

FOUR DIMENSIONS OF DEVELOPMENT

FRAMEWORK FOR ORGANIZING DEVELOPMENT TASKS

The indoor location-based services industry is currently in a ferment stage. There is a bottleneck effect slowing down the deployment of indoor LBS applications. While some amount of the current failure is a result of the immature infrastructure explored in previous chapters, it has become clear that the vision of technology and services is not the core problem. Instead, the prototyping and testing of envisioned services, concepts, early standards, and business models is.

Location-based services cannot be found just by cutting the wires of the existing Internet services. Instead, they must be designed from the ground up to gain added value from the inherent advantages of the infrastructure. Depending on the infrastructure types, different advantages can be realized. The problem, however, becomes one of generalizing the application to leverage multiple infrastructures. With respect to the communication and positioning infrastructures, these advantages include location sensitivity, context dependency, immediacy, and mobility, which emphasize the personal roles of the user. With respect to the mapping and software (service) infrastructures, the contents (i.e., pictures/MMS (camera phones), games, etc.) that users share with each other may well become the most significant driving force for the adoption of indoor LBS technologies.

Mobility also needs to be considered as a factor for the services infrastructure, with respect to seamless service offering and discovery. Service discovery is essential to indoor mobile users, providing the means to exploit services that are offered and to configure end devices automatically and to register on LBS applications with minimum (or zero) user intervention.

In developing services, the developer has to consider four dimensions:

1. Marketplace (what the user wants and needs)
2. Technical (infrastructure type, service development environments)
3. Organizational (internal and partners)
4. Economic (technology standard selection, business agreements)

In the process of developing a service, each of these four dimensions has to be analyzed and the different aspects taken into account. Any development project concerns much more than the pure technology selection and deployment; any technology is only as good as its surroundings — economical, organizational, and the marketplace for it. The four steps can be seen as a waterfall, but in practice, all four will be done simultaneously and interact during the development process.

Issues in the Marketplace Dimension: Finding out What the User Wants and Needs

As part of the marketplace dimension, the first step is to find out what the (potential) customers want. There are a few market surveys for outdoor location-based services that can be applicable for indoor location-based services. Much also depends on whether the system is developed to be sold as a packaged application, or as a custom development for a specific purpose. How you find out what the users want will vary significantly in the two cases.

In the custom development case, the system is developed for the purpose defined by the customer. This also determines the user group. It can be easily identified, interviewed about its expectations, and any constraints that may follow from the deployment to the specific user group can be identified. As part of the agreements by which the system is to be developed, there must be ample opportunity to conduct interviews and test it with the users.

If the system is to be sold as a package, the user group may not be as easily identified and approached. It may even be impossible to do so for commercial reasons, since that would destroy the potential market. In that case, there are other ways of finding out.

Apart from user characteristics, the marketplace dimension also captures the known social and behavioral implications of the service. Services that bring about fundamental behavioral changes in the way people live and work are usually referred to as social innovations. Very few systems can actually be called social innovations; the latest example is probably the World Wide Web. Location-based services in themselves will not be social innovations, but they will become the foundation for them.

Any process where you query users about their needs and desires includes uncertainties, such as the requirements definition due to the unknown user context, device diversity and ability to match user desires with technical capabilities (especially within the financial constraints given), delivering wrong value proposition, and the small mass market experience with indoor location-based services. The uncertain importance of privacy and security concerns and the widely varying estimates of future market size do not help.

This book is not about requirements management, so to manage the gathering and coordination of them, you have to find another source.

ISSUES IN THE TECHNICAL DIMENSION: DETERMINING INFRASTRUCTURE TYPE AND SERVICE DEVELOPMENT ENVIRONMENTS

The technical dimension captures the characteristics of the different infrastructure types: communication, positioning, mapping (location data), and software. For the communication infrastructure, the question it is about network-based vs. *ad hoc*-based types of service; the factors to be considered include coverage, bandwidth, cost structure, and ownership (e.g., network operator, business, or individual). These will affect the development of the system, since each of them will have an impact on what the most economic deployment is.

Measures of quality of service (QoS) in location services for communication systems include accuracy, periodicity, and response time as well as coverage area, availability, and deployment cost.

There are many positioning systems on the market, and most provide some kind of location capability. But different systems differ hugely in capability and cost. Ultimately it is the applications, and the benefits they provide in the short term and longer term, that count. The measure of a positioning system is the applications that it enables.

For instance, consider an application that sends a complete map with positions on it very frequently, vs. an application that updates only the position on a preloaded map rarely. You can either select the network and design the application to suit it, or select the method and select the network that fits. In the usual development process, the developer does not have much choice when it comes to the selection of the infrastructure; hence, it is not particularly useful to select protocols or their characteristics either. However, in the LBS case, you do have a choice. What is more, however, if you want the system to work seamlessly in different environments, you have to find ways to make it work in an equivalent fashion with each of the different networks and positioning methods you are going to use.

For the positioning infrastructure, the choice is first about whether to use only an indoor location system, and in that case which; having decided this, the choice is then about absolute vs. relative positioning systems. The parameters in the decision include coverage, accuracy, frequency of update, and positioning method. The location of a device can be determined either by the device itself or through the network. If the location determination is situated in the device, it has to communicate back its position to a server in the network for applications where several devices need to be coordinated; if the location determination is situated in the network, the location information can be made available to the server without communicating with the location-based application in the device. For example, radio frequency identification (RFID) systems may be the best choice when the area to be monitored is relatively small, and so can be instrumented with readers at moderate cost, or when the business requirement can be met by monitoring just a small number of key locations. For example, if the requirement is to know whether a given object is inside a space or not, only the entrances need to be monitored.

This distinction also impacts the ownership and use of the position information. The ownership of position information is a tricky legal issue, and the European Union has legislation that determines this. In the U.S., the situation is less clear, and in any case, agreements must be put in place to determine who actually owns the location information, and how it can be used. Here, the technical dimension interacts with the economical dimension. For wide-area roaming, the technologies can be divided into network based and handset based. Both have advantages and disadvantages, notably that handset-based technologies depend on the user having a suitable device, while network-based technologies depend on the user being within the limits of network coverage. The technologies that determine location are complex, especially the network-based technologies that perform very sophisticated interpolations of a handset's location based on the signals received at multiple base stations (e.g., APs). The good news for the application developer is that the technical complexity is hidden from the typical enterprise adopter with SCOTS systems like Ekahau.

For the mapping infrastructure the determining factor is what type of data model to select to model user location, as well as the features and content data of the indoor world: a symbolic or geometric data model, or a combination of both.

For the software and services infrastructure, the selection in the technical dimension is about simple or basic vs. location aware vs. navigational (or vs. context aware) types of services. It is also about the nature of the location-based service, which can be either positioning or navigation (something that in turn should be determined by the market dimension); it includes the servers, databases, and development platforms and characteristics (e.g., size, form factor, computing power, display size, battery life) of the different types of devices that can be used to deploy the service (PDAs, phones, and RF tags).

In terms of software, various models can be used in designing applications that do and do not require intercommunication, which include client–server and peer–peer. An application that does not need to communicate with other remote applications is considered to run locally. Most web applications are composed of two parts. The client runs one of the parts locally and communicates with the remote application, the server that runs the other part. Typically the client is more lightweight than the server in terms of its processing abilities. The brains of the application is typically part of the server. In many wireless LAN implementations, most of the clients are fairly lightweight and potentially will only have to measure signal strength values. The server would then calculate the location and display it to an administrator. Using a lightweight design for the client applications would allow the application to silently run as a background process and create the possibility of a client being monitored without user intervention. The third application development model, peer to peer, provides a client-to-client connection where the lines between a client and server are blurred. With this model the client and server are a single application, with all the connected clients having the same set of capabilities. A WAY implementation would be a type of peer-to-peer application. Also, there may be differences in the business models, and thus the deployment costs associated with different means of acquiring data. For example, handset-based hardware in the handset will usually be freely accessible to a local application, allowing a mobile client application to access the position data as frequently as desired and use it as it deems fit. In contrast, a network operator may require usage-based fees for access to network-based position data. In some cases, an application might not even be able to directly access location data. Instead, it may have to indirectly access mapping or navigation data from the network operator or its partner companies, paying a fee for each interaction. Directly accessing the location data may require a specific business relationship or partnership with the carrier itself.

In addition, there are a number of uncertainties in the technical dimension that have to be taken into account, such as the cost of the infrastructure (cost to build, or to rent or use); time to deployment; whether capabilities, reliability, and availability are proven; if the positioning capabilities and other infrastructure elements have been rolled out by communication infrastructure providers; if integration of communication networks (seamless communication handover) is possible; if integration of positioning types (seamless positioning (location) handover) is possible; if integration of mapping types (geometric + symbolic) is possible; and if integration within the software infrastructure (e.g., geographical databases, middleware, or legacy systems) is possible. Many of these interact directly with the economic dimension. They are not determined by the technology itself, but the agreements within which the technology is deployed.

The costs of RF tag location systems generally consist of three components: the tag costs, the infrastructure costs, and the installation costs. In terms of infrastructure costs, the IEEE 802.11 systems use an existing WLAN infrastructure and so could lead to very low incremental cost if the accuracy requirements could be achieved with existing infrastructure and the interference was not too great. However, in large buildings WLAN installations are likely to make use of leaky feeder antennas, which greatly reduce their value for location systems. For the other systems, cost is generally related to range from tag to receivers: the longer the range, the lower the cost. The Bluetooth and dedicated RF systems have indoor ranges of typically 50 to 100 m, while the ultrawide band (UWB) systems have shorter ranges because the higher operating frequency of these systems is more severely attenuated by building materials. Costs associated with Wi-Fi are minimal (if using existing hardware); medium for Bluetooth and dedicated RF; and relatively good for UWB.

Installation cost reflects the cost of installing each infrastructure element, and so it is generally proportional to the overall infrastructure cost. In terms of tag battery life, a long tag battery life is valuable, especially for applications such as asset tracking, when the tag is left in the field for an extended period. The maintenance of any location system is an often overlooked cost for long-term implementations. When a tag is attached to electrical equipment, a battery life of 6 months (or sometimes 1 year) can ensure that the battery is changed only when the equipment is periodically checked.

Proximity cards typically either are wholly passive or have a sealed battery with a life of many years that never requires changing during the lifetime of the tag. Wi-Fi and Bluetooth systems have relatively high battery consumption because the tags have to engage in protocols (IEEE 802.11 and IEEE 802.15.1, respectively) that offer relatively little support for power saving. Dedicated RF and UWB use various techniques to ensure long battery lifetime.

The following classes of properties for the assessment of sensors for the context element of location have been developed. The aim of the identification of these properties is to put criteria for the selection of suitable sensors at the disposal of location-aware applications' developers. They primarily assist in answering the question of which sensors to make use of in a particular application.

- Physical vs. symbolic location data: This criterion distinguishes between sensors that provide location data in the form of geographic coordinates and sensors providing symbolic location data.
- Absolute vs. relative location data: Absolute location data are based on a uniform representation system for all determined locations. In contrast to this there can be a separate reference system for every sensed location if the location data provided by a sensor are relative.
- Position-finding system part (infrastructure vs. tracked object): The process of determining an entity's location can be carried out by either the location detection infrastructure (e.g., a beacon) or the entity that is to be located itself (e.g., a GPS receiver).
- Accuracy and precision: A sensor's accuracy indicates the smallest area it can determine, while its precision gives information about the percentage of location detection processes achieving this accuracy.
- Scale: To assess the scale of a sensor, its covered area per unit of infrastructure as well as the number of locatable entities per unit of time are considered.
- Recognition: This property refers to a sensor's ability to recognize certain features of the tracked entity, such as its identity, color, shape, etc.
- Costs: There is a distinction between time costs, space costs, and capital costs a sensor causes. Time costs take a sensor's installation and administration into consideration, while space costs cover the infrastructure and hardware needed. The sensor's price and the incurred staff costs belong to capital costs.
- Limitations: Under this, property restrictions concerning a sensor's usability are subsumed as, for example, GPS sensors' limitation of being usable only outdoors.

ISSUES IN THE ORGANIZATIONAL DIMENSION: INTERNAL AND PARTNERS

The organizational dimension captures the internal competencies and resources of organizations planning to offer location-based services. It also captures the capabilities of partners necessary to offer location-based services. The term LBS is often used to describe a variety of positioning/location awareness technologies, although it is more appropriately used to describe the entire chain of technologies that turn raw position data into useful information, such as navigation directions or task efficiency. Such applications only begin with physical positions (i.e., coordinates); several more steps are required before such data can be incorporated into a decision that has business value.

The elements of the value chain subdivide the process into four basic groups. In principle, an enterprise could enter the chain at any point, from consuming raw location data (latitude, longitude, altitude) and developing custom applications on top of that data, all the way through to deploying a self-contained, off-the-shelf, remotely hosted application integrated with an existing back end. The decision about which components to build, which to buy, and which to access from a hosted service will depend on factors including the expertise of the enterprise, the degree of experience with LBS, and the size of deployment. Overall, uncertainties include the following:

1. Partners: Availability of partners with the right capabilities, competition from potential partners, incompatible strategies and visions, ability to reach acceptable revenue and cost-sharing arrangements.
2. Internal: Availability of relevant skill sets, ability for organization to communicate value propositions, ability to reach decision makers in target customer segments.

ISSUES IN THE ECONOMIC DIMENSION: TECHNOLOGY STANDARD SELECTION AND BUSINESS AGREEMENTS

The economic dimension captures the competitive landscape(s), privacy and security concerns, and, most importantly, technology standards.

The economic dimension of services development has two parts: the selection of the technical environment and the agreements that determine what the business model for the service will be.

As open-architecture approaches enable interoperability, this will allow for faster deployment of applications by the location information provider, service developer, or a third- party vendor. In order to achieve this interoperability, open standards such as the Open Mobile Alliance (OMA)/LIF MLP API and the OGIS OpenLS APIs are essential. In practice, these standards are the only examples of such standards that are implemented in systems on the market today.

Most developers of location-based applications have their own proprietary eXtensible Markup Language (XML) schemas. Closed architectures are a disadvantage for both telecom operators, who want to deploy services, and vendors of geographical information systems (GIS), who want to sell their Internet mapping technology to LBS application developers, who in turn build applications atop a GIS that they can then sell.

Uncertainties in the economic dimension include the possibility of economic growth, privacy and security regulations, commoditization of positioning capability via government-imposed interface standardization, legal liability for service failure, unanticipated social and behavioral impacts, unknown competitive threat from alternative technical solutions and other (possibly unknown) players, and complex value networks with many actors and diverse business models.

THREE STEPS TO LOCATION SERVICE DEVELOPMENT

Changes in one dimension can lead to changes in others. Technological innovations can make new services feasible (technology push), and lifestyle or organizational changes can create new opportunities (market pull). The deployment takes place in a value network, an interacting set of constraints on the actors that determine how the system can be deployed and what the business model will look like. These interactions create a complex environment for designing and managing all mobile services.

The design of indoor LBS presents a surprisingly complex dynamic environment for service design. This combined with the high level of uncertainty makes the design strategy formulation for these services a daunting challenge. Hence, this section on the steps and filters is dedicated as a guideline for managers in the basics of what is important to consider when developing indoor LBS. The last step deals with decision making regarding standards, what they are, why they are important, and what considerations are important when making the decision to adopt a standard.

Note that the paths are probably different, for example, for telecoms and niche players. Also, in either case, the contextual uncertainty and designers' understanding of the dimensions will inevitably change throughout implementation. There is a continued need to reassess the strategy in light of changes, market pressures, and uncertainties. Hence, designers should watch out for network externalities. Likewise, test beds and full-scale operational experience should be taken into consideration to support planning and execution of subsequent phases.

Development of indoor location-based services can be organized into four steps (with a number of substeps), which when applied will both create the appropriate environment and select the appropriate technology for the task at hand. For each step, a set of questions can be formulated — the filter — which allows us to better determine what is appropriate for this particular application and how it should be implemented.

STEP 1: DETERMINING THE BEST NICHE APPLICATION (MARKETPLACE DIMENSION)

The first step is to determine who are the potential users and what do they expect from the indoor location-based application under consideration, such as the Product Finder and Buddy Finder. First hand knowledge of customer needs gained from market surveys, focus groups, or service trials can expand the knowledge of social, economic, and technical constraints on services. If the service is developed as part of a consultancy agreement, this step should already have been taken by the orderer and be expressed as requirements. Surprisingly often, that is not the case, and additional analysis of the potential users may have to be done using the methods described here.

There is currently a number of practitioner or academic literature that helps in the design of location-based services based on what customers want, or in the definition of a coherent strategy to pursue location-based service opportunities based on infrastructural capability analysis. Anthropological methods, requirements analysis, and a number of other tools that are typically found in the usability or requirements analyst toolbox can be applied here.

Filter A: Filtering by Size of Opportunity

This filter seeks to weed out less promising applications by selecting the largest and easiest opportunities first, i.e., the "low-hanging fruit." This filter essentially provides high-level market and technical feasibility studies.

In the first filter, the applications with the largest revenue potential are selected. The analysis required to estimate revenue potential varies by the nature of the service, but includes a sizing of the target (niche) markets (e.g., industrial sector or consumer demographic), an assessment of the value proposition for these segments, and an analysis of their willingness and the ability of the customers to pay.

Network externalities should be considered where appropriate. Likely social and behavioral impacts (i.e., privacy) on the service use can help the designer understand the social constraints on the feasible problem, and thereby identify both commercially attractive applications and those that are not likely to succeed. Comparing the Buddy Finder to the Product Finder application, for example, privacy might be the main issue for the Buddy Finder application since its location-aware characteristic makes the control of the user's position data crucial to the service's adoption rate. In contrast, the Product Finder application can be designed as a simple or basic location-based service, not requiring the positioning and tracking of users.

The applications with the highest revenue potential are passed to the next stage for deeper analysis. The rejected applications are not completely cast aside, as the service designer should attempt to identify viable applications that can be integrated together that benefit from economies of scope. It makes sense to look at varying service bundles that target the same niche market and that use the same infrastructure types. For example, both the Product Finder and the Buddy Finder applications can function based on symbolic location (i.e., "Product Z is in Store ABC," "Bob is in the mall"), which means that the same positioning infrastructure (a local positioning system that provides relative positioning as opposed to absolute) can be leveraged.

Another option is a service bundle that leverages the service provider's strengths in some other way (i.e., provider offers complementary service, like Mobile 411 for the end customer, or in the case of a self-contained company, it offers end-to-end solutions for a location-based application developer).

In the case where the system is ordered by another company, and development is done under contract by the location-based system developer, there are other constraints. The revenue opportunity may not be a consideration, but the customer and user needs become more important. Having clear requirements becomes an absolute must; otherwise, the project is doomed to failure.

STEP 2: DETERMINING THE INFRASTRUCTURE TYPES (INFRASTRUCTURE DIMENSION)

The second step is to determine whether the service should start off as simple (no local positioning) or location aware. Determining a user's location might be a minor value-add to the core service (e.g., the MapQuest.com service offers directions to restaurants and other information without automatic position determination). In these cases, the positioning capability may be so marginal that it could be excluded, making the implementation easier.

The user's needs will affect the infrastructure directly, and hence it becomes important to carry over the requirements from step 1 to this step. Requirements determination is concerned with a move in its own solution space (i.e., existing resources such as the technical capabilities and infrastructures of a service provider). For example, if the existing communication provider has a Wi-Fi network, the solution space would entail leveraging this resource to enable a positioning infrastructure that is based on Wi-Fi cell ID positioning (either relative or absolute). The anomaly space, or the requirement space, on the other hand, contains all the potential problems that could conceivably be addressed for different stakeholders.

In the context of location-based service design, the large size of the solution space and the considerable uncertainty around its several dimensions make the discovery of the "determining an attractive" solution problematic. Overall, the concepts and requirements for application/service can come from a review of competitive offerings and plans, from a review of practitioner and academic literature, or from brainstorming by various groups from within the mobile industry or attractive customer segments. It becomes very important to form a deep understanding of the true needs of the users, so that this knowledge can be used to develop functional requirements, which in turn can be used to determine what the appropriate infrastructure (in terms of both positioning system and other equipment) may be.

Infrastructure Requirements and Potential Use of Standards

In the Buddy Finder application, seamless handover for both communication and positioning (location) from the mobile network or global positioning system is a requirement. That is, when entering the mall, the system automatically switches over to the indoor positioning system. The positioning accuracy for this use case is building scale. Note, however, that a Buddy Finder application, for example, might also cause a location trigger if the users (friends) are in the same room. This implies that positioning needs to be represented (modeled) on several scales to make the application/service scalable.

In terms of the mapping infrastructure, note that all of these location trigger-based applications do not really require a map to be displayed on the mobile device. However, location still needs to be represented (modeled) somehow. Here, the location is symbolic (i.e., "inside the mall"), which is linked to positioning that is relative of nature. Also, the modeling language requires not only a location component, but also event and time components. These requirements are seen when the user enters a conference room. The event here is that the user is walking from one zone (hallway) to another zone (room). The time component can be used as a restriction for the Buddy Finder service, to notify the user of his or her friends that are in the vicinity only between 1 and 4 P.M. on Fridays, for example. The syntax for such a *location + event + temporal* language could be prototyped as the following:

When/if *<object>* + is/are *<relationship>* *<location>* + *<temporal>* then *<action>* +

where *<relationship>* is "entering," "leaving," "in," "alone in," etc., and where *<location>* is one or more offices, conference rooms, open areas, or hallways (the selection of multiple locations is interpreted as a disjunction), or exactly one of the following special cases: "anywhere" or "nowhere."

In terms of the software (service) infrastructure, these types of trigger services (proactive) are required to push a response targeted at the user when his or her location changes. For this to work, an event service (server) is needed to enable this dynamic activity generation. Location-based events allow a user to register to receive alerts based on specific business-relevant events, for example, alerting an application when a user approaches a given destination or leaves a specific region. An event model is particularly attractive if location data are provided by an externally hosted service that charges per use. Instead of repeatedly querying the external service to check the user's location, the application simply waits to be notified when the relevant event occurs. The event service (ES) is a messaging service with the purpose of notifying mobile applications of events that are or will be occurring. Most LBS applications only offer location information to users, and potentially basic tour-type content and navigational directions. The ES will allow the presentation of more real-time, asynchronous information. The ES would pass messages to the client without the client's explicit request. ES will most likely be of the push type (publish–subscribe service). Standards have yet to appear in this market, making integration more complex, but in the future this appears to be a compelling candidate for the proposed web services eventing (WS eventing) mechanism. Web services eventing is a proposed standard mechanism unifying many kinds of event-based notifications between processes (http://msdn.microsoft.com/webservices/understanding/ specs/default. aspx?pull=/library/en-us/dnglobspec/html/ws-eventing.asp). Messages could be of different types and priorities; the client would react in different ways depending on the type and priority of the message. In addition, events could be based on context information such as time. For example, a message could be sent at noon reminding users of dining areas on campus for lunch. Events could also be based on location. This would require external communication between the event server and location server. The messages sent could carry URL information, linking to additional information regarding the event. This information could exist on external servers, or locally, on the client, in case of network inaccessibility.

The concept of tailored special offers from stores nearby also applies here, where stores could compete for customers with last-minute deals. Two constraints have to be noted, however. First, there have to be servers that capture the events and trigger the service based on them. And second, there are significant user privacy concerns.

Also, a virtual infrastructure can exist where a mobile user would use "phone-marks" to store a content of interest (i.e., a product) for later use (i.e., mobile commerce) on the web. Detailed georeferencing and the application of phonemarks opens the possibility for querying objects such as products in a shopping window or a painting in a museum. YDreams' FluidShopping was designed to reduce the anxieties of shoppers through the use of Internet-enabled mobile phones, helping find the product they want to buy and enabling its purchase after hours, from the shopping window.

In terms of the positioning infrastructure, positioning aids such as spatial code bar (SBC) or emitters are needed around a building. These have been used in various experimental settings, for instance, the Colombo Shopping Center in Lisbon, Portugal. Using Wi-Fi or Bluetooth access points (emitters) as anchors will allow users to automatically pinpoint their location without having to read or write any SBC.

Moreover, locating a user based on Wi-Fi cell ID might be adequate for navigating the user to a store, but might not be precise enough for a shopping experience inside the store. More precise positioning technologies could enable navigation to specific items within a store's aisle.

The mapping infrastructure is used for exploration and navigation of a place, which is usually associated with the use of symbolic representations, such as maps (of a shopping mall), that are then matched against the reality they try to symbolize. This matching is facilitated by the use of anchors (points of interest) and paths displayed in maps that are easily recognizable locations. These maps and their features (i.e., anchors, paths) can be modeled using either symbolic representations or geometric features.

If the service is using a geometric map, the user may optionally display distances along segments of the route. If the service is based on a symbolic model (spatial relationships are based on either associations, containment, or proximity), the directions will not be given precisely as to how many feet before making a right turn. In either case, geometric or symbolic, the directions should probably be given by means of symbolic reasoning (i.e., "When you reach store A, make a right turn," instead of "After 10 ft, make a right turn"), as people associate with this type of reasoning better. In the case of the Segway or an automatic wheelchair, directions need to be given in terms of geometric reasoning since the machine is not capable of symbolic reasoning.

Overall, indoor navigation is going to be more often the desire than the need until an infrastructure of positioning/navigating aids (i.e., anchors) is in place.

In terms of the software (service) infrastructure, the user may invoke a location-based concierge application that would determine his or her location, and he or she would subsequently receive navigation/routing details about getting to the destination. Optionally, the service is able to specify some waypoints (or these points have be specified by the user), and the user may specify route determination criteria. The criteria might be fastest, shortest, etc., and can also specify the preferred mode of transport of the user (elevator vs. staircase). The details of the returned information might include directions to the site and other relevant information (according to the user's profile, for instance). The route can also be optionally stored on the terminal or application server. The user may store it for as long as needed, thus requiring the means to also fetch a stored route. Regardless of how the endpoints and waypoints are established, the information is then sent to a route server that calculates the route.

Applying Open Geospatial Consortium's (OGC) OpenLS services, a subscriber would seek navigation advice from the service provider via a personal navigation service. This is an application service that utilizes OpenLS Core Services (gateway, directory, geocoder, presentation, and route determination).

Filter B: Filtering by Ease of Implementation

After the application and type of service have been determined, the second filtering stage involves a high- level assessment of the infrastructure and organizational dimensions, specifically the technical and organizational capabilities necessary to offer the applications. This corresponds to developing a hypothetical future solution space for each possible application, and thereby determining requirements for leveraging and transitioning from the current solution space. These requirements and related capabilities give an overall indication of the level of implementation complexity and what service requirements (the ones determined in step 2) can be achieved.

LBS functional requirements detail how the current solution space can be transformed into a future solution space. While there will be many possible internal and partnering solutions to providing the necessary capabilities, the filtering decision can be normally based on the simplest option. The nonfunctional requirements for the placement are defined by the constraints on the future requirements. For example, the LBS type switches from basic or simple to location aware. Or, the location-aware service requires a switch-over from relative positioning to absolute positioning to enable precise directions. The functional capability for location management can be broken down into a location positioning capability (positioning infrastructure) and location database (software (services) infrastructure) requirements.

Many requirements are dictated by the business context or consumer business model in which the location positioning capability is embedded. To support a wide range of applications, the service designer must also consider standardized interfaces to software components (i.e., databases, GIS servers, etc.) and build a capability for organizational and individual customers to instantiate their own location-sensitive databases and describe their own business logic.

Also, location-based services can also generate high volumes of data. Therefore, in addition to the network infrastructure to be used, service designers must carefully consider the type and quality of user interactions to avoid information overload, and to make sure the user gets a seamless service. The key is to understand the context of usage. The level of complexity faced by a service designer in implementing the service depends on internal and partner (i.e., service integrator) capabilities. Gaps in implementation capabilities may include service portal and personalization capability, availability and interoperability of software components, content/context management, and transaction processing.

In the first pass through the framework the service designer should probably focus on the simplest and easiest applications within a single market as a starting point. In subsequent passes, more challenging services or bundles of smaller services can be chosen for in-depth analysis. The requirements for location positioning capabilities also vary, something that may imply very different sets of functional and nonfunctional mappings. A positioning capability is important for many applications, but may be as simple as "in the room"/"not in the room." In this case, a relative LPS is sufficient. Symbolic reasoning (context?) aware processing is also often needed to provide useful descriptions of location and for navigation (i.e., "When you reach the ATM in the middle of the hall, Store ABC will be to your right").

For example, if step 2 for the Product Finder application determined that the service type is to be simple, the service designer can draw a conclusion that no LPS is necessary. However, the user's location can be determined in two ways: (1) user selects his or her location (i.e., shopping mall) based on an aggregation of the place (i.e., shopping mall) or (2) user enters his or her location in terms of an address, phone number, etc., for a particular place (i.e., store in a shopping mall).

For a location-aware type of an LBS, the service designer knows that a local positioning system is required to position (and track) the mobile user. The next step is to determine if this positioning is to be absolute or relative. Either a geometric or symbolic location data model is then needed for querying, storing, and modeling the user's position.

STEP 3: SERVICE PORTFOLIO AND STRATEGY SELECTION (ORGANIZATIONAL DIMENSION)

Having selected the most promising applications and the service type, as well as using the ease of an implementation filter to determine the infrastructure types to deploy, the service designer must decide on the overall development strategy associated with those services and infrastructures. It is important to note that this decision is carried out in a background of the uncertainties outlined above for each dimension and due to conflicting strategies and decisions of other players (e.g., partners, competitors, and regulators), uncertain market demand, instability of partnerships, and switch-over of technical feasibility.

Filter C: Filtering by Organizational Driving Forces

The risk and return for potential services are derived for each scenario from an understanding of the crucial driving forces behind the attractiveness of services, and the possible business relations that the network of actors behind a service implies. For each potential service and scenario combination, the designer needs to design a strategy that considers the following factors:

- Adoption rate
- Development of product offering stages (i.e., enhancing the simple or basic into a location-aware service, or service bundling with external service providers)

- Perspective on the complete LBS value chain:
  - Proportion of value chain that can be locked in
  - Commitment to the different types of infrastructure
- Level of commitment to reducing the sources of uncertainty surrounding the marketplace dimension (e.g., test beds) and infrastructure dimension (e.g., standards)

Analysis of strategic alternatives over a range of scenarios provides insight into the most important sources of uncertainty, and the mitigating actions most likely to increase return and reduce risk. Attractive services are likely to leverage existing infrastructure investments, internal and partner capabilities, and partnership arrangements.

Actions to reduce risk include the retention of flexibility where uncertainty is greatest, such as by making incremental investments (e.g., cost position, customer and partner relationships) that preserve a "right to play," and merely having an agreement in place.

The importance of IT standards has increased over the past 10 to 20 years. Increasingly, IT managers must make decisions regarding the adoption of standards for IT development, specifically about the particular standards that relate to current development projects. After reading about this filtering stage, an IT manager should have a solid grounding in the risks, factors, and challenges involved in evaluating the appropriateness of specific standards for specific development projects.

All standards are not created equal. On one end of the spectrum are the highly formalized, carefully crafted standards created by democratic international bodies such as the International Standards Organization (ISO). On the other end lie market-driven *de facto* standards, such as Windows, which derive their influence from widespread adoption. In between are standards developed by consortia like the OGC, which seek to maximize the benefits and minimize the detriments/disadvantages associated with both formal and *de facto* standards development processes. It is crucial to find out who is responsible for the creation of a standard, how they have gone about creating it, and what potential traps or benefits exist because of the creation process. The same procedure should be applied when selecting the products used in the service.

With this basic information the IT manager is ready to begin making the difficult decision about whether to adopt a given standard. The manager must look at the criteria for successful standards:

- Technical quality
- Timeliness
- Effectiveness
- Widespread adoption

In addition, many other factors must be weighed, including:

- The need/desirability for compatibility or interoperability with other systems

- The existence or cost of acquiring the necessary skills and expertise needed to implement the standard
- The strategic importance of adoption (product compatibility)
- The stability of the standard
- The consequences of not adopting the standard

In some cases, making this decision will be relatively painless, as in choosing a target operating system or web browser for a new system. On the other hand, choosing a programming language may pose more of a challenge. The size and influence of a company may also have an impact on the decision.

The following sections describe the main factors to be considered when assessing the pros and cons of using a standard.

COST-RELATED ISSUES

Using a system based on standards is typically cheaper than using a proprietary system. This is because the standards development is an open process, and the documentation of the standards becomes public and can be used by anyone. The trade-off is that there is no way to be certain of what knowledge the developer has, as there is no formal certification — something that is usually true for proprietary products.

There is also no certainty of the integration between the products implementing the standards. In a proprietary system, the vendor takes this responsibility; in a standards-based system, there are interfaces between components, and these are where the responsibility of the vendor ends.

TIMELINESS

A second major issue in the use of standards is the timeliness, with respect to market cycles, with which standards are developed. The dominant perception here is that consortia standards like the one of LIF and OGC are fast. Timeliness only becomes an issue when one is considering the creation of a new standard. In that case, managers need to make the decision about whether speed is more important than quality. In high-risk projects, it is crucial to build upon standards of the highest quality, since in the event of failure this helps to reduce liability damages.

LEGITIMACY

For all practical purposes, IT managers nowadays are free to adopt standards regardless of their origin. The legitimacy of standards is first tested in the market. At the same time, it is very important to understand that all standards are not created equally, and now more than ever managers must be careful to research the origins of any given standard and evaluate the likelihood of market success before deciding to adopt. Liabilities still stay with the vendor of the product used, but for open-source software, *caveat emptor* (buyer beware) probably applies since there is no responsible party.

Making the Decision

Now that we have a background in the standards process, this section will provide some realistic guidelines for deciding which standards to include in development projects. What are the elements of that decision?

There are a number of classes of considerations to be made. The importance of each will vary greatly depending on the particular development project in the particular organization. Those considerations are:

- Quality of the proposed standard (technical quality, timeliness, effectiveness, widespread adoption)
- Type/usage of the standard
- Compatibility/interoperability issues relating to one's current and future IT architecture
- Strategic importance
- Personnel/expertise issues
- Stability of the standard
- Consequences of nonadoption
- Capacity to participate in standards development or to wage a standards war

We will briefly describe each of these considerations and leave it up to managers to apply them as appropriate to their own organizations. There can be four qualities that describe good standards:

1. *High technical quality* of the standard is the first element to be examined. Hidden in this requirement is that the IT manager, or someone on the staff, must be competent to evaluate whether the specification is of high technical quality. Naturally, a standard that is not of high technical quality will result in end products that have flaws or other undesirable qualities.

2. *Timeliness* is the second quality of a good standard. Earlier we spent a good deal of discussion on this issue. Generally, the strength of the need for a standard dictates the speed with which it will be developed. Recently, consortia such as World Wide Web Consortium (W3C) have been spending considerable energy on predicting and even directing the future need for standards and planning to develop them in time to meet the demand. The hidden trap here is not to go with standards that exist before their time. Companies may build products to fit a standard whose need is never realized and be left with a warehouse full of gadgets no one wants to use.

3. *Effectiveness* is the third quality of a good standard. Effectiveness is different from technical quality in that a specification may be of high technical quality and still not effectively address the core problem — i.e., it is not an effective solution. Many standards exist that address similar issues or problems. IT managers must carefully choose the standard that is most appropriate to the given design problem. With respect to the LIF

and OGC standards, there is not much (if any) overlap with other standards from different standard bodies.

4. *Widespread adoption* is perhaps the most crucial success factor for a standard. Certainly this is easy to evaluate for established standards, but IT managers must beware of emerging standards that have not yet reached maturity. With respect to the Open GIS Web Map Server (WMS) standard, there are now hundreds of software products that are WMS compliant.

Using these four criteria, IT managers can make a good preliminary evaluation of a potential standard. Provided the standard meets these criteria, the next question is its appropriateness to the development project at hand.

How the standard will be used is very important. In our case, this consideration deals with interoperability and compatibility. There are numerous areas that need to be considered, such as backward compatibility with legacy systems, compatibility with the products the user currently owns, and forward compatibility with products a company and others will produce.

Especially in the area of information systems, interoperability with very different systems is increasingly an issue. Primarily the issue of compatibility can be seen in terms of switching costs: How much will it cost the company to switch to a new standard? Will users of the product also be willing to switch? Is there a general switching trend going on in the industry? Will the new standard lower switching costs in a way that can help or hurt the company?

The open-source movement and standards such as languages built on XML are bringing increasing focus on universal compatibility between radically different devices. The cost of adopting a standard might be prohibitively high for one compatibility reason or another, but in these times the costs need to be weighed carefully against the cost of *not* adopting the standard. There are more than compatibility issues at stake here, so these costs will be discussed last.

The other consideration is strategic and includes issues previously discussed, such as switching costs for users of the products. This can be done by asking five simple questions:

1. Can IT build barriers to entry into a given market?
2. Can IT build (or reduce) switching costs from your product to a competitor's?
3. Can IT change the basis of competition?
4. Can IT change the balance of power in supplier–consumer relations?
5. Can IT generate new products?

By their very nature and purpose, standards have four effects:

1. Reduce barriers to entry into a given market
2. Reduce the switching costs between competing products
3. Change the balance of power
4. Generate new product ideas

For example, standards in the frequencies at which remote control devices operate allow competing vendors to provide remote controls that will work with their competitors' products (e.g., TVs/VCRs), reducing the cost of switching to the consumer. Perhaps question 3 is the most crucial of all. Standards change the basis of competition. No longer is the base technology in question; the real challenge now becomes to focus on what features differentiate the product from its competitors. The same applies to software standards, such as the ones of LIF and OGC.

The existence of a standard reduces design and production time, but that in turn should free the design team to work on improvements, which also in turn may become standard some day. The real business profit lies in this marginal period lasting from when new and distinguishing features appear until they become standards themselves. Of course, such strategic considerations will apply more or less to different products, but they are important all the same. For example, when developing the Product Finder application, it is not necessary to make the service location aware, as the simple or basic version in most cases will be as effective. However, a good strategic approach is to consider that location awareness might become the norm or standard in all future LBS applications. As a result, adopting LIF's and OGC's standards early on might make sense. The same reasoning applies to joining the consortia.

Other factors can be addressed by the following question: Do you have the people with the experience and expertise necessary to utilize the standard? (If not, do you have the resources to develop them?) This is a basic question that applies to all the skills necessary to develop a product.

The next to last question to consider is extremely important: What will be the consequences of not adopting a given standard? Failure to adopt some standards can mean loss of profit, as noncompliance will result in the application's incapability to integrate with other applications. Therefore, IT managers need to work to make sure they are aware of the applicable standards in all cases, and to understand what will be the consequences of not adopting those standards.

The last issue we will discuss covers situations where either no standard exists or current standards are out of date or unsatisfactory for one reason or another. In these situations the IT manager has several options: develop a proprietary solution to the problem without regard for how others are approaching the problem, join/create an industry consortium to tackle the problem, or submit a request to a standard body (i.e., ISO) to handle the problem. Creating a proprietary solution is another possible approach. One example is the Japanese company ImaHima, which developed the Buddy Finder application, based on the Japanese proprietary location technologies. ImaHima was successful in locking in customers, which allowed the company to take advantage of a closed, proprietary standard to exploit the benefits of network externalities. Considering market expansion and opening up its standard, ImaHima might have considered submitting its standard to a standards body with the goal of making it an industry-accepted standard.

How to Use the Dimensions

The dimensions described earlier are general in nature. They could be used to describe the environment and options for a wide range of applications. In this section we describe the impacts of mobility and location awareness on service adoption from a number of different perspectives. These cover social and behavioral impacts of LBS, as well as user privacy and security concerns.

The social and behavioral impacts of LBS deal with both the positive and negative consequences of LBS and their implications for service adoption and diffusion. Current technology acceptance models, such as TAM, have many deficiencies for the range of contexts in which LBS are deployed. For example, current acceptance models cannot explain the stark differences in the uptakes of WAP and i-mode service in Europe/U.S. and Japan, respectively. In contrast, it is important to integrate the perspectives of the user as a consumer, a network member, and a technology user. This approach highlights the need to use specific technology adoption models for LBS at the individual level, but also raises the need to develop multiperspective frameworks for understanding adoption behaviors at the group and organization levels.

In addition, more research is needed to explore how the factors influencing LBS adoption differ by level of analysis in the consumer market (e.g., families and nonwork organizations) and how LBS adoption in the business segment impacts use in individuals' private lives, and vice versa. Services designers need to be able to understand the complex needs of all stakeholders in the overall social system, of which the LBS is only a part. There may be mobility implications that an LBS is not particularly well suited to address with its current capabilities. For example, while users frequently ask the OnStar service for directions to a particular store in a mall, the service can currently only provide the mall's street address.

Privacy concerns may be particularly sensitive as services allow colleagues, family members, or others to have real-time information on the location of individuals. Both privacy and security concerns could create resistance to LBS adoption. At the same time, a positioning capability is often used to increase security (e.g., tracking of children). In these applications location information provides a compelling value proposition. In both of these examples privacy and security are only maintained if access to the location information is restricted to authorized users. How location information will be managed when the positioning capability becomes ubiquitous or seamless across all the world (networks) is still uncertain.

In some applications, location and positioning are minor value-adds to the core service. In these cases the positioning capability may be so marginal that it could be excluded or made optional to counter privacy concerns. In short, applying these diverse perspectives on emerging indoor LBS applications provides insights into the types of users that are likely to adopt LBS, and major adoption barriers.

The dimensions, filters, and constraints discussed in this chapter should give the IT manager and service designer the background necessary to make the most appropriate decisions regarding the overall service deployment strategy. Each choice has its own set of costs, risks, rewards, and benefits. The manager needs to consider the resources available to the company and weigh them against the costs involved with each process. Following these guidelines should also allow the IT manager to make the most appropriate choice when it comes to evaluating standards for adoption in development projects.

REFERENCES

AT&T wireless, Find People Nearby, available at http://www.attwireless.com/personal/features/organization/findfriends.jhtml.

Campbell, B., Rosenberg, J., Schulzrinne, H., Huitema, C., and Gurle, D., SIP Extension for Instant Messaging, paper presented at IEEE Network Working Group RFC 3428, December 2002.

Dey, A., Futakawa, M., Salber, D., and Abowd, G., The Conference Assistant: combining context-awareness with wearable computing, in *Proceedings of the 3rd International Symposium on Wearable Computers (ISWC '99)*, San Francisco, October 1999, pp. 21–28.

Haeberlen, A., Flannery, E., Ladd, A.M., Rudys, A., Wallach, D.S., and Kavraki, L.E., Practical robust localization over large-scale 802.11 wireless networks, in *Proceedings of ACM MobiCom '04*, Philadelphia, PA, September 2004.

KDDI. GPS MAP, A Location Service for Mobile Phones, available at http://www.kddi.com/english/corporate/news_release/archive/2002/0718/.

NetLab, Wireless Information Delivery Environment, http://netlab.boun.edu.tr/~wide, June 2003.

NTT DoCoMo, iArea: Location Based Services, available at http://www.nttdocomo.com/corebiz/imode/services/iarea.html.

3 Infrastructure

THE INFRASTRUCTURE COMPONENTS

The infrastructure consists of the sensor system, the network, the software, the servers, which are the environment for the execution of the services (which often are large, central nodes, but in principle can be seen as terminals), and the terminals, which are used to provide the location-based services to the users (Figure 3.1). In this chapter, we will discuss the terminals, the servers, and the software. In the next chapter, we will cover sensor types and navigation algorithms.

The Role of the Infrastructure

In infrastructure-based systems, like Nexus, TEA, or the Context Toolkit, a specialized context infrastructure serves as a central access point for applications and sensors. These types of networks have nodes that are able to compute their physical location. The nodes compute their location by using ranging technologies (i.e., ultrasound) and location-sensing techniques (i.e., triangulation, proximity). The nodes can thus be used to locate objects. The advantages of a network node knowing its own position, and sharing this information with others, are becoming more and more evident as routing algorithms are becoming smarter and mobile-specific applications are being introduced at the user level. The infrastructure consists of the sensor system, the network, the software, and the terminals, which are used to provide the location-based services.

Applications access the infrastructure to retrieve context information. Sensors are linked to the infrastructure to provide it with their sensor information (which includes location, but may also include other types of information, such as temperature, light level, etc.). Using such an infrastructure allows applications to access context information that has been captured by sensors far away from their current position, but requires mobile devices and sensors to be connected to the infrastructure, as all communication takes place through the infrastructure. This can become costly, e.g., in terms of energy usage.

Context-aware computing relies on an available infrastructure that has a globally accessible data repository. This gives a well-known access point with a consistent information model, which enables distributed applications to interact with their environment according to the model and their location. Mobile nodes in such systems require access to the infrastructure in order to access stored model data.

The role of the infrastructure is crucial when deploying location-based applications. There are four types of infrastructure: the sensors (i.e., the ranging and triangulation devices), the communications network (through which the sensors provide their information to other parts of the system), the terminals and servers (i.e., the hardware on which the system runs), and the software (service) infrastruc-

65

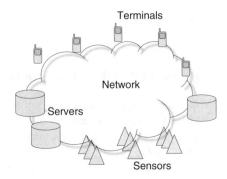

FIGURE 3.1 Conceptual layout of the system infrastructure.

ture. This might be the most important since it is not tied to the user directly. More specifically, this infrastructure determines what the user sees on his mobile device and how he or she interacts with it.

Most current applications are based on proprietary technologies, closed architectures, and full systems. The assumption is that everything should be bought from one vendor. While a good start, these stand-alone (stovepipe or silo) systems are unlikely to have a large impact on the marketplace, primarily because of the heterogeneity stemming from lack of interoperability, which prevents application integrations and service chaining. This means that there is no way for an application to share, access, or control the sensing resources without knowing the sensor and the network specifications. Moreover, there is no consensus on common standards for designing these applications, and as a result, there is no common software platform on which to build such applications that would enable easier deployment (Figure 3.2).

The reverse is already true in wide-area location-based service systems, where a standard application programming interface (API) to access location information exists, as well as standards for the representation of position in data. However, as with the outdoor world of location-based services, the stand-alone application approach currently present in the indoor world is likely to change as existing and

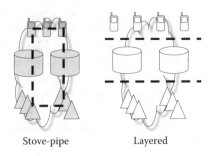

Stove-pipe   Layered

FIGURE 3.2 Stovepipe vs. layered system design.

emerging standards enable web proxies and servers to get location information directly from the indoor positioning system, in the same way as location-based systems in the wide area receive information from the cellular network. Moreover, technological driving forces such as web services will enable easier integration of the various software components and result in interoperability. There are two ways of providing location information: either computed by the network or provided by the individual nodes to the infrastructure (essentially reporting their own position). In such cases, the location information of the nodes has to be provided to the infrastructure, and thus they can retrieve location-dependent information. As a result, mobile devices have to obtain their geographic position. This implies mobile nodes with certain capabilities with respect to their communication, as well as computing power and sensor inputs (e.g., radio frequency identification (RFID) reader) for location determination.

An example of such an approach is the Nexus project, where the mobile device on which the application is running has a wireless connection to the infrastructure, e.g., 802.11 or GPRS. The mobile device also has some means to determine its current position, e.g., global positioning system (GPS) outdoors or an infrared-based system indoors (manual positioning by the user is another possibility). Given such an environment, an application can make use of an infrastructure-based platform such as the Nexus platform that provides context information.

THE SERVER AND TERMINAL COMPONENTS

Today, it is a foregone conclusion that the terminal and servers should be standard. The part of the system that processes information and presents it to the user would be uneconomical to build separately. That said, there are still a few choices that a systems designer has to consider.

The choices are fewest when it comes to the server. It has to be fast enough and have enough memory to execute the software provided. But today this can be solved with any off-the-shelf server computer. The main problem will be the cost of ownership, but the maintenance cost of a Linux computer today is lower than that of a Windows NT computer (due to less downtime after an admittedly more complex configuration), so the choice in practice is easy.

When it comes to terminals, the choice is somewhat more complex. There is no single criterion that determines which type of terminal to use. The alternatives include Windows-based PDAs, dedicated handheld computers, mobile phones with wireless local-area network (WLAN) functionality, and other types of terminals as well, since it is likely that in the future the terminal for the indoor location system will not be presenting information to a human user, but to a computer of some other type, for instance, a robot.

However, given the assumption that the information has to be presented to a person, all terminals have their drawbacks. The only common thing is an 802.11 interface and a screen with a web browser — probably with Java — so basing the user interaction on this is by far the easiest way of designing the system. It is the software infrastructure that will be the complicated part.

THE SOFTWARE INFRASTRUCTURE

The software infrastructure consists of three parts: the client component, the application server or middleware component, and the database component. In addition, the content can be regarded as a separate piece of software, since it — even though it will be part of the database — will require separate development. In the rest of this chapter, we will discuss each of these components and how they relate (Figure 3.3).

THE CLIENT COMPONENT

There are a number of issues on how to represent location information to the user. In the near future, the user may not be human, and its needs for location information may be dependent on the applications (e.g., a cleaning robot with navigation capabilities).

Other issues connected with the client side are the small screen, limited keyboard, memory, processing capacity, and limited bandwidth, which put a premium on anything heavy on network traffic or user interaction. This applies equally to WLAN-based systems, especially those with a variable peak load. We will discuss this further in a later section of this chapter.

In general, maps are the best means of depicting location-based information, and are hence an essential element of any location-based system. But while it is the overall best way, it is not necessarily the best way for all applications. Presenting mobile maps is important in the context of location-based systems, as the ability to display mapping information on mobile devices such as cell phones or handheld computers is limited. But alternative ways can be used; for instance, the location-based information may simply be text (i.e., point of interest), images, map images (i.e., area of interest), etc. And as we already mentioned, users may not be able to understand visual information, needing a stream of coordinate information to navigate, e.g., in the case of robots.

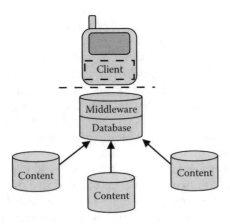

FIGURE 3.3 The software architecture.

The major concern is the limited display capability of mobile devices. Apart from the limited map features that can be displayed, the speed of data transmission to mobile devices is also slow in comparison to a wired network.

Mobile mapping requires standards that allow data content to be easily transferred and displayed across the wireless Internet to any one of a large variety of mobile devices. For displaying location-based data on the standard Internet, Scalable Vector Graphics (SVG) and Geographic Markup Language (GML) are two important standards. The data in these formats are delivered across the Internet as eXtensible Markup Language (XML).

To support legacy devices, you need to convert to their presentation formats. For example, the Oracle 9i Application Server Wireless Edition (Oracle 9iAS Wireless) converts content to XML and transforms the XML to any markup language supported by any device (HTML, WML, HDML, VoiceXML, VoxML, SMS, etc.). This may be required, since the previous generation of devices spoke a different wireless protocol and supported a variety of different wireless markup languages. For example, WAP-enabled cell phones support the Wireless Markup Language (WML). On the other hand, Palm Operating System devices support Tagged Text Markup Language (TTML), and voice-activated Internet applications support the VoiceXML and VoxML markup languages.

Since web-enabled devices are converging toward XHTML and CSS, and SVG is being implemented, this should soon be less of a problem except for legacy devices (which do not implement location systems anyway). In this book, we will assume that the development is done for terminals that have at least support for WAP 2, which implies XHTML and CSS, and probably SVG in the user interface.

Another general characteristic or requirement of local positioning systems is the reasonably high update rates, happening in real time. In general, one or more location updates per second are required for many location-based service (LBS) applications. This has implications in several ways: in terms of traffic and in terms of power consumption (since effect is radiated out through the antenna, which will mean that power has to be provided from the batteries). One of the keys to power management is to intelligently vary the update rates of individual objects. A fast-moving object may need to have its location updated multiple times a second, whereas a stationary one does not. Certain objects may need more frequent updates than others. An important aspect of a practical local positioning system is being able to intelligently vary update rates of individual clients in a dynamic fashion.

THE MIDDLEWARE COMPONENT

Between the database and the terminal is where the middleware is located. The origin of this term was a piece of software that delivered content through a web server, but the system has far outgrown this in complexity. The central task of the middleware is to deal with data management. Application servers are situated between the communication network, the databases, and the end user. These servers are essentially connection points for the different components of the software infrastructure. The most basic application server is the web server, using Hypertext Transfer Protocol (HTTP) to call the MPC and retrieve the position information. In a web environment, the application server requests a position by issuing an HTTP POST request toward the location server. The query is invoked by sending the request using HTTP POST to a URI, which is used to transport the data. XML is used to encapsulate the requests and responses and have functions to report the quality of the position. The web server usually acts as the front end to the application server, using HTTP and a markup language such as XHTML, HTML, WML, or SVG to deliver the presentation.

In addition to incorporating web servers, application servers also provide database interfaces (using CORBA, XML, EJB, JDBC, or ODBC interfaces). This means that a query by a user can combine multiple data requests, such as the subscriber's location, content information of multiple kinds, and mapping information of any kind. Some application servers provide for personalization beyond the use of position information, as well as the transcoding (i.e., using XSLT) of information. We will not discuss this in any depth, although it is clear that personalization services and location-based services are converging.

Moreover, the platform should consist of a middleware that will determine the positioning mechanism and the LPS that is appropriate to serve the location request. In a Wi-Fi network, for example, the gateway might be deployed in an environment where various local positioning systems and architectures might coexist, offering different types of service in terms of accuracy (relative vs. absolute positioning).

Another important point is to introduce abstraction of the components within the platform. It is not quite relevant to see them as layers, but the separation and interaction between them will be of the same kind as within a layered system. The middleware is a software platform that acts as a central system by gathering, integrating, and transmitting data between a variety of different components, including the actual positioning technologies, the network, database servers, billing and service management systems, geographic information system (GIS) servers, the end user terminal, and the actual location-based applications (which can be seen as a piece of logic that leverages all the services available). All of this is accomplished while implementing privacy rules that ensure proper and authorized use of the end user's location information.

In some legislatures (e.g., Europe), it is a requirement that consumers should be able to customize the settings and triggers of the services they subscribe to. For example, they may want to define trigger zones (location based) or set times (temporal based) when they want to be alerted of events and times when they do not. They will also want to determine who gets alerted to their presence.

The position information can be derived from the positioning infrastructure by using an API such as the LIF MLP API. The application programming interface (API) is a library of standardized functions that give application developers easy access to the system. For server-based applications, the Mobile Location Protocol (MLP) provides the most well-supported and well-deployed API; for terminal-based applications, JSR 179 provides such an API. However, these are lacking in that they do not provide quality-of-service parameters that reflect the diverse nature of mobile location data in today's wireless networks, or a flexible privacy model for supporting a wide range of service requirements.

One possibility is to build the location-based system as an event-driven instead of a query-driven system. An example of this is the Spatially Indexed Resource Identification and Tracking (SPIRIT) system, which was an event-driven middleware developed at AT&T Laboratories, Cambridge. The SPIRIT project designed and implemented a three-tier high-level software architecture over the raw coordinate positions generated by their proprietary Active Bat system. The software performs the various mechanics of two-dimensional spatial data handling and frees the application programmer from repeated reimplementation of such. Although the design goals included the ability to reason about a wide range of sensor technologies, the software authors admit they never came close to realizing this objective and SPIRIT is closely coupled to Active Bat. The most significant outcome of the project was the Programming with Space API. However, it is still not standardized, and can be regarded as more or less experimental.

One issue in this type of service is the tracking of users. Alert-based LBS involves continuously tracking mobile users. These services require dynamic time tracking of users along a complex location model. In addition, the moving users have relationships (or triggers) to each other, other moving points, and static objects (i.e., "Let me know when I am close to a friend"). Users may also have complex relationships with these points or other users. For example, in a date-match finder service, a date has certain characteristics (age, education, hobbies, etc.). Adding more subscribers causes an exponential growth in processing required to manage these services. This computational intensity will also escalate, and since the requests cannot be moved in time to any great extent, the system must be scaled to handle peak loads from the start.

Ideally, a platform like this should be able to connect to external information sources like the Internet using wireless communication and allow for seamless handover between different location-based systems. This handover can take place at different levels: communication between different types of networks (e.g., GPRS and Wi-Fi), and location information (e.g., from a WLAN to an RFID system, or a wide-area system). The platform needs to integrate the different LPS and hide their heterogeneity with respect to positioning methods.

Similar to the cellular network's approach with the GMLC/MPC standardized gateway for location servers, the indoor sensor network should have its own standardized component (the WLAN Location Center (WLC)) to hide the heterogenous functions of the indoor positioning methods and architectures (i.e., 802.11 Wi-Fi, Bluetooth, RFID, or other methods) that may occur. The WLC gateway, along with a LIF-like API, would unify a framework for retrieving position data of users from the LPS (as well as GPS and cellular networks when considering multiworlds and seamless handovers) (Figure 3.4).

The application server essentially works as a gateway between the MPC (or the WLC) and the mobile device. In theory, the mobile devices can connect to the application server by using any protocol: HTTP, DNS, SMTP, Session Initiation Protocol (SIP), or File Transfer Protocol (FTP). The application server will function as a gateway toward the LPS. With the LIF API, the application server interfaces to the mobile cellular network of the mobile operator (for network-based positioning). The API is used to call the position information, which means that developers do

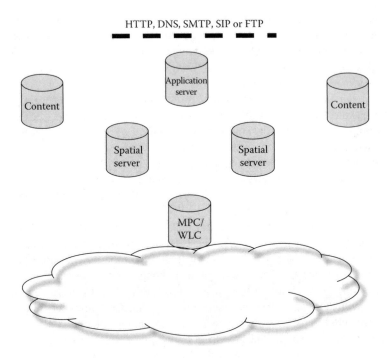

FIGURE 3.4 Middleware layout.

not need to know the back-end implementation details of the operator (nor do they have to write applications that interface directly to the system of the operator). Querying the proper MPC or WLC retrieves the user location; this may then use MLP, Parlay, or any other protocol to communicate with the location system.

THE DATABASE COMPONENT

Another part of the middleware is the spatial (GIS) servers, which provide data about static or mobile objects and users, and upon a request, return their location. They also provide the necessary geoprocessing functionalities (i.e., shortest-path calculation). Each GIS server stores information about spatial objects within a particular area (in essence, using the database component). For example, there could be a GIS server that stores two- and three-dimensional shapes of buildings. Distributed servers collaborate to provide a unified spatial view through a defined API to applications. In addition, it is important that the servers are able to describe locations at different levels of precision and scale.

A special kind of GIS server is a route server that computes a path through a path network, given two or more positions. A typical application of the route server is the ability to calculate and display the best or shortest route between two specified points on a path network. However, for the indoor world (where routes may include different floors), the calculation of a most efficient route will require different algorithms than in the wide-area environment.

The data flow is managed differently depending on the application server provider and architecture, and how it is deployed. Sometimes the application server is housed in a central location, like an Internet Service Provider (ISP), the service providers can house their applications in it, and it takes care of the interfaces to the network. Some of the network operators who have deployed LBS see themselves as ASPs; others plan to sell the data to companies that want to provide services. The services would then be provided in the same way as web service today — by companies that essentially are publishers but run their own infrastructure.

For an efficient communication within a network, the amount of data that has to be transferred between the web client and application servers must be minimized. The efficiency of data transmission can be increased by means of caching or hoarding techniques, and algorithms that split the processing in various ways. For this reason, the application server needs to preserve the states of its location-aware clients. To minimize location signaling overheads and to lower power consumption for mobile devices, the location cache would keep the location information of the recently tracked user. This may have implications on system design, e.g., determining the freshness of a session. The Nibble system, for example, introduced a predefined refresh period, which imposes lower bounds on the time-to-respond requirement.

To achieve the scalability necessary for a large-scale deployment of the platform, the GIS servers can be organized in a hierarchical fashion, similar to that of the Globe location system for software objects. Leaf servers in this hierarchy are responsible for managing the position and registration information for the mobile objects inside their disjunctive service areas, while the higher-level servers are responsible for forwarding queries and handovers.

As a fast processing of queries and especially updates concerning location information is of great importance, the location information can be managed in a special main memory data structure based on a quad tree, while the registration information is stored in a traditional database. The volatile position information, which will be out of date after a server failure anyway, can be recovered from the mobile objects.

The positioning information that is obtained from the local positioning systems is managed within databases, from which the application server will access the data for various location-based services. Databases are also used to handle the content and other information that is to be provided to the users. There are several types of database systems that can be used here, but relational database management systems (RDBMS) are generally used to store location data.

There are two problems with this: The first is that an RDBMS is designed only for transactions involving comparatively simple data types, such as characters and numeric data. Location data, especially data based on the geometric location model, are usually complex objects that require more than one data structure to describe them and their spatial relationships (i.e., topology).

The second problem is that the RDBMS inherently assumes that the data are located, or should be reached, from one single location. However, when there are multiple instances of data that may appear and disappear, having a central location to handle data becomes unsustainable. Just like the web does not really lend itself to a relational model, so too the distributed location data space.

Alternatives to RDBMS exist, notably in the form of object-oriented databases and XML databases (which may be adapted RDBMS). A hybrid is object-oriented. RDBMS merges the object-oriented management system which allows the storage of complex data as objects, and the relational database management system to offer the ability to manage the relationships between objects.

Another matter is querying the location space. In case there is a central engine that manages queries and tracks distributed data, the Structure Query Language (e.g., SQL2 or SQL3) can support all database management operations, as well as object-oriented data modeling. These enable users to store and manage complex location data, as an object, along with data from other sources, such as computer-aided design (CAD) and images in the same database. More importantly, ORDBMS allows spatial analysis to be performed in the database server using SQL commands instead of in the application. Oracle 8 Spatial and ESRI ArcSDE are examples of spatial databases that store geometric objects as abstract data types (ADTs), a user-defined data type, in feature-based tables, within the RDBMS.

Due to their wide deployment, it is likely that many, if not most, location-dependent applications will rely on RDBMS. However, there will undoubtedly be quite a few (and perhaps the most novel) applications that do not use RDBMS, but some alternate system.

From a GIS perspective, location-based services do not include many complex spatial analyses. However, it is the nature, completeness, and accuracy of the database content that impacts on the quality of the subsequent location-based service. For a certain service area, the database must include all the appropriate features, such as hallways and points of interest (POIs), e.g., ATMs, restaurants, stores, etc. In addition, digital maps of the area are needed. These can be a portfolio of raster maps (images), a vector map that can be created on the fly when requested, or archived photographs. All hallways and points of interest (and appropriate labels) must be shown and be georeferenced so that their locations on the map are correct. Nonetheless, the spatial data analysis might require geometric functions involving the computation of distance, area, volume, and directions.

Another feature of location-based systems is their real-time or near-real-time nature. Queries have to be served before the user moves out of the area, which, depending on the means of the locomotion, the accuracy of the system, and the size of the area, may mean within less than a few milliseconds. In addition, this will increase the traffic volume. As more devices and applications become available, the increase in location-sensitive data requests will skyrocket. Communication infra-structure providers (i.e., wireless carriers or wide-area network operators, Wi-Fi network providers) will need to support this increasing number of moving subscribers, which might be well beyond the amount of 20,000 customers per processor that a traditional relational database management system (RDBMS) scales to. Database vendors obviously have great experience in handling high volumes of short trans-actions, so the trend toward more real-time applications is likely to reinforce the growing role that the major database vendors have been playing in the GIS market. However, this trend also counteracts the trend toward RDBMS, since the real-time nature of queries makes them harder to serve using an RDBMS (which typically has an inherent latency in serving requests).

The traditional approach to analyzing real-time data is to continuously load the new information into a database management system and repeatedly run queries against the data, but there is significant overhead in indexing such dynamic results. In-memory spatial databases, more feasible now than a few years ago with the falling price of memory and the emergence of 64-bit processors, have started to emerge. But new approaches to handle queries are also emerging, such as the Apama method, which indexes the queries using the analytical model. Such queries can change over time, but they are much less dynamic than the data. Incoming data can be efficiently matched against relevant queries and acted on as appropriate. As update rates grow to thousands per second, potential niches for new technology open. This includes new query methods, main memory databases, and fully distributed systems. A further consequence is that the traditional query methods may not be relevant, since queries have to be served faster than a central SQL processor could handle.

THE CONTENT COMPONENT

System designers tend to forget that a database is just a container for content. Creating content that can be leveraged in location-based systems is not a matter of mechanical translation of blueprints; information has to be added, and added in such a way that it can be managed in the database system to create services that serve up different content in different locations.

There are two tool sets that are geared toward the content industry. To some extent they overlap the database component, but they are developed and handled independently. These are the content creation and content (or asset) management tools.

Content creation tools enable content creators to design content that can be made differently available at different locations. Traditional (if one can speak of tradition in an industry less than 10 years old) content creation tools are geared toward the formatting of content, possibly with an added component of device independency (or, more typically, device dependency, as it were). Here, content is formatted for use with specific devices, although this tends to be a process where designers overestimate the need and possibility of control.

Content creation tools are most often geared toward presenting web content. However, an important area of content creation is nonweb content, where content creation is geared toward content systems that vary the presentation and the content itself with the actions of the user (such as games). In many ways, location-dependent systems are more similar to games than web pages (and indeed, there are a few location-dependent games on the market).

Content management tools are geared toward the management of content objects, i.e., text, images, multimedia clips, and so on. The primary purpose is to connect the database storing the objects with the display system (typically the World Wide Web). But the content management system also makes sure that the objects are provided or retired according to their settings, e.g., whether they are fresh. In some sense, the content management system is a database application; in other ways, it overlaps the database orthogonally. Important to note is that the database in a content management system does not have to be an RDBMS, although that is typically what is used; instead, it can be a distributed system, an object store, or anything that allows the management of content.

One aspect of content management is mapping between different representations. A local positioning system providing an absolute position can usually be augmented to provide corresponding symbolic location information with additional information. For example, a web client application can access a separate database that contains the positions and geometric service regions of other objects to provide applications with symbolic information. Applications can thus link the physical position to determine a range of symbolic information, like getting to know the closest printer to the current position.

COMMUNICATION INFRASTRUCTURE

THE NETWORK COMPONENT

Deploying a positioning system that works well indoors is a challenge, because signals are reflected off walls, floors, and ceilings, which tend to confuse sensors. In addition, the coverage has to be complete in the areas where positioning is needed. Where existing networks can be leveraged, this will speed up deployment and simplify the creation of services. However, it is possible that the coverage of these networks needs to be complemented with additional base stations to give full coverage with full positioning detail in all the areas where positioning is required.

The proliferation of lightweight, portable computing devices and high-speed wireless local-area networks has enabled users to remain connected while moving about inside buildings. Because indoor settings have the disadvantage of absorbing and diffusing the radio frequencies of GPS and cellular network systems (i.e., GSM, CDMA/CDMA2000, and UMTS), their positioning mechanisms (i.e., TOF, OTR, AOA, etc.) are not appropriate to provide the location of a user inside buildings, at least not in cases where indoor base stations are not deployed. Nor is the resolution typically good enough from these systems. Usually, it is in the range of 10 m (25 to 30 ft).

Accuracy requirements for location-based applications are much higher. To pinpoint a shelf inside where a product is located, accuracy to within 30 cm (a foot or so) is required, specifically as (x, y) coordinates (absolute positioning). To achieve this level of accuracy, either a dedicated positioning system, such as the MIT Cricket, or a location function in a local-area network can be deployed to achieve this accuracy level.

The role of the network can be to provide the sensor system, such as where base stations are used to triangulate the terminal. But the network is also the communications interface between terminals and servers. In many environments of interest, like shopping malls, schools, convention centers, hospitals, etc., the communication infrastructure already exists to provide data-networking capability to mobile terminals. Many such locations are equipped with 802.11 standards-based wireless LANs. However, it is not necessarily sufficient just to have a hot spot in an area to deploy location-based services. The system requires a number of additions. These include, for instance, functions to compute positions based on the information from the network (either cell information, radio fingerprint, or triangulation in the same way as with wide-area networks). It is also likely that coverage has to be increased to cover all places where location-based services will be required. After the deployment of the other infrastructure types that are essential for indoor location-based services, such services will complement this data-networking capability of, for instance, wireless LANs (802.11 family of standards). This, in turn, will add value to such a network. Due to its wide deployment, WLAN is a strong candidate to handle the positioning centrally in an indoor system.

There are two main scenarios for indoor location-based services: use only within the network, or use with the wide-area network as well. In terms of seamless communication infrastructure, roaming between access points is already supported, and Wi-Fi networks, for example, can be extended to create "clouds of connectivity" inside the so-called hot spot (i.e., locations with high connection frequency, such as an office building). How to connect the hot spots to the wide-area network is not yet clear, however. Standardization is ongoing in 3GPP, the standards body that works on these issues, and it is likely that this will be solved during 2005, with products emerging in 2006.

There are two main philosophies of organizing the sensor network communication: either over a separate network or over the existing air and backbone network. For practical and economic reasons, it is likely that the second case will be the primarily deployed case. And for most purposes, this implies an 802.11-type network. This has implications on deployment. But it means that as a developer, you must be able to at least roughly model the traffic, to make sure that the network does not get congested just at the moment the information is needed.

Finally, there are two main ways to build a location system: passive or active. They can be combined, e.g., as in assisted GPS, but here we will look at the differences between them.

In passive systems, the location data are publicly offered and the terminal takes care of the processing to interpolate location. GPS is an example of a passive system. Satellites provide a timing signal and a known position; the terminal has to calculate its own position. Architectural challenges associated with passive systems include optimum formats for data, synchronization and broadcast intervals, and logical signal differentiation. From the perspective of limited devices there is also the problem of antenna effect and signal frequency, which will affect battery life. A passive positioning system provides an open infrastructure, anonymously broadcasting location information. Any privacy issues will be dependent on the terminal. A passive system, consisting, e.g., of RFID tags, does not communicate actively with the access points. Instead, it merely transmits back information when requested (through a radio trigger).

The active system uses terminals that actively communicate with the access points. This requires power, which has to be available for communication at the intervals prescribed by the system. This communication may need to occur anyway (as in the case of mobile telephone systems).

Location system implementations generally use one or more techniques to locate objects, people, or both. Most of the methods used to define a position are based on geometry computations such as triangulation (by measuring the bearings of an object from fixed points) and trilateration (by measuring the distance). In addition, scene analysis, proximity (detecting physical proximity), monitoring wireless cellular access points, and observing automatic ID systems can also be applied.

COMMUNICATION AMONG THE COMPONENTS

From a communications protocol perspective, LBS applications can be categorized as push or pull, which are two approaches to make the mobile device (or the service and network) aware of its current location. This assumes that the client is requesting the information, either explicitly or as part of a previously set up session. In a pull application, processing activity occurs only when it is initiated by the client. An example might be an application to find a nearby ATM where a user with a mobile device sends a query to a spatial server to return a map (or direction, or both) of the resulting ATM locations. The terminal may then add its location to the map. The location can be computed from a local database, or it may be requested (separately or as part of the query for the ATM information) from the location server (Figure 3.5).

In contrast, a push application will send potential customers a text message with a special discount offer when they are close to a particular store. For this type of application to occur, the positions of all users need to be updated frequently and automatically. This may be a real-time service, which may need to handle thousands of spatial updates per second. There is a trade-off between predictive applications (which poll the user's position less frequently, but instead compute his location from a direction vector) and applications that leverage the user's location. In cases where the terminal is computing the location, the system may easily be overloaded with location updates, but if the database can query a location server (which may be on the same local-area network), the traffic over the air will be negligible. Note that this (as usual in the telecoms world) assumes that the bandwidth in the fixed system is several magnitudes larger than that of the wireless system (which has been true so far, and which does not seem to change) (Figure 3.6).

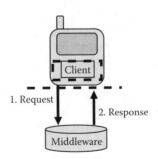

FIGURE 3.5 Pull protocol.

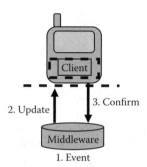

FIGURE 3.6 Push protocol. Note that the confirmation is optional.

Push applications can also be triggered by a state change in the location database where the client has registered, e.g., in the Buddy Finder case, where you register to get a notification when your friends move into a local area.

Note that the push and pull may occur at different levels. A system may be push or pull on the application level, and it may be push or pull on the location determination level (i.e., either changes in the location of the user are dynamically calculated and pushed to the positioning system and previously subscribed receivers, or the position is calculated when requested).

Other databases involved in transactions to compute the location of a user can be the user and security databases. A user database stores the information that is relative to the registered subscriber. Each user is assigned a unique ID. Just like with the cellular network's Home Location Register (HLR) and Visitor Location Register (VLR) databases, users can be classified as visitors or home (permanent) users. Moreover, this database is used for accounting and charging purposes, including post- and prepaid options. A security database holds all the required information that enables authentication and security to enforce the policy for the platform (WLC gateway).

If Mobile IP is used, it will automatically provide the visitor and home database functionalities, but it may not be an optimal (or even appropriate) protocol for the systems envisioned (among other things for performance reasons, since the additional traffic load from Mobile IP can be quite high).

An additional consideration, in both the push and pull cases, is that the user must be able to control the information retrieved from the system. The user must be able to determine his privacy, i.e., whether another user is allowed to retrieve the information requested. This can be done by using white lists or black lists, or by some more advanced means, such as the privacy-checking protocol used in the OMA MLP.

It is an advantage if the system uses open interoperability standards like the OSA/Parlay API and the LIF MLP API. The reason this is an advantage is that it makes testing of different systems (including clients against servers, terminals against base stations, etc.) against each other much easier. In many cases, test protocols exist that can be implemented or adapted for indoor location use.

Depending on the application and the way the system is deployed, the communications requirements are very different. In a system where the base stations communicate with the server through a backbone network, and the server computes the position, very little traffic (only the registration of a terminal with a base station) will be going through the network. In systems where the terminals themselves have to determine their location and communicate it to the server (and receive additional information from the server, such as the position of other terminals — for instance, in the Buddy Finder application), the requirements of communications over the air network will be much higher. This might mean that it becomes congested if a very large number of terminals are trying to position themselves at the same time, especially if there is a lot of other traffic going over the network. Much depends on how the application is structured. Tracking a list of member family/friends in order to be alerted if they are in the vicinity would require the system to determine every 5 to 15 min if the wandering event has occurred. This may work well with one single user in the system, but it does not scale, as the requests for information will escalate exponentially when other users are added to the system. This can quickly overload the communication infrastructure.

SERVICE DISCOVERY AND METADATA

Service discovery is essential to indoor mobile users, providing the means to exploit services that are offered, to configure end devices automatically, and to register in location-based applications with the minimum (or zero) user intervention. The mobile environment must also support ways of changing the collection of services available to the user as the user moves. Service discovery implies both finding the position information source and finding resources that can be used with positioned information. It also implies managing the authentication and authorization to use the services, although this is not necessarily a part of the discovery protocol.

Depending on the assumptions of how the system will work, different models can be applied. In a system where it is assumed that the network can appear and disappear with short notice, or be reorganized with an entirely different set of resources, for instance, as a new service node enters the system, it is necessary to provide for a distributed system of discovery. However, if we are assuming a relatively stable and permanent communications network underlying the positioning system, and that the servers providing the services do not move around (in theory, this could be possible, but so far both network constraints and computational constraints would force us to assume that the servers are not mobile, but probably placed on a backbone network in a computer room), this will have very different applications from the *ad hoc* networking case — luckily, making for a system that is much easier to build with the technology we have available today.

One assumption in a system with fixed servers and well-known access points is that the system can provide a directory, to which services can register and which can be used by clients to discover which services are available in the local area. The widest deployed technology for managing directory information and queries today is the LDAP model. An alternative model is to use the web services model, which has a looser association between the resource and the directory. In either case, the

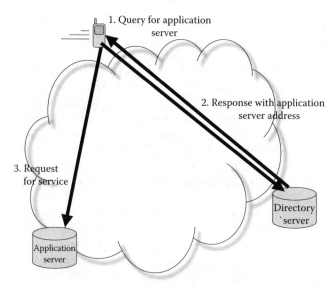

FIGURE 3.7 Service discovery with directory database.

end user should not be responsible for the configuration of which directory (or subdirectory) to query for the information; this should be handled automatically by the system.

The service discovery process will have multiple steps, from the perspectives of both the client and the server providing a service (thus becoming a resource).

In the client case, the mobile application must discover the existence of available services when it enters an area. A special case is when new services become available when a client is in an area and already has discovered the services available (Figure 3.7).

One solution is to use service discovery protocols, such as SLP, UPnP, Salutation, and Universal Description, Discovery, and Integration (UDDI). They attempt to move up from purely network-based addressing to higher-level descriptions of resources. They provide a bootstrap mechanism enabling the dynamic, spontaneous hookup of services and devices in ubiquitous computer environments. They use, in a centralized or distributed fashion, a generalized lookup service that may build upon and subsume the more specialized naming, trading, or directory services provided by underlying middleware and protocols.

Another solution is distributed directories, which also apply in the context of service directories. Distributed naming and directory services, such as the Internet Domain Name Service (DNS) or the X.500 directory service, offer scalable and fault-tolerant design to provide directory service to a very large number of clients. Mapping names to addresses is not much different from mapping names to locations. However, there are several constraints. Hierarchical directory services such as DNS do not cope very well with frequent updates. Nor is it extensible, even if there are several experimental RFCs describing how to include position records in the service records. LDAP (which is based on X.500) provides a more flexible system to manage directories, but is — despite the name — a quite heavy-weight protocol. As a result, the requirement for real-time information delivery is not met by those designs. On the other hand, service trading, a special kind of directory function, is required to cope with functionality changing information.

MIT Cricket, for example, handles the first case by allowing applications running on mobile and static nodes to learn about services in their vicinity via its Floorplan ActiveMap application that is sent from a map server application. It interacts with services by constructing queries for services at a required location. Once location information is obtained, services advertise themselves to a resource discovery service such as the MIT Intentional Naming System (INS), IETF Service Location Protocol, Berkeley Service Discovery Service, or Sun's Jini discovery service. Another example of such service discovery is Microsoft InConcert's lookup capabilities in the MS EasyLiving system.

Next, the mobile application must determine the newly found service's capabilities. Descriptions of services in the EasyLiving system are accomplished using a simple, open XML schema. In addition to ease of use, XML was chosen for two reasons. First, Extended Stylesheet Language — Transformations (XSLT) provides the ability to translate XML documents into multiple layouts. Second, it is straightforward to transform an XML-encoded description of a command into the XML-encoded command to be sent to the service. The service description schema is designed to support queries about available commands and their legal values. Additionally, the commands are associated with human-readable tags. While not a complete solution, this is a first step toward the automatic generation of user interfaces for different modalities.

As the number of services and catalogs available in an environment grow, there will be an increasing need for more sophisticated search engine-like tools that can consolidate, organize, and present information retrieved from various sources. Such tools may also provide interfaces through which users can pick services they need. Such tools can dispatch the users' requests to a variety of available catalogs, and then allow users to sort the results according to different criteria (e.g., store location, price of product, quality, or provider). As such, these tools will probably be similar to popular online price comparison sites (e.g., metaprices.com or simon.com), which allow users to pick a category of items to compare (e.g., CDs, books, electronics) and then return a list of items along with their prices, availability, special offers, and reviews from various online shopping websites.

An alternative is distributed queries. As the database grows in complexity and becomes more distributed, it is possible that it should be queried not by an SQL-like query to a central system, but a semantic query directed at the resource descriptions themselves — Google instead of DB2. An example of such a system is the Augmented World Querying Language (AWQL), an XML-based query language wrapped in a SOAP message and sent using the Hypertext Transfer Protocol (HTTP). An HTTP server passes the request to the AWQL servlet, which first uses an XML parser to extract the query from the SOAP message. The AWQL request is then passed to the AWQL parser component that ensures the validity of the inquiry and converts it to a sensor request. The sensor request is passed to the Sensor Worker component, which is able to access the sensors needed for answering the request. The AWML composer receives the sensor readings from the Sensor Worker and returns an AWML message wrapped in a SOAP envelope to the servlet. The servlet then passes the response to the requesting server.

Several alternative approaches exist, notably those based on the Resource Description Framework (RDF) from W3C.

In a web services model, application developers build location-based applications by using a suite of enabling technologies and APIs. They then publish these applications back to the directory system, and the provider of this then makes the applications available for subscriber use. This enables client applications to perform spatial queries without knowledge about the respective servers.

The indoor world can learn from the outdoor world and start applying the accepted standards for the appropriate software components as an early adopter. Of course, incremental growth needs to be considered and standards need to be studied first to see how they fit to specific needs. Trying to do too much coordination and utilizing standards to achieve interoperability too soon might not be economical (using cumbersome database management packages, handling many users or too much detail, worrying about editing and updating, trying to keep all the evolving interface standards in synch, etc.)

However, the service and network providers do recognize that sooner or later they would like to use their services in other settings. So, they have an interest in standards and integration actions that facilitate coherence and long-term goals while being easy to do and inexpensive (in time and money and performance) in the short run. In short, the indoor world might decide to apply the outdoor world standards, try to modify them, or design new ones. Of course, the preference would be to apply them.

Furthermore, as the market for indoor location-based services develops, many companies may want to deploy multiple vertical applications on top of a common software infrastructure. For example, a hospital may want applications for security, office productivity, and intelligent building infrastructure, in addition to specific health care applications. It is important that the overall platform can handle multiple applications developed by different vendors or organizations, running concurrently.

EXAMPLE: CRICKETS IN THE MIT FACILITIES ENVIRONMENT

MIT Crickets is an example of not only a position-sensing system, as we will see in the next chapter. It is also an example of a service discovery system, with its own directory system, maintained by the MIT Facitilites group, which is the building maintenance office of the university.

MIT Facilities has developed a mapping web site. The web site can map not only the building locations, but also the rooms and floors that have Wi-Fi coverage. The software architecture is set up in a way that users could enter their (x, y) coordinates and be displayed on the map in a fashion similar to that of Wi-Fi coverage mapping. Ideally, users would want the location-based service to locate them automatically without having to enter any input. The mapping web site could be utilized on mobile devices and made into a location-based service using MIT Cricket for automatic location positioning. However, MIT Cricket would need to be adopted, if it has not been already, to locate users according to the floor. Users would then be able to request such a location-based service to locate them and map the nearest bathroom location, for example. This service could also navigate the user to the bathroom of his or her choice, which would require either a route map result or text directions result.

MIT Cricket also has its own mapping software application, Floorplan ActiveMap, for a mobile device. As the user moves in a building, the navigation software running on the mobile device uses the listener API to update its current position. Then, by sending this information securely to a map server, it can obtain updates to the map displayed to the user.

In addition, mobile devices learn about services in their vicinity via the floor plan application that is sent from a map server application, and interact with services by constructing queries for services at a required location. Services appear as icons on the map that are a function of the user's current location. The services themselves learn their location information using their own listener devices, avoiding the need for any per-node configuration.

The relationship and interests between the players and why the need for possible coordination and deployment of standards are as follows. MIT Facilities has a campus-wide focus and current facilities management needs, so they have no current need for MIT Cricket-scale locations/activities. MIT Cricket, being currently a laboratory project, does not need to expand its deployment beyond their building to experiment with their location-based service ideas. Similarly to MIT Facilities, MIT IS also has a campus-wide focus and has no real motive (or business incentive) to deploy standards in order to ease coordination with the other two players.

Trying to do too much coordination too soon will slow down either project (i.e., using cumbersome database management packages, handling many users or too much detail, worrying about editing and updating, trying to keep all the evolving interface standards in synch, etc.).

However, MIT Facilities recognized that, for example, a common coordinate system is essential for doing facility management, especially in an environment such as MIT's, where one building is actually a series of 14 separate interconnected CAD plans that do not necessarily know where they lie on a local/global spatial reference system. For this reason, MIT Facilities invested in figuring out what the correct orientation is for each floor plan (i.e., a world file) to allow the construction of a room location model based on real-world coordinates. This is very important because MIT Facilities can now do horizontal and vertical adjacency studies not allowed with the CAD files as they were, and MIT Facilities can now overlay seemingly dissimilar data sets and perform analyses not possible before. A simple example for this would be to know where the closest fire hydrant is to a particular lab with a particular use. This is querying architectural information with something that is typically a civil/survey question. Even though this is not directly related to indoor location-based applications, MIT Facilities is aware that it does open the floodgates to such applications as indoor location services (i.e., way finding, emergency management, etc.).

Also, MIT Facilities recognized that in the future it would like to keep track of equipment in rooms and capabilities at a higher scale (higher granularity). This means that a positioning infrastructure will need to be deployed in every room, which would pose an incentive for MIT Cricket to expand. If MIT Cricket wants its position data of the users to be overlaid on top of the GIS maps from MIT Facilities, it needs to be in the same data format and same coordinate reference system. Use of GML for data format (for data display and integration) and LIF MLP API would take care of the same coordinate reference system.

Also, both MIT Facilities and MIT IS could benefit from a common data model. The development of detailed models is a costly task (especially when extending them to subdetailed world models). Therefore, different applications should be able to share the same model information. Having such a common model may increase interoperability between applications and make new classes of applications possible. The basic requirement for such an approach is a common language for describing and querying location information.

Similarly, the MIT Cricket project knows that its local positioning system must eventually be possible to integrate on top of the existing Wi-Fi networking (communication infrastructure) and web service (software infrastructure) to avoid stovepipe (or silo) implementations.

Overall, it is difficult to exchange information between all these players, as their systems are stand-alone and vertically integrated. Data conversion would be required for integration. Nor is it possible to notify the applications on one system about information changes based on the information sensed by another system. In order to make the exchange of information more interoperable, open standards from LIF for positioning and Open GIS for mapping, as well as Open GIS/OpenLS for software and services, could have been used.

All players have an interest in open interoperability standards that would ease integration and facilitate coherence and long-term goals while being easy to do and inexpensive (in time and money and performance) in the short run.

If all departments would follow OGC's WMS specification, for example, web-based maps from both would overlay without any need for conversations and transformations. Also, if MIT Cricket would use LIF's MLP API for positioning, then both the communication handover and positioning handover would be seamless if users would, for example, travel from the outdoor world, where the GSM cellular network is present, to the indoor world (where MIT Cricket's or MIT IS's Wi-Fi network is present). In short, MIT Facilities and MIT Cricket are good examples of how individual systems/projects start and develop over time.

THE SENSOR COMPONENT

The sensors are the focus for the next two chapters, but in general, the sensors may be a feature of the network (such as in mobile location determination systems based on GSM, UMTS, etc.), or it may use separate sensors (beacons). There may be hybrids, which use, e.g., an RFID tag to determine the area of the absolute position, and then a relative position to the base stations.

It is important to realize that different positioning systems express location in different ways — different measurements (geometric vs. symbolic), different spatial frame of reference, or different uncertainty. As a result, it is important to first determine which type of positioning (absolute vs. relative) the location-based service will require.

It is a challenging problem (to say the least) to build an ideal indoor positioning system that provides accurate and precise location at a high update rate. Different underlying sensors will give different results of accuracy and precision. Accuracy requirements for indoor location-based applications will vary significantly. For example, RFID tags are considered to be proximity sensors. These can detect a tag when it passes within a relatively short distance of a sensor (usually a few centimeters (1 foot or less). Other technologies can sense when a tag is within a room (relative positioning), but cannot identify its absolute location within the room.

Much depends on the requirements of the application. What is actually the need of the system when it comes to information accuracy? Is it sufficient to know you are in the right building, or do you have to know which room you are in? Which shelf in that room? This will determine how the sensor system should be organized to serve the application in the best way.

Most available local positioning systems are network based rather than terminal based (sensors are integrated into the mobile device), meaning that the network calculates the position of the mobile device, as opposed to a receiver inside the device calculating its own position and using the network to notify others of its location. Network-based positioning is especially common for local positioning systems that are typically based on small radio frequency or infrared cells, since the processing capacity in the terminals is too small for any calculations within any reasonable time. In fact, all of the local positioning systems that were found as part of the detailed survey done for this book were networked based.

POSITIONING INFRASTRUCTURE

There are, in the main, three ways to create a positioning system for indoor location: deploying a set of sensors or beacons independent of other networks, using existing infrastructure such as Wi-Fi networks, or using the existing wide-area networks, with their positioning capabilities.

Each of these approaches has its advantages and problems. In practice, since they will tend to overlap, there will be a need for the software system to manage handover of the location information (as previously discussed).

These are discussed in more detail in the next chapter.

4 Sensor Systems for Indoor Position Computation

POSITIONING SYSTEMS AND ALGORITHMS

The core of any positioning system is the means for measuring the whereabouts of a terminal, and the algorithm to compute those whereabouts in relation to some known location (whether absolute or relative). This can be done in the terminal or in the network. The nodes can be active, transmitting a signal themselves, or passive, just receiving the signal (leaving it to some other entity to compute the position). These two dimensions are complementary, not contradictory.

In addition, active systems transmit a signal themselves; passive systems just receive a signal. Most global positioning systems (GPS) are passive, simply receiving the signal (since it would not make much sense to communicate with the satellites anyway); most mobile phone systems (when used for positioning) are active, since the phone is required to send out a "keep alive" signal to the base stations now and then, to make sure the system does not remove it from the list of connected terminals. This also implies that in entirely passive systems, there is no way for the sender to know who uses the positioning signal.

In either case, you can further make the choice of basing the system on the existing communications network or a dedicated network that is only used to transmit positioning signals. If you base it on an existing network, there is the risk that the positioning signal crowds out other (useful) signals; if you base it on a dedicated network, you have to provide additional receivers in the terminals, which may bring an extra cost.

Moreover, active systems incorporate signal emitters, sensors, and landmarks placed in prepared and calibrated environments. Passive systems are completely self-contained, registering naturally occurring signals or physical phenomena. Examples include compasses sensing the Earth's magnetic field, inertial sensors measuring linear acceleration and angular motion, and vision systems sensing natural scene features. Most of the outdoor tracking is based on a sort of passive-target systems utilizing vision (Azuma, 1997). Vision methods can estimate camera position directly from the same imagery observed by the user. But vision systems suffer from a lack of robustness and high computational expense. Unfortunately, all tracking sensors used in passive systems have limitations. For example, poor lighting disturbs vision systems, close distance to ferrous material distorts magnetic measurements, and inertial sensors have noise and calibration error, resulting in a position and orientation drift. Hybrid systems attempt to compensate for the shortcomings of a single tech-

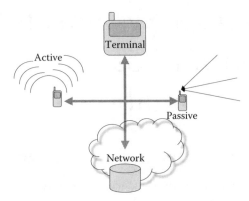

FIGURE 4.1 The two dimensions of positioning.

nology by using multiple sensor types. Among all other approaches, the most common is passive magnetic combined with a vision system because inertial gyroscope data can increase the computing efficiency of a vision system by providing a relative frame-to-frame estimate of camera orientation, and a vision system can correct for the accumulated drift of an inertial system.

CLASSIFYING POSITIONING SYSTEMS

There are a multitude of methods for determining the traveler's current location. These vary in the extent to which they require sensing of the environment or reception of signals provided by external positioning systems. At one extreme, there is inertial navigation, which requires no external sensing. At the other extreme are methods involving the matching of perspective video images of the environment to three-dimensional models stored in computer memory. In between are methods employing dead reckoning and a variety of local and positioning systems in which the navigator determines current position using signals from transmitters at known locations. Because of the high degree of accuracy of its position fixes, some of the local positioning systems (e.g., Ekahau) are the preferred choice for environments where obstructions and multipath distortion are accounted for. For environments in which indoor signals are only intermittently available, the positioning system of choice needs to be supplemented by inertial navigation or dead reckoning. When indoor signals are unavailable, a network of location identifiers will be needed to assist the user with navigation.

Figure 4.2 portrays a taxonomy of geolocation methods for local positioning systems with external sensing. Location-sensing techniques can be divided into three general categories: location fingerprinting (scene analysis), triangulation, and proximity. (All these are based on external sensing, in contrast to inertial navigation, which requires no external sensing.) These various techniques can be discussed in terms of being physical or symbolic and relative or absolute.

A physical technique generally results in a set of coordinates, such as longitude and latitude coordinates, whereas symbolic techniques provide more of an abstract description, such as location in terms of which building an object is in. The difference

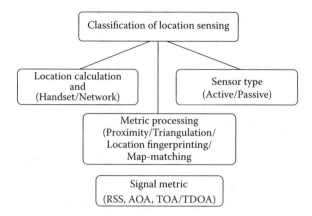

FIGURE 4.2 Classification of location sensing.

between absolute and relative systems is the frame of reference. In an absolute system, a frame of reference is used. A relative system, however, uses local references where physical and symbolic techniques may be combined. Signal strength is often used to determine proximity or sometimes range (through an attenuation model). As a proximity measurement, if a signal is received at several known locations, it is possible to intersect the coverage areas of that signal to determine a "containing" location area. If one knows the angle of bearing (relative to a sphere) and range (distance) from a known point to the target device, then the target location is precisely known in three dimensions. Similarly, if one knows the angle of bearing to a target from two known locations, then it is possible to triangulate to determine the (x, y) coordinates of the target. This can be extended to multiangulation for higher dimensions. In a similar fashion, if one knows the range from three known positions to a target, then using multilateration it is possible to determine the target location. Video cameras can be used to describe spatial relationships in scenes using image processing techniques, and thereby determine position. Proximity schemes are often enhanced by using received signal strength (RSS) to provide additional indications of the distance from the source. These schemes rely on propagation loss estimates often derived from a careful mapping of the environment characteristics.

With the proximity technique, in wireless local-area network (WLAN) location sensing we can measure the signal strength at an MC (receiver) and the signal strength at the transmitter. We could then use a radio wave propagation model to calculate the distance the signal has traveled. This approach provides the location of an MC relative to the known location of an access point (AP). Using only one measurement would allow us to place an MC in part of a building, as opposed to another part, but it would not be able to easily provide us with a coordinate for the location. This approach depends on the propagation model used and is hampered by interference caused by walls and furniture. An alternative proximity approach involves creating a table of measured signal strengths and then comparing the signal strength of an MC to the values in the table. Sampling and recording more values in the table would increase the resolution of the system. This approach is not as susceptible to error due to interference; however, it involves more work to set up the location system. An example of proximity-based localization is radio frequency identification (RFID) proximity cards (e.g., Bewator Cotag, Wavetrend). RFID proximity cards are in widespread use. They are active or passive devices that activate by proximity to a fixed reader. They are often used in access control systems. Location can be deduced by considering the last reader to see the card.

The location fingerprinting (scene analysis) technique examines a scene such as a room from a fixed vantage point. One such approach is adopted by the Microsoft Research group in the radar implementation. Radar measures signal strengths of mobile devices from the vantage point of a fixed base station. These measurements are then used to calculate the position of the devices on a two-dimensional coordinate system local to the building. Indoor positioning systems utilizing the location fingerprinting method require a user to record signal strengths of all APs in range from a point in a table. This is done for various points in the WLAN area. To find out the location of a mobile client, the signal strengths from all APs are measured and then the values are compared to the entries in the table. The closest entry in the table is the probable location of the user. The more points a user calibrates, the better the resolution of the software.

The third general technique is triangulation. This method derives its name from trigonometric calculations and can be done via lateration, which uses multiple distance measurements between known points, or via angulation, which measures an angle or bearing relative to points with known separation. These two techniques are also referred to as direction-based and distance-based techniques. Direction-based techniques measure the angle of arrival (AOA). Using directional antennas, the receiver must measure the direction of the signal from the transmitter with respect to a fixed direction, such as east or west. Two or more AOA measurements can tell us where paths between the transmitters and the receiver intersect, and using trigonometry we can calculate the location. Because this particular triangulation technique requires the use of special antennas, it would not be suitable for a WLAN location-sensing application that mandates the use of standard components. Distance-based techniques involve measurement and calculation of the distance between a receiver and one or more transmitters whose locations are known. These techniques involve using one or more of the following signal attributes: signal arrival time, signal strength, and signal phase.

The map-based positioning method does not need to log on to any of the Wi-Fi networks to determine the position, and some of the networks may well be password protected. But the Wi-Fi base station does send out an ID number when hit with a blast of radio frequency energy from the Wi-Fi card inside a laptop or handheld computer. And even if the signal is too weak for a laptop to connect to, it can still get the ID, which it then compares to Skyhook's database and plots on a map.

Classification Based on where Location Estimation Takes Place

Positioning systems can be terminal based or network based. In the first case, the terminal receives the information from the sensors or sensors (which may be a part of the network, as is the case when positioning is done using some 802.11-based technologies, or built into the terminal) and computes its own position. It then has to communicate that position to other entities that want to use it (for instance, servers that have location-based services). As wireless devices, the mobile devices are configured to communicate with a network through a wireless interface.

In the second case, no computation is done in the terminal, but the position is computed in the network and communicated to the terminal if required. Mostly, the location-based services are produced in the network as well, and the service access is transmitted to the terminal (for instance, through a web page or another type of client). Of course, hybrids exist (such as assisted GPS), but these are the two main types. Each has a different set of constraints and a different set of advantages.

Preferably, the mobile device needs no special client-side configuration, modules, or programs to be detected and tracked, since detection and tracking are preformed on the network side of the interface. The availability of applications and access to data may be selectively provided or inhibited as a function of the location of the mobile device and an identity of the mobile device or its user, or both.

Although network-based sensing is a compelling business case for operators, many systems have chosen to compute location on the handset because of the lack of standard methods for querying location from the network, and because this solution provides for simpler regulatory compliance in jurisdictions with strong location privacy legislation (i.e., private use of others' personal information is exempt from most data protection provisions). Computing location on the mobile device also respects the user's control for disclosure. A disadvantage of placing a network of location identifiers within the environment is the cost of installing and maintaining the network relative to the coverage achieved. One alternative is to use computer technology to locate the traveler and then make use of a spatial database of the environment to display to the traveler his or her location relative to the environment.

In some cases, such as a high-end portable laptop, it could be possible to compute position information, whereas on a portable handheld such as an iPaq, computing the position information could be cumbersome and utilize too much computational effort. In such cases it could suffice to report just the position with the highest accuracy, to just merge the two most accurate positions, or to use a less accurate merging technique. It is therefore important that the platform can be optimized for the underlying host characteristics and needs of the programs utilizing it.

As well as merging positioning information from local devices, the position can also be augmented with information from other nonlocal position platforms via limited range *ad hoc* networks such as a WaveLAN segment or Bluetooth networks.

Overall, sensor measurements are attained and then these measurements are collated to acquire the user's position. In most indoor systems the user carries either a transmitter or a receiver. If the user carries a transmitter that broadcasts signals, then sensors fitted in the room are used to detect the emitted signal. These signals may be radio frequency signals, infrared signals, or ultrasonic pulses. The measurements that are used to pinpoint the user may be either the signal strength of the signal detected by the sensor or the time-of-flight of the signal from the transmitter to the sensor. For example, the BAT system, which is one of AT&T's indoor location-based systems, can find the three-dimensional position of a user when given three or more time-of-flight measurements. All systems that exploit transmitters and receivers have their own pros and cons, but the fact that the users must carry either a transmitter or receiver is a universal problem for such systems.

Location determination methods are also classified according to the point of computation: server based or client based. The location methods can be further differentiated using the architecture of the implementation to emphasize autonomy. The architecture can vary from infrastructure dominated to infrastructure-less schemes. Infrastructure schemes can be active or passive. In active schemes, there are sensor servers that provide measurement and timing signals that are received by the clients (equipments). The location computations are performed by the clients. Sensors may be requested or they may be constantly generated. In passive infrastructure schemes, the sensors (or measurement signals) are generated by the clients and the locations are computed by one or more servers, as in cellular base stations performing pattern matching or angle-of-arrival schemes. The infrastructure-less class of systems, where location is performed by clients only, are either independent or cooperative. Independent schemes involve a single client operating from a known position with self-tracking capability, such as an inertial navigation (or variants such as star tracking, compass, etc.). Infrastructure-less, cooperative schemes, or *ad hoc* location-sensing schemes, involve measurements made by multiple clients that must be communicated among clients to determine position. The cooperative schemes cannot determine location individually and must self-organize to perform the communication necessary to share the measurement information.

Classification Based on Signal Metrics

The basic function of a wireless positioning system is to gather particular information about the position of a mobile device and process that position into a location estimate. The particular information could be one of the classical geolocation metrics for estimating the position:

- Received signal strength (RSS)
- Angle of arrival (AOA)
- Time of arrival (TOA)/time difference of arrival (TDOA)

Calculating direction using AOA requires additional antenna hardware that needs to be precisely calibrated; the systems based on the time of flight of electromagnetical waves require very accurate timing measurements, and thus high synchronicity.

In a networked environment of many nodes, cell site density, the distribution of nodes, the ability to share information, terrain, and physical obstruction play major roles in the accuracy derived by the system.

A positioning system may or may not be able to communicate and thus share data used to solve for the position. The systems that cannot communicate have to rely solely on their ability to detect an electromagnetic signal. They can employ methods based on signal strength, Doppler shift, and AOA measurements. If communication is available, data can be modulated onto the navigation signal or be available on a secondary communication link. Combinations of modulated data and secondary links, for example, differential GPS, are possible too. The data source of the modulated signal assists in deriving the GPS solution, the secondary communication link in correcting and refining the solution. TDOA and TOA measurements require a minimum of transmitted data consisting of at least a pulse shape, used for synchronization and thus measuring the time of flight of a signal, to be exchanged. Depending on the signal and the range anticipated, larger amounts of data can be transmitted. Hitachi (Uchida and Kinoshita, 2002) released TDOA-based location technology in October 2002. This system uses two types of access points, a Master AP and a Slave AP. Slave APs synchronize their clocks with that of a Master AP. Slave APs measure the arrival time from a mobile terminal, and the Master AP determines the location of the mobile terminal by using the time differences of arrival between the signal reception times at multiple Slave APs. Triangulation methods can be used in conjunction with angle and range data to solve for remote positions. Range vector data from range measurements such as TDOA, TOA, and received signal strength indicator (RSSI) or similar are considered the input data. The absolute position of a device/user position can be computed using range measurements (triangulation) to known positions. Multiple range measurements (TOA, TDOA, or RSSI) or angles (AOA) are necessary to determine the exact three-dimensional position on a remote navigation receiver.

Indoor Signal Propagation Characteristics

Before going into detail about the geolocation metrics, first we will talk about indoor signal propagation characteristics. Figure 4.3 illustrates a signal strength pattern around an access point with no environmental interference. Signal strength pattern represents an ideal, where signal strength alone would merely indicate proximity to the source. That is, ideally, the closer to the source, the stronger the signal and the higher the reading. The system draws coverage circles around each access point that hears the user. Each coverage circle reflects the border of the signal strength at which

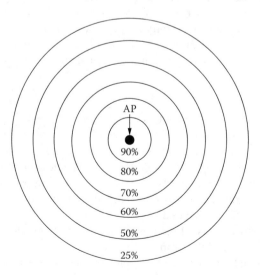

FIGURE 4.3 Signal strength pattern with no interference.

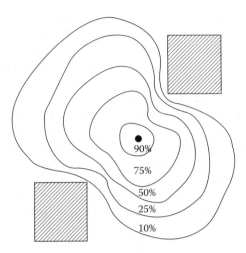

FIGURE 4.4 Signal strength pattern with interference.

the AP received the signal strength from the user/device. If an access point heard the device at –65 dBm, then the network management system would draw a circle at where it believes the –65 dBm demarcation is located on a coverage map. If another access point heard the device at –45 dBm, then the coverage circle around that AP might be considerably smaller (meaning the signal strength was higher), and it too is factored into locating the user.

However, in a practical deployment, a communications source or access point is likely to encounter some environmental impact, causing absorption, attenuation, reflection, or a combination of these factors on the communications medium in different areas throughout the defined space accessible by the signal, such as would occur in a digital map. Figure 4.4 shows a distorted signal strength pattern around the access point that is the result of environmental obstructions. The resulting nonuniformity provides an opportunity for local position detection since, depending on the environment, locales (or other locations of interest within the defined space) may have highly distinguishable nonuniform signal strength profiles.

Radio Propagation

Radio wave propagation is defined as the transfer of energy by electromagnetic radiation at radio frequencies (SMI, 1998). Radio propagation studies look at how radio waves travel through a given medium such as air. Radio waves encounter various outdoor objects ranging from buildings and plants, and indoor objects, such as walls and furniture. These objects obstruct radio wave propagation and affect the amount of time it takes for the wave to reach the receiver from the transmitter. In addition to the interference caused by obstructions, other factors, such as terrain, wave frequency, and velocity of the transmitter and receiver, all impact radio propagation. Since 802.11 works over the 2.4-GHz frequency, there is interference from microwaves, Bluetooth devices, cordless phones, and other similar devices. Multipath fading, where a signal reaches the receiver through different paths, each having its own phase and amplitude, is another common problem faced by radio waves. Even environmental changes such as humidity and temperature affect the signal strength. At a fixed location, the signal strength received from an access point varies with time and its physical surroundings.

The wavelength of the radio waves used by a WLAN is significantly smaller than the obstructions that the radio waves encounter; therefore, we can simplify the study of these waves by treating them as rays traveling in straight lines. The shortest path that a wave can take is the unobstructed path, or the line of sight (LOS). When obstructions are encountered, the signal has to take multiple paths to travel from the transmitter to the receiver. This behavior, called multipath delay spread (more on multipath below), introduces a delay in the transmission time when compared to LOS. Usually, non-line-of-sight (NLOS) land mobile radio channels have a characteristic random multiple-path radio propagation, with the principal parameters of multipath fading, shadowing, and path loss. Traditionally, the transmission link of such a system is modeled by an elevated base station antenna, a relatively short LOS propagation path followed by many long NLOS reflected propagation paths and an antenna on a mobile transceiver.

Variations in the channel occur due to changes in the static environment, moving interferers, and motion of the user/transceiver. When natural and constructed obstacles, and especially movement in the environment or a transmitter, result in multiple signal paths, the situation is referred to as multipath propagation, influenced by the following factors: The rate of variations of a signal between moving objects is referred to as Doppler spread. This is then a time-selective or time-variable signal phase on the received signal.

Particularly relevant to location sensing are the radio propagation behaviors that cause the signal transmission to take a longer path or be delayed. One approach involves measuring the time it takes for a signal to reach its destination and, based on that measurement, calculating the distance traveled. Any radio propagation delays can ultimately introduce errors into the distance calculations. Signal attenuation is perhaps the most fundamental technical problem to be addressed, because no radio positioning system can operate when the tracking signals are too weak to be received. Any receiver is essentially a device that separates signals from noise and recovers the information content of those signals. Usually, there are only one or a few sources of the desired signal, but there are many sources of noise in a typical environment. In general, given typical wood or reinforced concrete buildings, attenuation is an increasing function of frequency. This explains why GPS does not work indoors; its microwave carrier signal at 1.5 GHz undergoes severe attenuation in the course of passing through even the thinnest roofing materials. Commercially available indoor radio positioning systems use the industrial, scientific, and medical (ISM) allocation at 2.4 to 2.45 GHz. These frequencies are attenuated even more severely by passage through materials as common as reinforced concrete, people, and plant leaves. It is not uncommon for a 2.4-GHz signal to be attenuated by more than 20 dB when a single reinforced concrete wall appears in an otherwise free-space signal path. In general, sources of error for signal propagation arise from the receiver's equipment, noise and interference, propagation channel (multipath and non-line of sight (NLOS), which is a range/direction estimation error — most important impact on location accuracy), and nonlinear algorithm (positioning error).

In most cases, signal propagation in indoor environments is usually affected by multipath. In wireless networks, the straightest path (direct line of sight, or DLOS) between objects is often difficult to ascertain. The strongest signal (the one the mobile device is using to bind to the AP) may have bounced off a nearby wall, while the DLOS path (the shortest route) may have been drowned by RF interference. Multipaths are present due to reflections, diffraction around sharp corners, or scattering from walls, ceilings, or floor surfaces.

Wireless signals are also attenuated with increasing distance and obstacles in their paths. If the signal strength is below some threshold value, the receiving antenna will encounter bit errors when decoding the signal, which gets worse under significant RF interference from other sources. Encountering such bit errors, the receiver does not send an acknowledge signal back to the sender, causing the sender to retransmit the frame. These retransmissions decrease the data rate of the system. So as the distance between the AP and the mobile increases, the data rate of the system decreases. Thus, in order to increase the data rate, the coverage area should be decreased, which is the premise for IEEE 802.11a systems operating at 54 Mbps and IEEE 802.11b systems operating at 11 Mbps. Since the coverage area decreases to about 30% of that of the 802.11b standard because of the increased path loss, the 802.11a standard requires the deployment of more APs and more overall infrastructure investment.

If the signal strength falls below the receiver sensitivity, the mobile is disconnected from the particular access point. The receiver sensitivity depends on the wireless technology used and the data rate of communication. The receiver sensitivity is lower for higher data rates, and also determines acceptable path-loss measures. Also, dead spots are present where there is no received signal due to these effects.

Therefore, before deploying a WLAN, an RF site survey should be performed for producing signal propagation models, enabling the determination of optimal locations of AP to provide the best possible coverage. A site survey is also essential for the purpose of location fingerprinting, which is the most accurate method of location tracking. Both site survey and location fingerprinting are discussed in detail later in this chapter.

Many studies (including the ones of Place Lab at Intel) also compare the effects of density of calibration data, noise in calibration, and age of data on the centroid

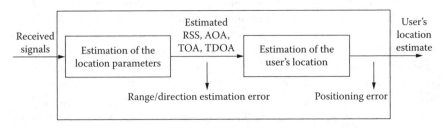

FIGURE 4.5

algorithm, particle filter, and fingerprinting algorithms. An algorithm combining these techniques is feasible, and it may prove to be both robust and accurate. Moreover, one could incorporate additional environmental data, such as, for example, constrained GIS maps of hallways (streets) or navigation applications, to improve the positioning accuracy. There are a variety of ways to determine or improve a position estimate through the sharing of information with other clients. One of the simplest forms of cooperative location, communication proximity, is for a node to request the positions of other entities in the current communication neighborhood, using a broadcast request message. Depending on the transmission range, a set of regions (ellipsoids) can be collected by the source and used to estimate its own location. The accuracy depends on the density of responding devices, the environmental characteristics (e.g., obstructions), and the radio channel (e.g., multipath effects and noise). For short-range radios, such as Bluetooth (e.g., 10 m) or the infrared (IR) medium, knowledge of even a single neighbor can provide significant accuracy. Variations of this approach for *ad hoc* networks are described in Bulusu et al. (2000) and Doherty et al. (2001), in which neighbors can be used to define an average location value. In general, the more devices that can respond with their position, the more accurately the source position can be determined.

Two concepts are key in radio propagation: transmission power and signal strength. Radio propagation is the transfer of energy and is measured in terms of units of power, or watts. This power is measured at the transmitter (transmission power) and also at the receiver; the signal strength is the total amount of power measured at the receiver. Due to the nature of radio wave propagation, the latter measurement is less than the former because the signal loses power as it moves through the air in the form of radio waves.

More on Multipath

Multipath is common in radio channels. At the receiver, multipath signals are multiple copies of the transmitted signal, each with a different delay. This results in the correlator output having multiple correlation peaks of various sizes corresponding to the transmitted code word. A receiver can take advantage of multipath and combine all of the multipath-related correlation peaks out of each code word correlator, and then do the comparison among all such code word correlators. This receiver would perform better than a conventional receiver, which would only look at the correlator output at a particular instant in time. This type of receiver can be implemented by changing the basis for comparision in comparator. The multipath phenomenon has been recognized in pagers and mobile telephones. In this context, the combination of wireless signals may cancel each other out, making reception difficult. This is sometimes referred to as multipath interference. Efforts have been made to reduce multipath interference in order to improve the signal-to-noise ratio or bit error rate of the received signal. For example, in one technique, two or more signals with relatively noncoherent amplitudes may be transmitted using space, frequency, time, or antenna polarity diversity. Multipath interference is avoided by selecting the strongest received signal for detection and demodulation. According to this method, the particular path or paths taken by the strongest signal is not important.

Some receivers provide an example of time diversity reception in direct-sequence spread-spectrum communications. The receiver de-correlates the received signal by applying several time-delayed versions of the known pseudo-random sequence used by the transmitter. The signal from the direct path (if there is one) and the strongest echoes may be de-correlated and combined to generate a signal having a lower bit error rate than can be obtained from the signal from any one of the paths.

Where the wireless signals are digitally modulated, an adaptive equalizer may be deployed at the receiver. Adaptive equalizers pass the receive signal through a tapped delay line. The tap take-off parameters are adaptively adjusted to cancel out echoes. As a further alternative, directional antennas may be used at one of or both the transmitter and receiver. At the transmitter, a directional antenna limits the number of paths that the transmitted wireless signal may take. At the receiver, a directional antenna reduces the number of paths from which wireless signals can be received. In either case, the direct signal path can be strengthened in relation to the reflected or echo signals. However, directional antennas are inconvenient and have limited use because they must be oriented to direct or receive the wireless signals.

In wireless distance measurements, the problem is not to increase the signal-to-noise ratio or bit error rate irrespective of signal propagation time or distance. The object is to identify the direct path signal as closely as possible. If reflected signals contribute significantly to the measurement, the resulting distance will be inaccurate.

There are ways that enable the number, position, and relative strength of multipath signals to be identified. In the context of distance measurement, some solutions (e.g., AeroScout) permit the direct path between two wireless devices to be identified from among other paths taken by the transmission signal. Some solutions may be used to correct wireless distance measurements for inaccuracies caused by the multipath phenomenon. Moreover, some solutions are capable of determining the distance between two wireless devices using only the two wireless devices, as described in further detail below. To achieve this result, a first wireless device transmits a forward path signal to a second wireless device. The second wireless device generates a reverse-path RF signal sequence using the forward path signal such that the forward and reverse path signals are coherent. The reverse-path RF signal sequence includes different frequencies that have the same or substantially the same multipath characteristics.

A wireless receiver of the first wireless device receives and generates amplitude and phase information using the reverse-path RF signals and the forward path signal. A curve drawn from the received amplitude information and frequencies will be periodic in frequency, with the period depending on the relative amplitude and delays of the reflected or echo signals. The first wireless device converts, for example, by inverse Fourier transform, the amplitude, phase, and frequency information into time domain data. The time domain data indicate the number, time delay, and relative amplitude of the echoes composing the received signal. For distance measurement application, the time domain data are sufficient to distinguish the direct path signal from the echo signals and correct the measured distance from effects of multipath distortion. After the direct path signal is distinguished and all other paths are filtered out, a Fourier transform can be applied, resulting in a flat frequency response and a linear phase shift. This information is sufficient to determine the measured distance. Then a direct reading of the propagation delay of the direct path may also be determined.

Following is a summary of the main effects of multipath on phase–slope distance measuring systems and the measures to take to counter it:

- Multipath signals will distort the distance reading to a degree depending on the strength and delay time of the echoes.
- The apparent delay in the presence of multipath can be corrected by analyzing the passband amplitude and phase behavior to find relative echo strength and delay.
- The degree of resolution of echoes depends on the number of frequency hops and total bandwidth of the system.

RSS Positioning Technique

This approach makes use of the relationship between the received signal strength and the distance. Theoretically, there exists an inverse proportional relationship between the received signal and the distance from the receiving station that can be represented linearly. The distance between the mobile and the AP can be determined from the RSS values either at the mobile device end or at the AP end. There are two ways to determine the distance. The first is to map the path loss of the received signal to the distance traveled by the signal from the transmitter to the mobile device. Second, with the knowledge of the RSS from at least three APs, triangulation is used to locate the receiver. Geometrically, if a mobile is at distance di from APi, possible locations of the mobile are represented with a circle with center at APi and radius di.

There is no database search, and the positioning delay is just related to the communication and computation. However, inside a building, the variation of the RSS with distance (the inaccuracy of the path-loss model) is significant due to obstructions and multipath fading effects. As such, it is usually not reliable to use the RSS in this manner. This method is used in some systems along with training and interpolation to improve accuracy.

Use the path-loss models to compile a location fingerprinting database of RSS values used for database correlation/pattern-matching algorithms (see more on location fingerprinting below in the "Positioning Algorithms" section). The measured RSS values are then compared with this database to obtain the mobile device's location. This seems counterintuitive, as it retains the inaccuracy of the path-loss models and the delay associated with the database search. In order to employ RSS-based techniques with greater accuracy, this second approach uses the RSS in a fingerprinting scheme (see more below on location fingerprinting).

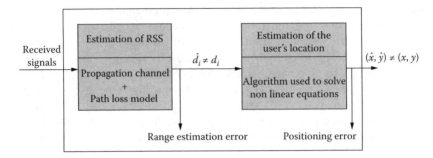

FIGURE 4.6

Path-loss models measure received signal strength as a function of distance. Signal strength measurements are affected by various radio wave propagation mechanisms (PAH, 2002). The models become more complex as each of these mechanisms and the previously discussed sources of delay are taken into account. There are three ways that signal strength can be affected: reflection, diffraction, and scattering.

Radio waves typically encounter obstructions larger than the wavelength, and depending on the wave frequency and the angle at which they hit the obstruction, the rays are reflected away from the obstruction. Reflection is an important consideration in indoor applications. The radio waves are diffracted when they encounter edges, such as when they come into contact with the edge of a building or a wall. As a result of diffraction, the waves are able to propagate away from the edge and reach places that are not directly within LOS. The third mechanism that needs to be considered is called scattering. Irregular objects such as walls, furniture, and leaves on a tree can cause the rays to scatter in all directions. Scattering is observed when the object dimensions are close to the radio wavelength, and becomes a significant issue if the transmitter or receiver is located in a highly cluttered area (PAH, 2002).

Signals bounce around, and it is documented that the strongest signal a device receives — the one that is used to bind with the wireless networking appliance — is not always the DLOS signal from the transmitter. However, signal strength may be the easiest metric to record on an ordinary mobile device, since it is directly applicable to a device physically gaining access to a wireless network. Although the DLOS path between the mobile device and the transmitter is not determinable by RSS, this technique is suitable for propagation modeling and location fingerprinting.

Figure 4.6 shows that received signals at three base stations and a path-loss model provide distance or range estimates (d1, d2, d3) between the mobile device and base station. Each estimated range gives a circle centered at the receiver (base station) on which the transmitter (mobile device) must lie. Range estimation errors are due to multipath, NLOS, and local shadowing. Position estimation error is due to the algorithm used to solve nonlinear equations.

Many radio-based positioning systems use the observed signal strength as an indicator of distance from a radio source. In practice, this works only as well as the radio beacon's signal strength decays predictably with distance and is not overly attenuated by factors such as the number of walls crossed, the composition of those walls, and multipath effects. For instance, buildings with brick walls attenuate radio signals by a different amount than buildings made of wood or glass. In addition to fixed obstructions in the environment, people, vehicles, and other moving objects can cause the attenuation to vary in a given place over time.

Advantages of the RSS metric are the following:

- Easiest metric to obtain (does not require additional signal processing, as is the case with AOA and TOA/TDAO and multipath); an exception is when multipath characteristics are needed for location fingerprinting.
- Does not require usage of additional hardware.
- Use of existing infrastructure with minimum additional devices.

Disadvantages include the following:

- Various phenomena like multipath fading and shadowing make it impossible to establish a precise relationship. As a result, the RSS metric is not very accurate and can vary by up to tens of decibels over a distance of only a fraction of wavelength.
- Path loss predicates how the signal power decays with distance from a base station. As a result, most solutions use the averaged values of signals rather than their instantaneous values.

AOA Positioning Technique

This technique computes the angle of arrival of the signals from the MS to more than one base station, making use of directional antennas and using simple geometric rules to calculate the distance from the angle measurements. Two APs are enough to accurately estimate the mobile's location. Each estimated angle gives a line between the AP and the mobile station. The lines of position are straight lines whose intersection provides the location of the MS. The accuracy of AOA diminishes with increasing distance between mobile station and base station due to the scattering environment. Figure 4.7 shows that the received signals at two base stations (array antennas) and the estimation algorithm provide angle or direction estimates between the mobile device and base stations.

Disadvantages include the following:

- Need of extra hardware — antenna arrays at the AP. The performance of this system highly depends on the accuracy of the antennas used for the angle measurement.
- Need for line-of-sight (LOS) conditions for accuracy.
- Changing scattering characteristics and multipath signals hinder the performance. One way of reducing the scattering characteristic and the multipath issues is to elevate the antenna to an appropriate height (which makes these systems almost impractical for micro-cell-based networks).

Overall, AOA is not a very suitable metric in indoor environments.

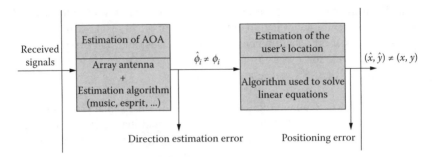

FIGURE 4.7 Received signal angle estimates.

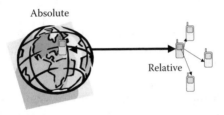

FIGURE 4.8 Angle of arrival.

TOA/TDOA Positioning Technique

The distance between the mobile and AP can be calculated using TOA/TDOA information as well. The TOA technique is based on estimating the time of arrival of a signal transmitted by the mobile device and received at the minimum of three base stations. The TDOA technique is based on the time difference of arrival of a signal received at multiple pairs of base stations. Knowing the distances (having obtained the TOA information, the distance between the mobile and the AP is just the speed of light times the travel time) to each AP, location can be calculated using the triangulation method.

The usage of TDOA information is very similar to that of TOA, but this time the difference between the distances from the mobile to each AP is calculated. The lines of position are hyperbolas (instead of circles as with AOA) whose intersection provides the location of mobile station.

The use of time of flight of signals to measure distance is not a new concept. GPS uses the one-way delay of radio waves from satellites to estimate distance, while collision avoidance mechanisms used in robotics determine the distance to obstacles by measuring the time of flight of an ultrasonic signal being bounced off them.

The major concern with this type of technique is the preciseness of time measurement, and unlike other techniques, the error factor decreases as the distance between the mobile station and the base station increases. Since the time delay to be measured in microcellular technologies is very small, technologies like Bluetooth do not easily support this technique.

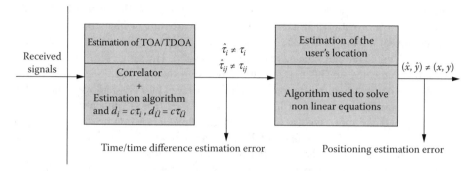

FIGURE 4.9 Received signal time estimates.

Figure 4.9 shows the received signals at three base stations and estimation algorithms provide time/time difference estimates between the mobile device and base stations (pairs of base stations).

Advantages include the following:

- Compared with RSS and AOA, TOA metrics perform better in terms of accuracy. (The synchronization of clocks is very important since 1 nsec of timing error will cause 0.3 m of location error.)
- It is almost totally independent of the preferred wireless technology of the mobile device. TDOA technology therefore can be applied to a wide range of wireless networks (primarily TDMA, CDMA, and FDMA based).

Disadvantages include the following:

- The network infrastructure needs to be extremely well synchronized (at the MS and the transmitter ends), which in turn implies higher costs.

BER Positioning Technique

Bit error rate (BER) is part of the signal transmission and can be a good indicator of the distance between two devices. Like RSS, BER is directly proportional to the distance between the mobile device and the base station. This technique is found in Bluetooth-based positioning systems.

Deploying Sensors and Sensors

In a system that uses dedicated sensors for the positioning signal, you may still need a network to communicate with the sensors. This depends on how intelligent (and energy efficient) you want to make the network of sensors. It is quite possible to keep sensors switched off until a terminal is about to enter their coverage range (from the reach of another beacon). This may not be economical, if the effect radiated is low. In the MIT Cricket case, the system uses the existing wireless communication network installed in the building of the Institute of Computer Science (and could do the same for a wider deployment across the campus). However, MIT Cricket requires extra investment in the positioning infrastructure, which consists of ultrasound sensors. This makes the positioning infrastructure investment more costly (i.e., installing, configuring, and maintaining sensors). The same is true in any system that uses dedicated sensors to transmit position information. The trade-off for higher cost is higher positioning accuracy. The more sensors that are deployed, the higher the accuracy. For many indoor location-based service (LBS) applications, accuracy requirements are more precise than one might think. For example, for most applications it is essential to be able to tell with certainty which room a person is in, or which side of a partition he is on. For this reason, accuracy to within 30 cm (a foot or so) is required for many of the indoor LBS applications. In case the coordinates required are three-dimensional, for instance, when finding a product somewhere on a shelf, the requirement is the same but in three dimensions. This requires additional sensors to cover the third dimension — and it may not be sufficient just placing a row of them along the floor.

In most cases a simpler approach (i.e., a single tag and sensor per room, like Active Badge, instead of multiple sensors per room, like MIT Cricket) might suffice. It will also be significantly cheaper for some basic indoor location-based applications. However, it is important to think in terms of the long-term deployment of the positioning infrastructure, enabling seamless extensions in the future. There are additional location-based applications that might become interesting, which can only be implemented with a more precise local positioning system. Hence, it is important to consider whether the simpler approach will suffice for all indoor location-based applications that may be needed in the long term. If not, it makes sense (due to the cost of any positioning infrastructure) to implement one single positioning infrastructure, which can be extended in the future.

Known Network Positions vs. *Ad Hoc* Networks

Some existing infrastructure-based indoor positioning research systems include Active Badge, Active Bat, and PinPoint, which require the user to wear a transmitter that periodically emits a pulse picked up by a grid of receivers (the infrastructure), whose positions are known. The receivers compute the transmission time to determine the position of the user:

- Following the earlier classification, this makes them active network-based systems.
- Nodes need a wireless network connection and location sensor.
- Nodes need to obtain their geographic position from the network, which computes their location.
- The location model is stored on the servers.

These systems require that the position of the sensor nodes is fixed and well known, so they can compute the position of the mobile terminals. A possible alternative is using *ad hoc* networks, which rely on mobile devices (users) for the exchange of location information. This includes systems based on Bluetooth, which

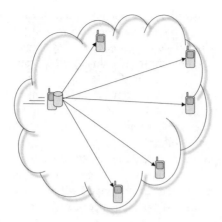

FIGURE 4.10 Absolute vs. relative positioning.

can connect up to nine terminals that do not have a fixed position, and which determine their proximity to each other. However, to resolve the relation to other things in the physical environment (and not just the network), a system that can provide at least one absolute position in the network of terminals is required.

The computational capacity of the terminals will also be a factor. One advantage of the Active Badge-type systems is that the terminals do not have any computational capacity whatsoever. All computing is done in the network. A system that is based on *ad hoc* networks, however, cannot be this simple, since the terminals need to compute the routing paths for packets. Due to the unpredictable mobility of these mobile devices (or nodes), the topology of an *ad hoc* network is unstable and likely to change frequently. Routing in such conditions becomes a challenging task. This means that this type of system always will be more costly than systems with passive sensors. A trade-off is if terminals are already used for some computing-intensive application, such as displaying data to the user, however. Then they will probably have capacity to spare.

When one attempts to implement a localization algorithm in real-world systems, significant problems arise. The first problem is that *physical obstacles* that obstruct line-of-sight connectivity between neighboring nodes prevent pairwise node distances from being obtained. This problem arises inside many buildings, where it is often hard to obtain line-of sight connectivity between rooms and open spaces. Moreover, physically realizable ranging hardware is often not omni-directional; for example, in an ultrasound-based location infrastructure such as Cricket, in which ceiling- and wall-mounted *sensors* broadcast spatial and ranging information to mobile *listeners*, the ultrasonic transmitters point toward the floor, making it less likely that a given pair of sensors will be able to measure their mutual distance.

Infrastructure-Based vs. *Ad Hoc*-Based (Self-Configuring) Systems

An additional problematic factor is the mobility of the terminals. This affects both the computational capacity required and the capacity required in the network. If terminals by and large remain in a fixed position, or move very slowly, it is possible to determine their position with less frequency, and hence less load on networks and less requirement for calculation and memory capacity in terminals — a factor that has to be considered when calculating the cost of systems. For instance, a system based on RFID cannot require terminals to do any significant amount of calculation, or they will not be as cheap as required.

With the development of wireless devices, a large amount of research has been (and is being) conducted in mobile and wireless communications. This research focuses in particular on how to use and how to deal simultaneously with heterogeneous devices and how to organize them into *ad hoc*, self-configuring networks that do not require a preestablished infrastructure. Such networks are assumed to be formed by mobile devices users carry, e.g., cell phones or personal digital assistants (PDAs). These devices are equipped with short-range wireless communication, such as Bluetooth or Wi-Fi in peer-to-peer mode, and can obtain the location of other devices.

These autonomous networks are called *ad hoc* networks. To make these networks autonomous, each node has to collaborate with its neighbors in order to exchange information. Thus, nodes can behave at the same time as routers or end systems. Hence, in contrast to the infrastructure-based network, in an *ad hoc*-based network the context information is retrieved directly from autonomous sensors in the vicinity and stored on the mobile devices.

To obtain context information that is not available locally, mobile devices have to exchange their stored sensor information with other mobile devices. As a result, it is likely that applications will only gain access to context in an infrastructure and the use within an *ad hoc*-based system.

- Nodes build *ad hoc* network
- Location model stored on devices

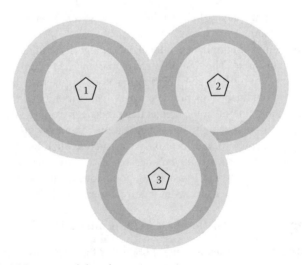

FIGURE 4.11 *Ad hoc* network location.

The scalability issue in *ad hoc* networks appears to be an important matter when the number of mobile nodes increases. Indeed, IP addresses are still used in many *ad hoc* networks in order to allow for the use of existing applications and to provide the system with authentication facilities. Routing cannot be based on IP addresses' network identifiers, however (although there is plenty of research around this as well). An alternative is used in Bluetooth, where each node has a unique address and the routing table is maintained by the master nodes.

In systems using IP addresses, such as Mobile IP, each mobile node joining an *ad h*oc network will keep its home IP address in order to maintain communication while moving. The network ID will not reflect its new network attachment. So the mobility issue complicates the routing process. However, there is plenty of research around this, and as long as the terminal can be identified in a locally unambiguous way, it can be positioned. We will not go deeper into the *ad hoc* routing algorithms.

Ad hoc systems use a set of sensors (nodes with known positions) that are spatially distributed throughout the areas of interest, so that mobile devices (nodes whose positions are known) can localize themselves by listening to nearby sensors. Most other positioning systems require an underlying infrastructure of sensors and have two advantages over positioning systems without sensors, in which mobile devices must establish a coordinate system and locate themselves in that coordinate system solely by communicating with each other. First, having sensors spatially distributed throughout the area lets devices compute their position efficiently in a scalable, decentralized manner without loss of accuracy. Second, even when the application permits offline, centralized position estimation algorithms, both the convergence and estimation accuracy can be significantly improved.

There also are *ad hoc* systems where the localization does not rely on each node being able to obtain its distances to the nodes near it (e.g., within radio range). In the absence of an external coordinate reference, this assignment can be unique only up to an arbitrary rotation, translation, and possible reflection, but its scale is determined by the measured ranges. For example, if three nodes are placed such that the pairwise distances between them are 3, 4, and 5 units, a correct coordinate assignment would be $(0, 0)$, $(3, 0)$, and $(0, 4)$. Because manually configuring each reference node with its position is cumbersome and error-prone, such systems benefit from a method to automatically localize the reference nodes. Given these pairwise distances, a distributed or centralized algorithm then computes a coordinate assignment for all nodes that is consistent with the measured pairwise node distances.

The following explains how *ad hoc* positioning works. Each mobile device measures the signal strength of both the neighboring mobile device and neighboring access points. The mobile device notifies the location server of the list of neighbors' IDs and their signal strength values. Each access point also measures the signal strengths of its neighbors and notifies the location server of the list of neighbors' IDs and their signal strength values. In addition, access points also notify the location server of their absolute locations. The location server calculates locations of mobile terminals using this information.

The process of position calculation on the server is described as follows: The list of signal strength values is received from mobile devices and access points (it is presumed that the positions of APs are known). Here, each mobile device and access point sends a packet containing a list of the signal strengths of neighboring mobile terminals and access points to the server. In this case, *neighbor* means that the terminal is within radio communication range of another terminal. Each access point adds its own location to the packet. Following, the signal strength is converted into the distance by the server and the initial positions of mobile devices are determined. The server calculates the initial positions of mobile terminals by using the positions of neighboring access points and neighboring initialized terminals. The initial positions of mobile terminals are determined roughly. Rough initialization at this step can avoid large errors between the final estimated position and the real position, and reduce the iteration in the following step.

An iteration process takes place until the modification of the positions of mobile devices reaches convergence. When the server starts to initialize the position of mobile devices, the server finds a set of the neighboring mobile devices and a set of the neighboring access points. The set of the neighboring mobile devices is subdivided into a set of initialized devices and a set of noninitialized devices. Lastly, mobile devices receive their position information from the server.

Applications using an *ad hoc* network rely on their localized location model. Inconsistencies do not occur — at least from the point of view of an application — since decisions based on the model are based on the locally stored data. *Ad hoc* networks do not allow the access of services from every position in the network due to partitioning of the network. As a result, applications relying on model data have to store them locally. Different applications on different devices can rely on potentially inconsistent states of the same model.

Relying on locally accessible data helps optimize the power consumption of devices. Second, applications on such devices are typically related to the user. Hence, the discovery of services in the proximity of the user can guide her through a smart environment. However, service discovery in *ad hoc* networks brings a number of additional problems, as we noted in the previous chapter.

Mobile *ad hoc* networks are constituted through mobile nodes without any *a priori* known topology, which is highly dynamic. Nodes can be devices, including cell phones, PDAs, etc. Sensing object locations with no fixed infrastructure represents a highly scalable and low-cost approach. In the future, infrastructural systems could incorporate *ad hoc* concepts to increase accuracy or reduce cost. For example, it might be possible for a system like Active Bat to use a sparser ceiling-mounted ultrasound receiver grid if Bats could also accurately measure their distance from other Bats and share this information with the infrastructure.

Examples of systems that attempt to implement such a combination include SpotON and HP Locus. The SpotON system implements *ad hoc* lateration with low-cost tags. SpotON tags use radio frequency signal attenuation to estimate intertag distance. They exploit the density of tags and correlation of multiple measurements to improve both accuracy and precision.

The *ad hoc*-based access to sensors makes context-aware applications possible even on devices that do not have any integrated sensors. The limited communication range of the devices involved ensures that the requesting device is in the vicinity of the node.

In many cases, this allows the assumption that the environmental context of the device is similar to the node context. Additionally, the *ad hoc* mode of the node can be used to make the context information available to devices that are in its vicinity but too far away to communicate directly. For such situations, it has been shown that a flooding-based information dissemination process can effectively distribute information in an *ad hoc* network. An application that makes use of such mechanisms is the Usenet on the fly. Here information that is grouped according to topics is replicated on devices in the spatial vicinity of the source.

Ad hoc sensor networking has a large community investigating many issues from distributed computation to cryptography to *ad hoc* routing protocols to data dissemination in low-power wireless networks. A primary driver for this work is the DARPA SensIT program, which seeks to create "cheap, pervasive platforms that combine multiple sensor types, embedded processors, positioning ability and wireless communication."

CLASSIFICATION BY SENSOR TYPES

In addition to the classification above (terminal vs. network, active vs. passive), the systems can be classified according to the sensor types used. In active systems, the signal is used for ranging, i.e., determining the distance of the beacon (which is usually worn) from the sensor (which is usually fixed). Since the position of the sensor is known, the position of the sensor can be computed.

Sensor Types: Active Sensors

Active sensors can be of several types, either based on the network (such as 802.11-type networks) or dedicated sensors (using other wavelengths of the electromagnetic spectrum).

Infrared Sensors

Infrared radiation (IR) is commonly used in remote controls for TVs and other consumer electronic products. IR-based positioning systems use short-range transmissions of modulated IR light to transmit the identity of a mobile device (tagged object) to a fixed receiver in a particular known location. Typically, a receiver is placed in every location in which a mobile device might be found. Several IR transmitters, which can automatically send their own IDs, are attached to places throughout a building. Mobile device emits IR light. If it is in the same room as a receiver, its position is known. A mobile device with an IR receiver uses these signals to determine its current position. In augmented reality (and to a lesser extent, virtual reality) applications there is a need for very high update rate, low latency, and high accuracy. To prevent motion sickness, more than a 60-Hz update rate and latency less than 10 msec are needed. In addition, registration errors between an augmented reality scene and the user's local environment must be minimized at all costs since they can contribute to user disorientation.

Because the IR light transmission range is limited to a few meters and is restricted to line-of-sight optical propagation, reception of a beacon message from a mobile device by a fixed base station is sufficient to establish proximity of the mobile device.

	Indoor GPS	Wi-Fi	Bluetooth	RFID	TV	UWB	IR	IrDA	Home RF
Network		1–11 Mbits/sec	1 Mbit/sec	N/A		40–60 Mbits/sec		16 Mbits/sec	1 Mbit/sec
Exchange of position/same communication network (infrastructure)		Yes	Yes	No			No	No/yes (limited)	
Security		Very good?	Very good	Good				Good	
Range		2–5 m	100 m/20 m/10 m	0–20 m		30 ft	10–30 cm	1–3 m	
Accuracy		Up to 100 m	Range	Range (triangulation)	Room level	6 in.	Room level	Range (triangulation)	
Power consumption	Low?		Low	Low				Low	
Device modification?									

The granularity of IR-based systems is low; it is not possible to determine the location of the badge at a higher resolution than the known locations of the base stations.

Systems using IR technology usually suffer from the limited range of IR, require significant installation costs, and perform poorly under direct sunlight or high ambient heat, which is likely to be a problem in rooms with windows or that hold machinery that emits heat (this may be a problem even if the ambient heat is not high, but if the system is placed close to a system with high infrared radiation characteristics, such as a printer). Usually, IR sensors are installed all over the indoor environment to pick up the signals emitted by the wearable badge from the mobile device.

The main incarnation of the infrared location-tracking technology is the Active Badge system, where it consists of a network of fixed IR transmitters and receivers (badge sensors) and a number of mobile IR computers (or badges). While IR components are now quite cheap (and the Infrared Data Association (IrDA) standard more resilient than its predecessors), the cost of deploying the transceivers is still quite high. There are commercial systems based on this technology (for instance, using IR instead of RF in a type of RFID tags), but so far they have not received wide deployment.

IR-based systems need not be expensive to produce, since they consist of only an inexpensive microcontroller and a few support components, and their batteries last about 1 year due to their 10-sec beacon repeat timing. The back-end processing infrastructure is also inexpensive, but since the mobile device low-power IR light signals travel only about 30 m in line of sight, frequent replication of this infrastructure is necessary to ensure that a user's beacon signal is received by the infrastructure. This leads to a relatively high total installation and maintenance expense, since a typical indoor environment is full of obstructing walls and corridors. As intervening objects can easily block infrared signals, RF-based positioning has emerged as a more attractive alternative.

The HiBall tracker uses a matrix of 3000 infrared light-emitting diodes (LEDs) mounted on the ceiling, which illuminate a wire-tethered tracking device consisting of six lateral-effect photodiodes. The LEDs are flashed in a predetermined sequence, and the tracking unit uses this pattern to estimate its position and orientation. Welch and Bishop claim a 70-Hz position update rate with RMS accuracies of about 0.2 mm (position) and 0.03° (orientation) over several meters of measurement distance. This is excellent performance, but must be weighed against the very high cost and complexity of the system. Because of the extreme infrastructure requirements and the tethered nature of the tracking device, the system will probably remain confined to research applications in augmented reality for which no other solution will work.

Advantages include:

- Does not penetrate walls or other opaque materials.
- Accurate location information due to short range and line of sight.
- Since it is the oldest and most installed positioning technology, there are a lot of vendor solutions available.
- No need for a host mobile computer in the object (or held by the person) to be located.

Disadvantages include:

- Scales poorly due to the limited range of IR.
- Incurs significant installation, reconfiguration, and maintenance costs.
- Receiver must be present in every room; connected using special wiring.
- Signal gets blocked and performs poorly in the presence of direct sunlight.

IrDA

Infrared Data Association (IrDA) is a communication system based on IR. It is commonly used in mobile devices for inexpensive point-to-point communication (IrDA is possibly the most common wireless link (100 million ports installed, but how common is its use is unknown). Mobile phones (in addition to digital cameras and laptops) use IrDA for wireless communication.

Similar to Bluetooth, IrDA is able to share the position information with other devices. IrDA can create networks with bypassing devices, where they take a role of either a master or slave. The roles of master and slave are dynamic in IrDA, though some IrDA devices may never be able to take the role of master because of its simplicity. Through the created networks they can easily share the position information they have. IrDA has automatic service discovery, which checks if the positioning service is present in the discovered device.

IrDA is a direct line-of-sight narrow angle communication system working at the IR frequency spectrum. IrDA has classified its products in two different standards, IrDA data and IrDA control. The IrDA data standard is usually used for high-speed, short-range applications, whereas IrDA control is usually recommended for low-speed applications, like keyboard, joystick, and mouse controls. IrDA has its own set of protocols covering all layers of data transfer and some optional protocols for handling network management. The working range for IrDA is 1 to 2 m and supports bidirectional communication with data transfer rates of 9600 bps to 4 Mbps.

Among Bluetooth, IR, and RFID, Bluetooth was found the most suitable communication system based on range, speed, and services.

Ultrasound Sensors

Ultrasound technology is mostly used in positioning systems for improving the accuracy, as it provides the finest granularity. Among the several technologies for indoor ranging, ultrasonic time of sight (TOS) provides the finest granularity, since it has the smallest minimum unit of distance that can be measured accurately. This is the technology of choice for the best-known indoor localization systems in research. Two of the best-known systems, the Active Bat and the Cricket systems, use ultrasound and radio frequency measurements. The MIT Cricket and Active Bat location systems are two primary examples that use the ultrasonic technology. Normally, these systems use an ultrasound time-of-flight measurement technique to provide location information. Most of them share a significant advantage, which is the overall accuracy. Cricket, for example, can accurately delineate 4×4 ft^2 regions within a room, while Active Bat can locate Bats to within 9 cm of their true position for 95% of the measurements. However, the use of ultrasonic technology this way requires a great deal of infrastructure in order to be highly effective and accurate, yet the cost is so exorbitant that it is inaccessible to most users. Position detection by means of ultrasound calculates an entity's three-dimensional position based on the distance between ultrasonic senders and receivers. Either the sender or the receiver can take on the fixed part of the system; it can, for example, be installed in the ceiling of a building like in the Bat system.

However, ultrasound has the same disadvantage as the infrared systems discussed earlier. These systems usually require excessive manual intervention, such as *a priori* knowledge of the position of the nodes that compute the location of a node transmitting the signal. Similar with the Wi-Fi systems that achieve high accuracy due to site surveys that point out the AP nodes, the location of the beacons has to be entered manually in a database or programmed into the beacons themselves. In addition, the placement of beacons to minimize interference between the ultrasound transmissions can be a constraint.

The Active Bat system requires *a priori* knowledge of the position of the nodes that compute the location of a node transmitting the radio frequency and ultrasound pulses. In addition, triangulation computations to locate the transmitter node are done on a central computer. Hence, this system is centralized, not scalable, and requires manual entry of the receiver node position. In the MIT Cricket system, the location of the beacons has to be entered manually in a database or programmed into the beacons themselves. In addition, this system has specific constraints on the beacons' placement to minimize interference between the ultrasound transmissions.

MIT Cricket comprises a fixed infrastructure of ultrasound emitters used by mobile nodes to determine their positions by trilateration. There is no centralized authority providing spatial indexing. Instead, mobile nodes are assumed to be capable of moderate-bandwidth wireless communication with a backbone network, and to carry sufficient battery power for the cost of communicating to be negligible. Nodes discover and bind to local spatial services in accordance with their own agendas. Services independently choose indexing algorithms as best optimize their application. Unfortunately, the model is best suited to indexing static data sets with the changing positions of inquiring nodes. Determining the positions of and interactions between moving objects requires each to push its location to a central service at regular intervals.

Precise distance measurements require sensitive ultrasonic sensors, but such sensors react to ambient ultrasonic noise and high-energy sound pulses. In particular, malfunctioning fluorescent lights, people jangling keys, and loud noises caused by slamming doors cause the listener to record bad distance samples. For this reason, MIT Cricket uses good outlier detection algorithms to filter out many such sources of error.

Radio-Based Sensors: Bluetooth, UWB, RFID

Bluetooth

Bluetooth is a short-range radio standard that allows various wireless equipment (mobile phones, mobile computers, etc.) to communicate over relatively short ranges of about 100 m, enabling point-to-multipoint voice and data transfer. Bluetooth operates in the industrial, scientific, and medical (ISM) 2.4-GHz region. Bluetooth operates in the free ISM band. This band is applied by various other technologies, like 802.11b and g and microwave ovens. The Bluetooth protocol uses a half-duplex, frequency-hopping scheme operating in the 2.4-GHz band. Bluetooth devices hop through 1600 frequency channels per second, with 800 transmit and 800 receive channels. The channels span 79 MHz, with a 1-MHz spacing between adjacent channels. Bluetooth facilitates high-quality voice and data transmissions. It is a low-cost and low-power technology, and it makes wireless communication between many different types of devices possible. Bluetooth is designed to be fully functional even in very noisy radio frequency environments.

Both Wi-Fi and Bluetooth operate in the 2.400- to 2.485-GHz spectrum in the ISM band. Bluetooth jumps between the channels 1600 times per second, whereas 802.11b sticks to a predefined channel.

Bluetooth technology is getting popular and cheaper, and almost all the handheld devices and cellular phones (all GPRS and UMTS phones sold today have Bluetooth installed) now come with Bluetooth. Bluetooth application-specific integrated circuit (ASIC) chips are under production, and once mass produced, would bring the cost down to $5 or less. By integrating Bluetooth positioning technology with GPRS/UMTS as content bearer, BLIP Systems can be used to provide localized messaging and web content to anybody with a Bluetooth-equipped mobile phone.

Bluetooth networks provide an underpinning for sharing information among devices located within 3 m of each other. This can be used to allow multiple devices to share one 802.11b network connection, and determine pico positioning around a device with known location attributes.

One of the target applications for Bluetooth networks is to provide more accurate and lower-cost indoor positioning. Due to the expected popularity of indoor local positioning with Bluetooth, there has been a special interest group (SIG) to define the related positioning standard. The first draft was released in July 2001 and focused entirely on developing the underlying software architecture to maintain and exchange location information between Bluetooth-enabled devices.

The local positioning profile defines the protocols and procedures that shall be used by Bluetooth devices implementing the following usage models: position determination and location awareness. Moreover, the specification defines the data transfer mechanism and related data formats. The minimum mandatory requirement is for LP-compatible clients to locate, select, and request or exchange LP data (position coordinates plus associated uncertainty and confidence) with LP-compatible servers. If the server does not know its position accurately, it may increase the uncertainty or it may degrade its confidence to 0% (position undefined). Optional data may also be exchanged.

Because Bluetooth uses a relatively short-range radio, devices at unknown positions can estimate their positions by proximity to devices with known positions. Bluetooth is able to share the position information with other devices. It can create networks with bypassing devices, where they take a role of either a master or slave. The roles of master and slave are dynamic. Through the created networks, Bluetooth can easily share the position information they have. Also, Bluetooth has automatic service discovery, which checks if the positioning service is present in the discovered device.

With the Bluetooth 1.2 specification, the adaptive frequency-hopping (AFH) feature was introduced in Bluetooth. The goal of AFH is to allow Bluetooth to coexist with other non-frequency-hopping technologies and interferers (such as IEEE 802.11b wireless LAN devices or microwave ovens) in the ISM band. This is accomplished by adapting the Bluetooth hopping sequence to avoid those frequencies occupied by other users of the ISM band. The Bluetooth devices keep statistic over bad and good channels. Bluetooth channels located at the same frequencies as an 802.11 channel will over time be marked as bad, due to the increased retransmission rate. Both the master and the slave can identify bad channels. Adaptive frequency hopping is good for reallocation of Bluetooth when frequencies are in continuous use by other technologies. A fully successful reallocation requires that it is possible to find at least 20 MHz of available spectrum. AFH involves exchange of channel maps between the master and the slave. Such maps can only be exchanged upon connection establishment. This implies that AFH has no effect on the connection establish procedures such as inquiry and paging. A train of ID packets are transmitted continuously during inquiry and paging procedures.

Normally paging and inquiry are only performed occasionally when a link is established. BlipNet may, dependent on the use case, perform periodic inquiries and pagings, e.g., to detect moving devices to ensure handover between APs. If a BlipNode is applied for such use cases, it is recommended to locate the BlipNode at least 2 m from the 802.11 access point.

It is important to check if there is no need to do a Bluetooth qualification. The BlipNet product, for example, is already qualified for a set of profiles. There is no requirement for qualification of applications when using the BlipNet API. (Before a Bluetooth product can be marketed as Bluetooth enabled, it must be qualified by the Bluetooth SIG. An independent test house must verify that the product is implemented according to the specifications. The qualification process involves a lot of work and can be quite expensive.) The Bluetooth web site (http://qualweb.bluetooth.org/Template2.cfm?LinkQualified=QualifiedProducts) contains a complete list of all qualified devices.

The Bluetooth SIG has defined a set of use cases. For each use case, a profile has been defined. The profiles describe a reference architecture, including the communication protocols that should be applied, and they list the mandatory and optional procedures that should be supported.

In terms of establishing a basic link, when a paging is made the Bluetooth address of the remote device must be known. If the initiator of the paging has knowledge about the value of the clock in the remote device, the paging can be completed very quickly. The clock is obtained as a part of an inquiry procedure. Normally such clock data cannot be shared between Bluetooth access points.

Two different positioning methods are applicable with Bluetooth:

- Triangulation
- *Ad hoc*

With triangulation, received RSS levels are measured to obtain distance estimates to multiple known location Bluetooth devices and triangulate the user device position. One Bluetooth-based positioning system's range estimation is based on an approximation of the relation between the RSSI and the associated distance between sender and receiver. The actual location estimation is carried out by using the triangulation method.

The Bluetooth specification (Bluetooth Special Interest Group, 2001) makes it optional for the device manufacturers to give RSS information for a particular link. (This means that the device will need to be manufactured for positioning services.) It does not provide a means for software developers for the exact measurement of the RSS (as in the case of Wi-Fi) to enable easy positioning calculations (specifically, it is optional to implement the host controller interface command Read_RSSI for reading RSSI). A Bluetooth device only needs to be able to tell whether the RSS is acceptable, too strong, or too weak. (This granularity is not enough for developing a positioning system.)

Bluetooth performance can be measured in terms of file transfer time, which points to the effective bandwidth for a Bluetooth link. (Note: A separate working group called the Local Positioning Working Group has been created with an aim to develop a Bluetooth profile that describes the type and format of messages allowing Bluetooth devices to exchange position information and also the algorithm to compute the position.)

Because Bluetooth works in a relatively short range, the system is relatively limited in accuracy and may not be sufficient for certain applications. Fortunately, more than one metric is available for positioning calculation, and it is expected that these metrics can be combined in order to improve accuracy.

The expectation is that a set of heuristics can be created to allow the Get_Link_Quality and Read_RSSI (or BER) commands to be used to perform relative distance calculations between two Bluetooth devices. The Bluetooth protocol allows two devices to dynamically adjust their transmit powers in order to maintain a reasonable quality of service. Thus, the Read_Transmit_Power_Level command must be used to allow the positioning service to dynamically adjust its calculations based on the current transmit power level.

To increase the accuracy, the Bluetooth standard on the local positioning says that "measuring RSSI and/or BER can be used to enhance the accuracy of the proximity estimation." To measure the RSSI or BER, a client–server application can be developed using open-source Bluetooth kits. The changes in RSSI and BER with the distance can be measured using protocol analyzers. When connected to a Bluetooth piconet, these analyzers monitor and record network activity and display information about the recorded packets. BER measurements are very low if the Bluetooth devices are close by, and as the distance between devices increases, the error probability is mainly governed by the failure in synchronization (communication failed) rather than by the error probability in the payload of the packets.

Once a connection to the service is established, the device offering the service will periodically determine the current transmit power level for the connection with the remote device. This information and the current known position will be shared with the remote device. The remote device can dynamically adjust its distance calculations (whether using link quality or RSS as a metric) in order to determine the relative distance from each other and, hence, its relative distance from a known position.

The other method works based on mapping known Bluetooth device addresses to location information. The basic concept in Bluetooth communications is the piconet, a collection of Bluetooth devices that can discover and communicate with each other by sharing a common channel.

Bluetooth devices form these mini-cells, which is similar to cellular networks. If enough cells are installed, the position of a transmitter can be surmised by knowing the cell with which a device is communicating. Or the greater the number of cells, the smaller the size of each, and that translates into greater accuracy. Further, these two methods can be combined into a hybrid system.

A piconet is a star-shaped wireless network in which one device plays the role of the master and every other device plays the role of the slave. Master and slaves of a piconet are synchronized according to a channel-hopping sequence that is a function of the master's BD_ADDR and clock.

A Bluetooth piconet can have 7 active slaves and up to 200 inactive devices in parked mode. For positioning application, it is enough to see another Bluetooth device. Also, it does not matter if the other device is a master or a slave. The device should be within the range to be detected. This implies that a Bluetooth sensor would be able to detect up to 200 other Bluetooth devices.

The microcellular structure of Bluetooth networks can be used to determine the position of Bluetooth devices with an accuracy determined by the size of the piconet. A positioning system can consider each room of the building as a granule of location information. A room can be defined as a space that can fit into a circle of 10 m radius since this is the dimension of the greatest coverage area of a Bluetooth picocell. This is in the order of 10 m for low-power (0-dBm) chips or 100 m for high-power (20-dBm) chips.

If two Bluetooth devices can communicate with each other, they are nominally within 10 m of one another (for standard, class 2 Bluetooth radios), though in an unobstructed environment, the distance may be much greater, up to 100 m. The simplest approach would be to assume the client's position is the same as that of the most recently contacted server.

A nearby Bluetooth device needs to be discovered. (This process is called an inquiry; when a Bluetooth device is in the inquiry state, it continuously transmits inquiry packets carrying an access code and hops frequencies 3200 times a second, twice as fast as the normal connection mode.) A device that allows itself to be discovered regularly enters the inquiry scan state to respond to inquiry messages, hopping frequencies only once every 1.28 sec. When it receives an inquiry packet, it first waits for a random period of time. Then it listens for another inquiry packet and responds to it with a frequency hop synchronization packet. This packet essentially carries all necessary information for establishing a connection between the devices. The reason for the random delay is to avoid all nearby devices responding simultaneously to the first inquiry packet.

In order to discover all discoverable devices in an error-free environment, the device must spend at least 10.24 sec in the inquiry mode when the maximum number of slaves in the piconet reaches 10; on average the master succeeds in discovering about 90% of these slaves in the first 1 sec. Only in the second operational cycle does it discover 100% of the slaves. When the number of slaves increases by 5 (i.e., 15 to 20), usually the slaves are all discovered in two cycles.

Some mobile devices (e.g., SymbianOS) support Bluetooth device discovery (support it with its RHostResolver and TInquirySockAddr classes). By default, inquiry results are cached for about 2 min. This cache can be disabled so that there will not be false positive position fixes 2 min after the user has left the vicinity of a Bluetooth sensor. Other devices (e.g., Nokia 3650) present a more significant challenge where the inquiry state is terminated after about 5 sec. Because of this time-out, there is only about a 50% chance for a device to be discovered during an inquiry period.

When a Bluetooth sensor is implemented in a mobile device, the device periodically scans for nearby Bluetooth devices. Positioning is realized with either of the following two approaches:

- Run the Bluetooth software on an end user's mobile phone. Use dummy Bluetooth tag-like devices in known locations as reference points to be discovered. Location can be derived based on the addresses of discovered tags. If multiple tags have been discovered, positioning accuracy can be improved by triangulation.
- Run the Bluetooth software on a suitable device (e.g., a mobile phone or a PDA) in a known location. End-user mobile phones are devices to be discovered, and their device addresses map to user accounts.

The time taken by a Bluetooth device to detect another Bluetooth device in its vicinity is not short enough to track a moving object. The device needs to be powered on and initialized. Initialization may require providing a PIN for the creation of a link key, device authentication, and data encryption. In addition, many of the Bluetooth products available today have their refresh time fixed to a certain value. For example, in the 3COM Universal Serial Bus (USB) adapter, the smallest refresh time is 5 min, and it increases in multiples of 5. Future devices should have an option so that this time is much shorter and configurable. This would mean that other Bluetooth devices would be detected with very little delay and a system for tracking moving objects can be designed.

A feature mentioned in the Bluetooth specification but not implemented in any of the recent Bluetooth products is to give the signal strength information for a particular link. If this information is made available, it can greatly aid in accurate location sensing. In the current Bluetooth specification, there is provision for three power levels. The minimum power level gives a range of 10 m. If future implementations can give a shorter range, then location sensing can be made more accurate. Since Bluetooth devices are expected to be cheaper in the future, we can use more sensors covering a very short area, thus improving accuracy.

Some of the limitations of Bluetooth are the following: The Bluetooth positioning system needs to be made more reliable. To achieve this, the inquiry time-out should be made longer. This would make the positioning latency longer but more predictable. To shorten the latency the Bluetooth sensor should not wait for the inquiry to time out before sending the device addresses of found devices, but send them as soon as they are discovered. Guessing user location based on his or her previous locations could be another possibility.

Also, because a Bluetooth network is typically used primarily for purposes other than positioning, and might well be constrained for power consumption, devices may only be available for positioning part-time. This introduces latency and reduces the accuracy and availability in environments where devices are moving. As time passes since the information was obtained, its uncertainty increases and its confidence decreases according to how likely it is that the device has moved.

There is 802.11b interference. Some studies propose the use of 802.11a as an alternative. Thus, there will not be any chance of interference between Bluetooth and an 802.11a device.

In general, there are no problems in operating multiple Bluetooth piconets within the same area. The throughput in piconet degrades gracefully as the number of piconets within an area is increased. Devices within a piconet are synchronized. So it is the number of co-located piconets that determines the number of collisions and not the number of devices. AFH cannot be applied to prevent interference from other Bluetooth devices, because of the hopping nature of the disturbance. Disturbance from other Bluetooth devices is only really relevant in case Bluetooth 1.1 SCO (voice) links are applied. Users of SCO links operated in areas with other Bluetooth piconets may notice a popcorn effect when SCO packets are lost due to collisions. This problem is solved with the Bluetooth 1.2 specification, where the extended SCO link is introduced. SCO links are not using retransmissions; extended SCO are able to retransmit voice. For example, BlipNet allows definition of channel maps via the BlipManager; this makes it feasible to control where piconets are located in the ISM band.

Interconnected piconets are called *scatternets*, and their aim is to allow more than eight active Bluetooth devices in the same network while augmenting their range by bridging. However, scatternet formation and operation algorithms are not part of the Bluetooth specifications yet. In the frame of our work we try to develop new methods for optimizing communications in scatternets, taking advantage of localization information that we can gather from the mobile devices. While the Bluetooth specification is strict with respect to the operation of the piconet, the creation and maintenance of a Bluetooth scatternet are open issues. Each piconet is based on a star topology, with a master and bounded number of slaves. The connection with the other piconets is achieved by using shared piconet members — *bridge nodes*. For example, a bridge node can be a slave in two neighboring piconets, and it participates in both piconets on a time-sharing basis. For efficient functioning of devices connected into a scatternet, several problems should be addressed. Scheduling schemes should be defined for bridge nodes that provide relevant time multiplexing of these nodes between different piconets. These schedules can be based on usage of low-power modes introduced in Bluetooth specification. Furthermore, some performance degradation is expected in a scatternet because of time multiplexing of bridge nodes and since a scatternet is a multihop network.

In broadcast-based wireless LANs such as 802.11, the network topology is determined by the physical distance between nodes. In Bluetooth, an explicit topology formation process is required since nearby devices need to discover each other and explicitly establish a point-to-point link. During the link formation process, the two Bluetooth nodes synchronize the frequency-hopping sequence and gather necessary clock information. The essential *ad hoc* discovery process could be lengthy, and clever solutions are required to quickly form a network topology that spans across all nodes within the transmission proximity.

UWB

Overall, positioning and location are hard problems, especially in buildings where multipath signals tend to ruin the position accuracy. One possible solution is ultra-wide band (UWB). UWB achieves higher accuracy due to its wide bandwidth, solving the problem of multipath in an indoor environment. UWB works in the unlicensed frequency of the RF spectrum commonly used for garage door openers, portable telephones, and baby monitors.

UWB operates by emitting a series of extremely short pulses across a very high band of frequencies simultaneously. Also, UWB technology is built around sending short (~1 nsec) discrete pulses (instead of using short RF pulses of high bandwidth), which enables accurate TOA measurements with good multipath tolerance. Such pulses are only 30 cm wide and, as a result, can be distinguished more than 1 ft apart. For these reasons, UWB systems are robust to multipath delays of more than one pulse width and overcome the potential for self-interference caused by the reflection of RF signals between readers and tags within a building. This is because UWB transmits short-duration radio waves that are finished before they can reflect off walls and ceilings, thus eliminating the opportunity for cancellation. Positioning using UWB can also be more accurate than other technologies. Most location-based positioning systems that use radio frequency-based technologies (i.e., Wi-Fi or Bluetooth) can pinpoint objects within 3 to 4 m (10 to 16 ft) without any additional computations. Systems that use UWB (i.e., UbiSense) can pinpoint and track to an accuracy of approximately 2 m (6 ft). UWB's very wide bandwidth allows transceivers to avoid the multipath problem, which is common in indoor positioning systems and reduces the position accuracy. Ubisense UbiTags use unidirectional UWB with a conventional bidirectional time division multiple access (TDMA) control channel Tags transmit UWB signals to networked receivers and are located using angle of arrival (AOA) and differential time of arrival (DTOA). There are other UWB systems to date that have been fielded (Fontana, 2000; Fontana and Gunderson, 2002). The soldier tracking system was the first to be developed and fielded, and was designed to track personnel and vehicles without the use of GPS over areas exceeding a few square kilometers. The system was tested at the Ft. Benning, GA, McKenna military operations in urban terrain (MOUT) site in 1997 and demonstrated the ability to achieve foot-type resolutions over a 4-km area. A smaller version of this system was subsequently developed in 1998 to perform indoor mapping, wherein the UWB tracking system was used to correlate position information with video still imagery to construct a three-dimensional AutoCAD model of the inside of a facility. A further size reduction and improvement in performance resulted in development in 2002 of the precision asset location system (PALS) (Fontana and Gunderson, 2002), which was used for tracking of ISO containers inside a Navy ship, a particularly severe multipath environment with all metal floors, walls, and ceilings. Each of these systems operated at band, and hence could not be used commercially.

UWB ranging makes use of nonsinusoidal electromagnetic signals such as impulses. Since there is no carrier frequency, these are sometimes called time domain, carrierless, or baseband signals.

The advantage of the UWB paradigm is the improved ability to reject multipath signals. If precise and robust electromagnetic ranging in indoor environments ever happens, it will be based on UWB impulses. Logically, this ought to be doable with much simpler electronics than those required to demodulate a complicated spread-spectrum signal. If this turns out to be true in practice, and if the Federal Communications Commission (FCC) develops a policy that allows UWB transmissions without too many restrictions, then this may eventually become a preferable method of ranging in VE motion-tracking systems. Compared with radio frequency identification (RFID), UWB has better range, which makes it better for positioning or tracking. RFID tags generally must be within 1.5 to 2 m (6 to 8 ft) from scanners, while UWB tags (e.g., Ubisense's UbiTags) have demonstrated its UWB system at over 50 m (150 ft) (though with reduced accuracy).

However, there are problems in deploying UWB systems, in particular since they are seen as noise by other systems in the licensed bands that the system uses. This actually makes it illegal to deploy them in some countries. In addition, the systems are still expensive, since they have not yet reached wide volumes; this may make it cheaper to add triangulation to Bluetooth, which would allow it to address the same type of problems.

Today's UWB-enabled chips have a range of only around 30 ft, but that is due to a cap on the amount of power they can transmit. The FCC may change that cap, which would allow telecoms to replace their code division multiple-access (CDMA) cellular towers with the cheaper, more powerful UWB chips.

In addition, UWB's data throughput potential, and its ability to support a piconet (an *ad hoc* network of devices using the Bluetooth networking standard), means it has the potential to displace technologies used in local-area networks.

UWB's impact could be dramatic on technologies used in local-area networks and, eventually, even the CDMA cellular networking standard that is deployed by many U.S. cellular carriers. A number of manufacturers plan to make UWB-enabled cell phones in order to provide the same kinds of short-range networking features and functions that Bluetooth can provide.

Some of the advantages of UWB are the following:

- The wide-bandwidth communication protocol allows a device to consume very little transmit energy while transmitting large amounts of data. Transmitted radio signals behave similarly to radar, which allows accurate location of the transmitters.
- UWB is superior to infrared tracking, ultrasonic tracking, and other radio-based positioning systems in terms of accuracy. (Most systems using radio technologies such as Bluetooth or Wi-Fi can only pinpoint an item within about 10 to 16 ft (3 ft with site surveying) in an indoor environment. With UWB, Ubisense is able to track to an accuracy of 6 in.).

- Compared to RFID, UWB has better range. RFID tags generally must be within 6 to 8 ft from scanners, while one existing UWB system requires 150 ft (though with reduced accuracy).
- Multipath immunity — UWB's very wide bandwidth diminishes the multipath problem in an indoor environment.
- In addition, the use of short-pulse RF waveforms provides inherent precision for TOA/TDOA measurements (three-dimensional positions possible).
- High-speed data transit capabilities of 40 to 60 Mbits/sec, in some cases nearly 10 times as fast as Wi-Fi; low-power requirements; its ability to penetrate walls, and its use GPRS information.

RFID

RFID is a less sophisticated form of positioning offered by such manufacturers as WhereNet and CheckPoint. A company can place an RFID tag (which is a small database accessible through an RFID scanner) on a variety of items. With the installation of RFID scanners through a facility (such as a warehouse), the system can keep track of the items. When an item containing an RFID tag passes a scanner, the scanner retrieves the identification code and sends it to the server. The server keeps track of all items using this approach.

RFID tags have many applications, but they are typically used as a substitute for bar codes. RFID tags can also be used for location determination. RFID technology has been successfully used in a wide range of markets and is expected to play a primary role in future mobile location applications since it enables the automated data collection and tracking of objects as they move across a limited geographical area. In such systems, RFID tags are placed at key points and the relation between tag ID and location is entered into a database. When a user moves the tag reader closer to a tag, the reader reports the location of the tag to the user. Some manufacturers are already attempting to push RFID into the mainstream with the launch of RFID-enabled mobile phones. The MIT AutoID Laboratory, for example, is looking for a global system that enables the efficient tracking of the RFID information across suppliers and retailers. It is building a device-neutral middleware framework for managing spatially separate device networks, which may include RFID tags and other sensors.

The accuracy of an RFID LBS depends on the number and placement of RFID scanners. For example, the system may only identify the location of an item within a few hundred feet if scanners are far apart. Some RFID scanners, though, can determine the exact location of an item as it passes the scanner. As a result, RFID LBS is very common in mail processing plants and airport baggage transport systems, where it is necessary to identify a particular item and route it a certain direction on a conveyor belt.

RFID is used for identifying and tracking objects within a few square meters. Through RFID labels, almost any device or item can be cost-effectively transformed into a smart information source. In conjunction with the electronic product code infrastructure (EPC Network), smart items will be capable of wirelessly telling who and where they are, what status they have, and which services they offer.

A basic RFID-based system consists of three components: (1) the transmitter, (2) a transceiver with decoder, and (3) an RFID tag electronically programmed with unique information. An RFID system integrates a transmitter with electronic circuitry to form a transponder that, when polled by a remote interrogator, will bounce back an identification number.

The transmitter emits radio signals to activate the tag and read and write data to it. Transmitters are the link between the tag and the transceiver, which controls the system's data acquisition and communication. The electromagnetic field produced by an antenna can be constantly present when multiple tags are expected continually. If constant interrogation is not required, a sensor device can activate the field. Often the transmitter is packaged with the transceiver and decoder to become a reader, which can be configured as either a handheld or a fixed-mount device. The reader emits radio waves in ranges of anywhere from an inch to about 100 ft or more, depending upon its power output and the radio frequency used. When an RFID tag passes through the electromagnetic zone, it detects the reader's activation signal. The reader decodes the data encoded in the tag's integrated circuit and the data are passed to the location-sensing computer for processing.

RFID readers have a range of 10 ft, assuming the tags are oriented correctly. If the tags, often manually applied to pallets, are oriented incorrectly or in the wrong place, readers will not be able to pick them up. Tags may cost as little as $0.23 each in volume, but the readers range from $800 for a PC card device to about $6000 for a top-of-the-line model (2004 figures).

Each reader needs to connect to the Ethernet network; that means wiring drops and Ethernet switches in the venue's ceiling. Although you can associate a tagged palette with the last reader that scanned it, the cost of deploying enough readers in a warehouse to know where the palette is all the time can be prohibitive.

One well-known location-sensing system using the RFID technology is SpotON. SpotON uses an aggregation algorithm for three-dimensional location sensing based on radio signal strength analysis. SpotON researchers have designed and built hardware that will serve as object location tags. In the SpotON approach, objects are located by homogenous sensor nodes without central control. SpotON tags use received radio signal strength information as a sensor measurement for estimating intertag distance. However, a complete system has not been made available as of yet.

Mobile robotics use RFID technology for robot navigation. Kantor and Singh (2002) used RFID tags for robot localization and mapping. Once the positions of the RFID tags are known, their system uses time-of-arrival type of information to estimate the distance from detected tags. For instance, Tsukiyama (2003) developed a navigation system for mobile robots using RFID tags. The National Institute of Standards and Technology's (NIST) RFID-Assisted Localization and Communication for First Responders project, for example, determines the feasibility of using RFID-assisted localization in combination with an *ad hoc* wireless communication network to provide reliable tracking of first responders in stressed indoor RF environments, where GPS-based localization and links to external communication systems are known to be unreliable.

The research also considers the means and potential for embedding critical building/occupant information in specific on-site RFID tags to enhance the safety and efficiency of the first responders' mission as well as to minimize dependence upon communication with external building databases.

Active vs. Passive RFID Tags. RFID tags are categorized as either *active* or *passive*. Active RFID tags are used in RTLSs (real-time location systems). RTLS tags cost about $55 each and have a range of approximately 350 ft. An RTLS reader costs about $3000 (2004 figure). An active RFID tag is powered by an internal battery and is typically read/write. An active tag's memory size varies according to application requirements. The battery-supplied power of an active tag generally gives it a longer read range. The trade-off is greater size, greater cost, and a limited operational life.

Passive RFID tags can only be located by associating them with the reader that reads them. At a 10-ft range, you know the item is within those 10 ft. At 150 ft, send out a search party. They operate without a separate external power source and obtain operating power generated from the reader. An antenna transmits a radio signal to the RFID tag, which disturbs the signal in a known predictable way and bounces the modified signal back to the transmitter. As a result, passive tags are much lighter than active tags, less expensive, and offer a virtually unlimited operational lifetime. The trade-off is that they have shorter read ranges than active tags and require a higher-powered reader. Read-only tags are typically passive and are programmed with a unique set of data (can hold small amounts of data and the reader can change these data) that cannot be modified. Read-only tags can be thought of as a replacement to bar codes.

Their frequency ranges also distinguish RFID systems. Low-frequency (30 to 500 KHz) systems have short reading ranges and lower system costs. They are most commonly used in security access, asset tracking, and animal identification applications. High-frequency (850 to 950 MHz and 2.4 to 2.5 GHz) systems, offering long read ranges (greater than 90 ft) and high reading speeds, are used for such applications as railroad car tracking and automated toll collection. However, the higher performance of high-frequency RFID systems incurs higher system costs.

Some of the advantages of using RFID for positioning are the following:

- There is no contact, non-line-of-sight nature of the technology.
- Tags can be read through a variety of substances and environmentally challenging conditions, where bar codes or other optically read technologies would be useless. RFID tags can also be read in challenging circumstances at remarkable speeds, in some cases responding in less than 100 msec.
- The read/write capability of an active RFID system is also a significant advantage in interactive applications such as work in process or maintenance tracking.
- Though it is a more costly technology (compared with bar codes), RFID has become indispensable for a wide range of automated data collection and identification applications that would not be possible otherwise.
- Identification at a distance, hands-free operation, versatile memory and processing requirements, and high accuracy due to the very short operating range.

- Low tag cost of $0.50 (passive tags only).
- Active tags can transmit longer distances with smaller antennas (2 to 3 m).

Some of the disadvantages are the following:

- RFID reader devices do not have a communication network that supports exchange of positioning information.
- RF tags are manually programmed and cannot change the positioning information without reprogramming — unless their ID number, along with the position, is stored in a location server and changed there. However, RFID has no means of providing a network connection, so the lookup would have to go through some other communication network.
- Passive and active tags: Read distance is limited (up to several inches for passive; 2 to 3 m for active). Can be increased to several feet by making the antennas larger (however, this makes the antenna cost go up).
- Antenna needed at every door to achieve any localization.
- Close proximity needed between the tagged device and reader so tag can be read.

ZigBee

The evolving open standard has a radio frequency range of 30 to 225 ft and uses very little power. The initial review of companies at the ZigBee Alliance Meeting in 2005 revealed only Motorola (via spin-off FreeScale) was engaged in developing a ZigBee solution for positioning, based on the NeuRFon Work Initiative within Motorola Research. Other companies, such as Ember and Millenial, were not working on this. The consensus from over 16 companies surveyed was that position finding with ZigBee would be difficult unless a portal strategy is used, i.e., restricted physical portals set up through which assets must pass. Participants estimated the likely accuracy for any other solution using ZigBee to be on the order of 3 m, with an update frequency of perhaps 1 to 3 min.

Sensor Types: Passive Sensing

Video-Based Sensors

Location information does not have to be derived from a signal. There are other ways of determining what is around the terminal. This information can also be derived from analysis of data such as video images, as in the MIT Smart Rooms project. Vision is a natural sensing modality for tracking the location, because it does not require that the room's occupants carry or wear any special devices. Also, vision can resolve the location of people in the room well enough to infer a person's intent based on his or her position. Vision can even tell when either person stands or sits. Vision can maintain the identity of people in the room, allowing the room's devices to react to a specific person's personal preferences. Vision can be used to find objects in the room. For instance, cameras are used to locate the wireless keyboard on the coffee table. Vision can also be used to make a geometric model of the room. Images from the people-tracking cameras are used to make a floor plan of the room.

While vision has unique advantages over other sensors for tracking people, it also presents unique challenges. The systems have the same line-of-sight problem as IR and work well with only a small number of persons in a room with nonfrequent occlusions. Accurate object locations can be determined in this way using relatively cheap hardware, but large amounts of computer processing are required, and all images of objects that are to be located need to be fed into the system. Furthermore, current image analysis techniques can only deal with simple scenes in which extensive features are tracked, making them unsuitable for locating many objects in cluttered indoor environments. It is particularly difficult to track people. A person's appearance in an image varies significantly due to posture, facing direction, distance from the camera, and occlusions. It can be particularly difficult to keep track of multiple people in a room as they move around and occlude each other. Although a variety of algorithms can overcome these difficulties, the final solution must also work fast enough to make the system responsive to the room's occupants. A low-cost alternative is to affix bar codes on known locations, but this requires that the mobile system can identify and locate these, and that there is a way of resolving their position. The current vision system, using two sets of stereo cameras mounted high on the room's walls, successfully tracks the location and identity of people in the room with an update rate of about 3.5 Hz.

With computer vision the room is fitted with cameras that gather color and depth information, and it is these color and depth measurements that are interpreted to locate the user. Microsoft's EasyLiving system is an example of a location system that utilizes computer vision. Regardless of the technique used to locate the user, the important thing is that this system is scalable. That is, it must be able to cover a relatively large area, and in addition, it must be able to locate a specified number of users within a particular distance in a specified time. Computer vision solves the universal problem of users needing to carry a transmitter or a receiver. Systems based on computer vision suffer from the typical problems of computer vision, including great complexity, the need for clear lines of sight around the room, and slow performance even on very fast computing hardware. Vision-based solutions are also generally prohibitive for portable systems for reasons of cost, size, and power consumption.

There is a large body of existing work in location tracking in support of virtual reality and animation motion capture. Technically, many of these systems can provide valuable insight into developing similar systems for ubiquitous computing. For example, it has been shown that CDMA-like radio technology can be used for precise position tracking (on the order of 2-mm grain size) for virtual environments. However, three important issues separate location sensing for invisible computing from most of these systems. First, these systems are often quite expensive, and thus not readily deployable in the ubiquitous sense. But more important than cost, many of these systems are not designed to be scalable even to a building-wide level — they are designed to capture position well in a single-room immersive environment. Hence, we will not discuss them further, as they do not provide realistic alternatives for the deployment of location-sensing technologies.

Electromagnetic and Optical Sensors

Other systems that have been used in research include active and passive electro-magnetic and optical trackers. For virtual reality applications, specifically the capture of human motion for the purpose of animating computer-generated characters, some prior work has been done using the amplitude measurement of very low frequency magnetic fields.

Electromagnetic trackers can determine object locations and orientations to a high accuracy and resolution (around 1 mm in position and 0.2° in orientation), but are expensive and require tethers to control units. Furthermore, electromagnetic trackers have a short range (generally only a few meters) and are sensitive to the presence of metallic objects.

These systems consist of a single transmitter, an antenna made from two orthog-onal coils, and a large number of small receiving antennas, which are placed on the clothing of the person whose motion is to be recorded. These receiving antennas have little onboard circuitry; all of the signal processing is done in a large box containing the transmitter. The attached computer system can then keep track of the motion of as many as 50 points on the person's body. The person being tracked is therefore wired with a rather large and unwieldy umbilical tether that connects her to a big metal box. The maximum range of this system depends on the size and placement of the transmitting antenna, but typical ranges of 15 m are claimed, with a measurement precision of 5 to 7 cm. The transmitting antenna has a fixed position and possesses field coils that induce currents in the coils of the receiver. Since these currents vary with the relative position and orientation of the receiving antenna, they are used to calculate position and orientation information. Apart from these types of position trackers, several other systems originally implemented for a different purpose can serve as sensors for the context element of location as well. Some examples of such derived positioning systems are the following:

- Chip card readers or other systems for admission control or registration of working time that provide information about the stay of people or objects in buildings.
- Electronically administered reservations of rooms, official or rental cars or other means of transport, and other shared equipment.
- Electronically administered (business) trip application forms and plans.
- Electronic agendas and schedules containing appointments along with the locations they take place at.
- Indices providing the positions of fixed entities, such as machinery or other equipment. This includes plans of buildings or factories, maps, directory services, and so on.

The system claims to provide position updates at rates of 30 to 50 measurements per second. While this performance is quite sufficient for the system's intended use in motion capture from a performer on a stage, it is not sufficient for many other applications in which the performer must be free to move over a wide distance. This system, because all signal processing is done centrally at the transmitter, is not suited to spatially tiling with other such systems to cover a wide operational space.

Systems using pulsed direct current (DC) magnetic fields can be used to determine user orientation and have been used in combination with systems that use ultrasound signals to determine user location. While these technologies and systems are very interesting, they generally suffer the same drawbacks as their IR and RF tag counterparts. Their specialized hardware is generally targeted at niche markets, tending to make the system cost prohibitive, range limited, and unsuitable for large-scale deployment.

In optical systems, detectors such as cameras or photodiodes observing the entities that are to be located are employed. Like infrared-based systems, optical positioning requires a line of sight between the detector and the entity. Usually the reliability of currently available optical systems suffers from an increase in the number of entities within the observed area. Optical location detection, for example, is made use of in Microsoft's EasyLiving project. Optical trackers (for example, the HiBall system) are very robust and can achieve levels of accuracy and resolution similar to those of electromagnetic tracking systems. However, they are most useful in well-constrained environments and tend to be expensive and mechanically complex. Examples of this class of positioning device are a head tracker for augmented reality systems and a laser-scanning system for tracking human body motion (for example, the Ascension MotionStar system).

Some systems use motion detectors and reed switches (which monitor movement of people and the positions of doors). Much of this information can be provided using a single, low-powered, and untethered device, thus simplifying the physical and computing infrastructure required to support the interactive environment. Interactive Office, for example, gathers information about the activity of the occupants this way.

Accelerometers

The raw motion data collected by motion sensors decodes by using motion algorithms, the building blocks for smart motion-sensing systems. The processing hardware includes analog filters, discrete logic circuits, and embedded digital microcontrollers. Caveo's proprietary technology includes using very compact code, so the system can manage the algorithms with very small microcontrollers to provide small size and low power consumption.

Inertial Navigation

An inertial navigation system (INS) provides the user's position using a dead-reckoning algorithm. That is, relative to a known reference it calculates the deviation of the position. The system consists of accelerometers for determining the distance traveled and gyroscopes/magnetic compasses for direction information. Inertial navigation is often used for complementing a GPS, e.g., by using the INS when GPS signals are not available. For dead reckoning (DR) with vehicles, the traditional approach consists in using a triad of accelerometers and gyroscopes generating signals, which are integrated to obtain the relative displacement. However, the classic GPS/IMU approach used for vehicle navigation does not adapt well to pedestrian navigation. This concept is difficult to implement with a low-cost system. The principal reason is that the sensors' noise blurs the signals that correspond to the displacement of a person. The Geodetic Engineering Laboratory of EPFL has been involved in the development of a Pedestrian Navigation Module (PNM) fulfilling these requirements. This project results from a close collaboration between the EPFL and Leica Vectronix AG (Ladetto and Bertrand, 2002). The PNM consists of a GPS receiver, a digital magnetic compass, a gyroscope, a barometer, and embedded DR algorithms. All the sensors are placed in a small box. The weight of the device is about 150 g, and it works generally if clipped at the belt. Hence, it can be worn without discomfort.

The first area of interest consists in the determination of the physiological parameters necessary to quantify the speed of walk and the step length. While the variations in the signals of the accelerometers are good speedometers, the frequency of the steps improves the robustness of such models.

The various technologies have been integrated to build an autonomous pedestrian navigation system. During periods with a fair reception of the satellite signals, the embedded GPS receiver allows calibratation of the different sensor parameters and physiological models. An initialization phase has been developed to individualize the parameters of the walk and to adapt them from the general model. The consideration of several phenomena, specific to the displacement of humans, brings artificial intelligence into pedestrian navigation.

Spread-Spectrum Microwave Systems

These systems are used for the tracking of large portable objects, like hospital beds containing critically ill patients or very expensive packages. They intend to leverage the proliferation of cellular telephones and wireless networking equipment by adapting the components used for those purposes to the problem of positioning. Several prototype systems were developed that use direct-sequence spread-spectrum (DSSS) techniques in the 2.4-GHz industrial, scientific, and medical (ISM) band set aside by the FCC for such purposes.

While it is a step in the right direction, this approach has several fundamental limitations. First, the 2.4-GHz microwave signal is strongly attenuated by building materials, so a complete set of fixed station units must be replicated in each room to be served by the system. Second, the many fixed units must be connected by a data network of some kind, leading to wiring, antenna mounting, and network management issues. Third, the FCC's bandwidth limitations for the 2.4-GHz band, along with the strong multipath interference caused by even small conductive objects in the room, limit the precision with which one may make DSSS ranging calculations. The prototypes claim an RMS error of about 3 to 5 m. Finally, even though microwave electronics have become much cheaper, easier to use, and less power hungry, these issues still render DSSS a less than adequate solution for many uses.

Pressure Sensors

Systems using pressurized sensors (for example, the SmartFloor system) identify persons by their footstep force profiles. Though the accuracy of identifying a moving user is only around 90%, it provides a very unobtrusive way for users to provide their location information to the system. However, it works only for people, not for other objects, such as mobile devices, since these do not have footstep force profiles. Even if they do — such as the Sony Aibo or Honda Asimo robots — the force profile will be exactly identical between different robots.

Leveraging Existing Infrastructure

From an economic standpoint, it is preferable to employ and leverage an existing communications infrastructure (i.e., Wi-Fi networks) to determine the position of users. This will decrease the deployment costs of indoor location-based services. However, one size does not automatically fit all. Different indoor location-based applications have different requirements for accuracy and timing.

Determining location using a Wi-Fi network does not provide the best positioning accuracy possible, but users can easily be positioned to room-scale accuracy, or even more accurate (with additional measures such as triangulation, location fingerprinting/pattern matching, or adding another sensor type like ultrasound to compensate for the low precision of Wi-Fi-based positioning); the precision can be brought down to the centimeter (one third of an inch) range, which is more than adequate for many indoor location-based applications. This addition does not have to be pervasive — it may be quite sufficient to deploy it in areas where the more accurate positioning is required. A wide variety of network technologies exist in most building environments (i.e., Ethernet, wireless LANs, cellular, IR, etc.).

The communication infrastructure and positioning infrastructure systems may be combined or independent. For example, the Active Badge and ParcTab systems use the same wireless IR link for both tracking and data transfer, and the radio frequency link of the GUIDE and radar systems has the same role. Most of the other systems reviewed use separate channels.

Using 802.11 (Wi-Fi)

RF-based systems mainly use IEEE 802.11 wireless LAN networks and attempt to deal with the noisy characteristics of wireless radio; multipath fading is the major reason for this noise. Most new mobile devices, such as laptops, PDAs equipped with WLAN devices, and cellular phones, should be equipped with WLAN devices to provide Voice-over Internet Protocol (VoIP) function in the near future. Slightly more expensive than RFID tags at $100 apiece, Wi-Fi tags have a range of about 150 ft. The cost of the reader is very reasonable; in many cases vendors are practically giving away Wi-Fi access points. A Wi-Fi tag vendor example is Ekahau, which sells location software called Ekahau Positioning Engine, which we discuss in Chapter 5.

The 802.11 family of standards is the basis for a wireless access network technology, based on Ethernet, which today allows for up to 2-Gbps connections (although this may not work with all systems). The standard was developed by the Institute of Electrical and Electronic Engineers (IEEE), an international organization that develops standards for various electronic and electrical technologies. The IEEE 802 committee develops standards for local- and wide-area networks (LANs and WANs). For example, the 802.3 committee develops standards for Ethernet-based wired networks, the 802.15 group develops standards for personal-area networks, and the 802.11 committee develops standards for wireless local-area networks (WLANs). A WLAN is conceptually very similar to both a cellular network and a LAN. The major differences between cellular and WLAN are in how they are implemented and include the method of delivery of data to users, data rate limitations, and frequency band regulations. While a cellular network was designed to serve similar functions to that of traditional telephone networks, a WLAN was designed to serve the functions of a wired LAN. There are two main categories of WLANs: infrastructure networks in which there is a backbone and *ad hoc* networks with no backbone. These different networks could still be further interconnected and provide a truly ubiquitous network.

802.11 has a number of substandards. 802.11b, or Wi-Fi (which is a trademark of the Wi-Fi Alliance, which certifies networking products), is a standard for wireless LANs operating in the 2.4-GHz spectrum with a bandwidth of 11 Mbps. 802.11a is a different standard for wireless LANs operating in the 5-GHz frequency range with a maximum data rate of 54 Mbps. Another draft standard, 802.11g, is for WLANs operating in the 2.4-GHz frequency but with a maximum data rate of 54 Mbps. Other task groups are working on enhanced security (802.11i), spectrum and power control management (802.11h), quality of service (802.11e), etc. When we talk about WLANs, 802.11, or Wi-Fi in this book, we usually mean 802.11b, since this is the standard for which the measurement technology that enables positioning has been developed. However, in principle, the positioning technologies should work for all 802.11 technologies.

802.11 networks can be configured as peers, or using an access point that bridges to the fixed network (these usually use Ethernet). Access points (APs) transmitting radio signals can be used to estimate the location of the mobile host. The strengths of the radio signals arriving from more APs can be related to the position of the mobile terminal and can be used to derive the location of the user. The accuracy of the system might be limited by the (possibly large) cell size. Roaming between APs is supported, and Wi-Fi networks can be extended to create "clouds of connectivity" inside so-called hot spots, i.e., locations such as office buildings.

As in the Global System for Mobile Communications (GSM), devices have a unique identity, derived in the same way as in the Ethernet, from a hierarchic number series.

The IEEE 802.11 standard is used for Wireless LAN MAC and PHY. IEEE 802.11b and g operate in the 2.4 GHz ISM (industrial, scientific, and medical) band. 802.11b has two types of PHY, called direct-sequence spread spectrum (DSSS) and frequency-hopping spread spectrum (FHSS), the latter of which is used only in the low-data-rate mode. PHY has data rates of 1 to 11 Mbits/sec, but 802.11g uses orthogonal frequency division multiplexing (OFDM) instead of spread spectrum (as of June 2003) to extend the data rate of 802.11b to 54 Mbits/sec. 802.11a operates in the 5.4-GHz band and uses OFDM with a data rate up to 54 Mbits/sec. 802.11a and g have the same data rate, but 802.11a achieves a better effective data rate than 802.11g. As a medium-sharing mechanism of the MAC layer, carrier sense multiple access with collision avoidance (CSMA/CA) with ACK is employed instead of CSMA/CD on wired LANs. RTS and CTS packets are supported as an option to resolve hidden terminal problems.

No additional hardware or battery power is required to support it. Therefore, many current algorithms for positioning in 802.11 networks concentrate on using the current signal strength or noise signature (location fingerprint) to discern location. However, the potential cost-effectiveness of using 802.11 signal strengths can come at the expense of accuracy, because 802.11 operates in the 2.5-GHz radio band, where signals are attenuated by line-of-site obstructions and sometimes reflected. It is also a well-known problem that the technology does not allow for the traversal of walls (especially if they contain metal). Depending on accuracy, the access point identity to which the user is connected can be used instead of the attenuation. However, this still requires careful preparation of the network: the access points have to be measured to make sure they do not overlap, and their position has to be measured. Each access point in a WLAN network has a range of roughly 300 m in open space, and interference between different stations is dealt with by automatically selecting different channels and by a CSMA/CA access protocol. Terminals are connected by WLAN interfaces (in PCs, normally a PC card) that communicate with the access point that gives the strongest signal. However, various technologies allow for the enhancement of the system using software that aggregates the position. One such example is Ekahau's server software, which uses a database of the network to enhance the position accuracy to 1 m (3 to 5 ft) on average. This positioning algorithm was originally intended for enhancing the GSM network solution. It is not actually dependent on network technology since it calculates the location from base station signals and compares them to prerecorded signal patterns and to the map of the location. The received signal strength of a WLAN fluctuates in an indoor environment because of signal reflection, diffraction, and scattering. An indirect path is mainly generated by reflection, and the transmitted signal reaches the receiver via both direct and indirect paths. This multipath phenomenon, or multipath fading, strongly influences the signal strength.

Several indoor positioning systems perform detection and location within a defined local area using a LAN comprised of a set of access points (APs). The APs are communication ports for wireless devices, wherein the communication occurs across an "air link" between the wireless device and the APs. That is, APs pass messages received from the wireless device across the LAN to other servers, computers, applications, and subsystems or systems, as appropriate. The APs are bidirectional, and so are also configured to transmit to the wireless device. Typically, the APs are coupled to one or more network servers, which manage the message traffic flow. Application servers may be coupled to or accessed via the network servers, to provide data or typical application functionality to the wireless device. In such systems, the process of defining the local area (e.g., room layouts, ground layouts, and so on) to the network is often referred to as training the area or system, where the area is divided up into spaces, which wireless devices transition between as they migrate through the trained area. The location and detection within the trained area are typically determined as a function of the signal strength from the wireless device with respect to one or more APs. The resulting database (training data set) is composed of a sequence of measurements: each measurement contains, for example, a coordinate and a Wi-Fi scan composed of a sequence of readings, one per access point heard during the scan. The APs are configured to determine the signal strength and pass it on to a back-end subsystem for processing. Each reading records the multiple-access control (MAC) address of the access point and its signal strength. Location and detection are typically determined as a function of received signal strength indicator (RSSI) values obtained from the communications between the wireless device and the LAN. As a general rule, the higher the signal strength, the closer a transmitting wireless device is presumed to be to an AP. Changes in the signal strength as the wireless device moves about the trained area allow for tracking. If there are at least three APs that receive the signals from the wireless device, trilateration can be used to determine the location of the device within the trained area.

Furthermore, while detection and tracking are desired for substantially all wireless devices within the trained area, they are much more difficult to achieve, since the many types of wireless devices may all have different configurations. It is possible to track the positions of multiple locations in the defined space using a single access point, as detected by the mobile device. However, this technique is equally valid when using multiple sources, or communications access points. Properly placed communications access points can provide another independent signal profile, greatly improving the accuracy of the position detection and motion tracking that can resolve ambiguities arising from symmetries or similar signal strength distributions from a single source. In the limit, the addition of multiple access points can be thought of as a multidimensional system, whose coordinate indices begin to uniquely and accurately define more and smaller locations in the space, as indicated in Figure 4.12, where each access point A, B, C, and D has its own signal strength pattern, wherein X is located at about A = 50%, B = 60%, C = 25%, and D = 50%.

The simplest and least accurate location method determines the AP that is closest to the device by monitoring signal strength. It is possible to use software ("sniffers") to infer the location of a Wi-Fi-enabled mobile device by analyzing the RSS (or signal-to-noise ratio) of the Wi-Fi AP with respect to that mobile device. There are several commercial products that can be used to find ("sniff") the coverage areas of all available APs. By using such software to infer location, no additional hardware needs to be attached to a mobile device beyond a wireless network PC card. In some cases a specialized directional antenna is deployed and

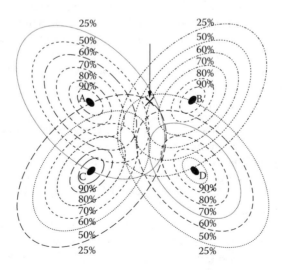

FIGURE 4.12 Signal strength pattern.

broadcast strength on APs is adjusted to reduce the leakage of the RF traffic. But this solution falls short of addressing any of the threats, such as rogue access points, that are not configured by the network administrator and neighboring WLANs that leak into the space. The algorithmic approach in these systems is mainly based in nearest-neighbor, error-minimization techniques or use of Bayesian networks to infer location from signal strengths.

The chief difficulty in localization is predicting RSS due to multipath effects and environmental effects (i.e., building geometry, traffic and atmospheric conditions). To overcome the indoor propagation problems, Wi-Fi positioning systems tabulate this function by sampling it at selected locations in the area of interest. This sampling results in a so-called radio map, which captures the signal signature of each AP at certain points in the area of interest. Given that APs can have coverage areas with a radius well over 100 ft, including up and down, nearest access point level of granularity will point toward the likely area within 30,000 ft² on multiple floors. This does not provide sufficient information to determine if devices are inside or outside the space. The range of communication between the AP and the mobile typically depends on the product type, propagation factors (i.e., range, multipath), and number of users.[1] A typical coverage range changes from 100 to 300 ft. Using multiple APs, coverage can be extended using roaming and microcells. The use of external antennas is another solution for extending the coverage range.

Positioning results indicate an accuracy of 2 m with a 90% probability for static devices, and for moving (walking) devices, an accuracy of 5 m with a 90% probability. With Ekahau, you can currently achieve 5 to 7 ft of accuracy in areas having very good signal strength and coverage. With this accuracy, however, you certainly would not want to use the system for positioning robots, but several feet of accuracy satisfies the needs of many applications (such as the one at IIT). It is possible to obtain accuracy with the Ekahau solution of 3 ft throughout a large area. However, this would likely require at least 10 access points within range of the client device at all times. In most wireless LANs, it is unlikely that users will be within range of more than four or five access points as they roam throughout the facility. In fact, a higher density of access points causes a significant amount of inter-access-point interference, which decreases performance capacity. See Chapter 5 for a description of the Ekahau positioning system.

Using Bluetooth and Wireless USB

The Bluetooth network, which is a low-power network originally intended for cable replacement (but currently used for micronetworks), and the wireless USB system both work in principle in the same way when connected to a fixed network. They form a hierarchical subnetwork with the master connecting to the fixed network. Bluetooth networks, though, can be organized *ad hoc* (without the need for a fixed connection anywhere).

In the following, we will be talking about WLAN or Wi-Fi; this should be taken to include all networks that use access points for the connection of terminals to a larger network. If a network of access points is deployed in such a way that the

---

[1] Note that total bandwidth available is shared among all the users connected to a particular AP.

access points do not move around, they can be provisioned with absolute positions (irrespective of whether they are connected to a larger network or not), and this can be used for absolute positioning (since the access points, being reference points, do not move). A Bluetooth *ad hoc* network, however, cannot be provisioned with an absolute position in this way, since the terminals may move around and the reference thus would be worthless. This is probably a special case, however, and when we need to single out Bluetooth, we will mention it specifically.

Using TV

TV signals are 10,000 times stronger than GPS signals. That means tracking through TV signals is much easier and quicker than via satellite. TV synchronization signals (part of the standard for TV set forth by the American Television Standards Committee) can be used for positioning mobile devices, including cell phones. Compared to GPS, TV-based positioning gives better accuracy, reliability, and rapid acquisition time. Rosum TV-GPS is a robust location technology based on analog and digital commercial broadcast television signals. TV towers are deployed globally around centers of population and commerce, they often broadcast multiple channels per tower at high power, and they utilize frequencies that easily penetrate buildings and other urban structures. As a result, Rosum's TV-based positioning system provides reliability and accuracy indoors and in urban canyons where GPS and related technologies fail. Rosum has integrated its TV-based positioning with GPS to provide reliable and accurate position fixes across all environments. Rosum's TV-GPS, which is based on existing standards for digital TV synchronization signals, gives mean positioning errors ranging from 3.2 to 23.3 m in tests.

This technique does not require changes to the TV broadcast stations. A general requirement is the need for TV coverage, which is good throughout urban areas. Also, the effects of multipath are significantly diminished. TV signal-based positioning can track signals that are below the noise floor and for which a conventional TV signal receiver would be unable to acquire timing information. These techniques extract timing information in a manner far more precise than a typical TV receiver. These techniques also accommodate all the variable characteristics of the TV signal, such that these variations do not affect the precision of position location. Moreover, TV-GPS hybrid positioning is foreseen in the future that will allow seamless indoor–outdoor positioning capability.

Various methods can be used to measure TV signals to use in position location. These include the following:

1. A location server tells a mobile station of the best TV signals to range from.
2. Mobile station exchanges messages with the location server by way of base station.
3. Mobile station selects TV signal to monitor based on the identity of base station and a stored table correlating base stations and TV signals.
4. Mobile station accepts a location input from the user that gives a general indication of the area, such as the name of the nearest city, and uses this information to select TV signals for processing.

5. Location fingerprinting: Mobile station scans available TV channels to assemble a fingerprint of the location based on power levels of the available TV signals. Mobile station compares this fingerprint to a stored table that matches known fingerprints with known locations to determine the location and to select TV signals for processing.
6. Using pseudo-ranges: Mobile station takes pseudo-range measurements on all of the available signals and communicates these pseudo-ranges to the location server, which determines the position of the mobile station based on the locations of the TV transmitters. Then the mobile station determines a pseudo-range between the mobile station and each TV transmitter. Each pseudo-range represents the time difference (or equivalent distance) between a time of transmission from a transmitter of a GCR signal burst within a broadcast TV signal and a time of reception at the mobile station of the GCR signal burst, as well as a clock offset between the mobile station and the monitor units, and a clock offset in the time of transmission of the component of the TV broadcast signal. Following, the mobile station transmits the pseudo-ranges to the location server.
   The location server can be implemented in the following ways:
   a. As a general-purpose computer executing software designed to perform the operations tied with pseudo-range measurements
   b. As an application-specific integrated circuit (ASIC)
   c. Implemented within or near a base station.
7. Using monitor units: The TV signals can also be received by a plurality of monitor units.

There currently are three basic types of commercial digital TV (DTV) broadcast systems: terrestrial, cable, and satellite. Satellite TV systems are all digital, whereas cable and terrestrial systems are converting to digital from their mostly analog predecessors. These digital TV systems use frame structures that are made up of encoded data packets. These data packets are also referred to as data segments. The receiver is meant to refer to any object capable of implementing the low-data-rate channel described below. Receivers may also be capable of receiving the television programming data carried on the DTV, signals but this is not required. Receivers may be stationary or mobile. Virtually any object that could include a chip or software implementing the robust low-data-rate channel described below could be a receiver. Thus, DTVs, pagers, mobile telephones, PDAs, computers, cars, and other vehicles would all be examples of receivers if properly equipped.

See Chapter 5 for a description of the Rosum TV-based positioning system.

Home RF

Home RF supports both *ad hoc* and infrastructure network technologies. It is a simple home networking technology aiming to make use of the existing home PC industry infrastructure. Working in the frequency range of 2.45 GHz and covering up to 150 ft, Home RF supports the equivalent maximum of six full duplex voice connections.

FM Radio

The use of commercial FM radio signal strengths for location was explored in Krumm et al. (2003), which showed accuracies down to the suburb level.

USING WIDE-AREA NETWORK TECHNOLOGIES

Passive Sensors: GPS and Indoor GPS

To determine a user's precise location, traditional positioning technology based on the GPS requires visual contact to special satellites circling the Earth in geostationary orbits at an altitude of approximately 20,000 km (12,000 mi). To determine the position, a locating device must receive the signals from at least three of these satellites. With a traditional GPS receiver it can sometimes take several minutes to collect all the satellite navigation data and compute the precise location.

GPS and LORAN are very successful in the wide area, but are ineffective in buildings due to loss of line of sight as well as signal blockage, fading, and shadowing. Spread-spectrum signals are susceptible to multipath that is less than one chip width away from a direct path ray. For a GPS C/A code, the chip length is about 300 m. Most indoor signal reflection delays would be significantly shorter than that distance.

Some server-based GPS already claim some indoor positioning capabilities. However, such systems are in general only accurate to within a few tens of meters. Although some experimental setups show decent navigation performance, there is a question of whether GPS should be used for such applications in the first place. GPS position determination can be aided by the transmission of data that make GPS more robust to attenuation and multipath. Examples of this type of data include the current positions of the GPS satellites and clock correction information based on atomic standards. Additional examples include Doppler information, information about the Earth's atmosphere, including Ionospheric effects that may alter or disrupt the satellite signals. Further examples include a list of satellites that should be visible given the location of the user, spatial coordinate data for those selected satellites, and predicted Doppler data for those selected satellites. Examples might also include the data that are modulated onto the C/A GPS code signals. These data might also include information that aids a device in tracking and determining the precise distance to GPS satellites. These types of data are referred to herein as GPS aiding information. The energy efficiency is an issue, as the terminal will need to constantly scan for pseudolites. The cost of pseudolites is also an issue, especially in situations where the area to be covered is not a single open space, but where there are multiple walls and ceilings that interrupt the signal — like in an office building.

GPS pseudolites have been deployed for indoor use. These are devices that generate a GPS-like positioning signal. As in the GPS, at least four pseudolites have to be visible for navigation, unless additional means, such as altitude aiding, are used. The signal generated by the pseudolites is monitored by a number of reference receivers. Latest innovations in this area have delivered receivers to the size of a stamp. In 1999, the Seoul National University GPS Lab (SNUGL) developed a centimeter-accuracy indoor GPS navigation system using asynchronous pseudolites.

Devices can use the low-data-rate GPS aiding information to integrate GPS satellite signals more efficiently or for longer periods according to conventional, well-known techniques. These techniques are generally referred to as assisted GPS, aided GPS, or A-GPS. Devices using GPS aiding information are more robust to attenuation caused by obstructions; further, the GPS aiding information makes it possible for these devices to obtain severely attenuated GPS signals from far below the noise floor. As a result, GPS position determination can be improved by the use of this GPS aiding information, especially when the GPS pseudo-noise signals are attenuated.

Indoor GPS focus on exploiting the advantages of GPS for developing a positioning system for indoor environments. Although some deployments show decent indoor positioning, there is a question of whether GPS should be used for such applications in the first place. Finding GPS signals in the traditional manner is impractical because the receiver cannot dwell long enough at each search location to achieve the required sensitivity.

In order to detect highly attenuated GPS signals indoors, dwell times must be increased by two to three orders of magnitude. This is done using two primary techniques: assisted GPS (A-GPS), which provides additional information through a communications data link, and massively increasing the number of correlators available for acquiring the encoded GPS signals. With A-GPS, the mobile device receives information about the satellites' orbits, frequencies, and functionalities over the wireless network. As a result, it can detect and analyze even weak satellite signals at lightning speed. The A-GPS technology uses the radio link between base station and mobile devices to transmit this assisted satellite data within a matter of seconds, thus saving time and battery power compared to traditional GPS. Even under difficult reception conditions, it takes the unit only a few seconds rather than minutes to display the correct coordinates.

A-GPS provides half of the answer to indoor use of GPS for mobile devices (e.g., cell phones). Indoor GPS solutions are mostly applicable to wide space areas where no significant barriers exist to signal propagation.

Nonetheless, there are still some advantages of using GPS:

- Takes into account the low power consumption and small size requirements of mobile phones and handheld computers.
- Provides a relatively low barrier to entry (although an external card and antenna are still additional costs over current commodity hardware).
- The signal is designed to be similar to the GPS signal in order to allow pseudolite-compatible receivers to be built with minimal modifications to existing GPS receivers.

Furthermore, GPS receivers normally determine their position by computing relative times of arrival of signals transmitted simultaneously from a multiplicity of GPS (or NAVSTAR) satellites. These satellites transmit, as part of their message, both satellite positioning data, so-called ephemeris data, as well as data on clock timing. This information is sent in a channel that broadcasts at 50 bps. In other words, it is possible for a device to receive GPS aiding information directly from the GPS satellite itself. However, the process of searching for and acquiring GPS signals, reading the ephemeris data for a multiplicity of satellites, and computing the location of the device from these data is time-consuming, often requiring several minutes. In many cases, this lengthy processing time is unacceptable and, furthermore, greatly limits battery life in portable applications. This processing time also makes it very difficult to perform real-time position fixes for applications that require constant position updates.

GALILEO, Europe's satellite positioning and navigation system, will offer greater accuracy than the U.S.'s GPS — down to a meter and less; greater penetration — in urban centers, inside buildings, and under trees; and a faster fix. That is, local components will improve mitigation of thermal noise, multipath and narrow band interference with local data distribution via terrestrial radio links or existing communication networks; these components will be deployed to provide extra accuracy, integrity, or extended coverage around airports and in urban areas, and to extend navigation services to indoor users. The range of GALILEO services is designed to meet practical objectives and expectations, from improving the coverage of open-access services in urban environments (to cover 95% of urban districts compared with the 50% currently covered by GPS alone). Devices that will make use of both GPS and GALILEO will be able to improve positioning accuracy up to 10 cm.

Cellular Mobile Network Technologies

Cellular networks have fast developed into an extensive wireless communication infrastructure providing wireless voice and data communications with almost worldwide coverage. Use of cellular phones is greatly increasing worldwide, and the number of cell phone subscribers has quadrupled to over half a billion in the past 5 years (STO, 2002). A cellular network is a wireless communication service area subdivided into hexagonal areas termed cells. These cells vary in size from a few kilometers in diameter, in modern digital networks, to around a hundred kilometers in older analog networks. Each of these cells has a base station (cellular tower) associated with it. Figure 4.13 shows a simplified cellular architecture.

Each base station has a certain range of radio frequency channels associated with it. To avoid overlap and radio interference, these channels are different from channels associated with all of its neighboring cells. Channels can be reused in cells

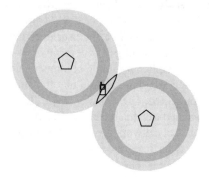

FIGURE 4.13 Cellular network.

that are far enough so that no interference occurs. These cells are grouped together as clusters for a required coverage area.

All the base stations (BSs) are connected to a mobile switching center (MSC) using fixed links. Each MSC of a cluster is then connected to the MSCs of other clusters and a public switched telephone network (PSTN) switching center. The MSC stores information about the subscribers located within the cluster and is responsible for directing calls to them.

One of the most important issues in cellular networks is tracking of a mobile client when it is moving through a network. To counter this, cellular networks use location management techniques. Location management essentially involves two processes, location update and paging. Location update is the information provided by the mobile device to the network about its current location. Paging, on the other hand, is done by the network, where it actively queries the mobile device to find out what cell it is located in, so that the incoming call can be routed correctly to the appropriate BS. The performance of a network greatly depends on the multiple-access control (MAC) techniques used for sharing the available frequencies among multiple subscribers. The three major MAC techniques in use currently are frequency division multiple access (FDMA), time division multiple access (TDMA), and code division multiple access (CDMA). FDMA divides the available channel bandwidth into equal subchannels among the mobile users. Each user is assigned a subchannel for transmission. Interference between these channels is reduced by inserting guard bands between adjacent subchannels. This technology is used by most analog cellular services.

TDMA works on a logic similar to that of FDMA, but it also divides each subchannel into a number of time slots in order to increase the amount of data that can be carried. A mobile user is assigned the entire channel for a period of time. In this case, guard bands are required in between frequency channels and time slots as well. Global System for Mobile Communications (GSM) and the North American digital standard IS-136 both work on the principles of TDMA. CDMA uses a spread-spectrum technique to scatter a radio signal across a wide range of frequencies. The frequency channel (1.23 MHz in IS-95) is shared between all the users of the system. A signal to be broadcast on the channel is first spread out over the entire bandwidth using a unique PN code. Since each signal on the channel is then unique, it can be distinguished from the other signals and can be recovered at the receiving end using the unique PN code.

GPS is a passive technology, because the user does not communicate with the satellite (or pseudolite). The terminal is just a receiver. It is different in wide-area cellular technologies. In principle, all mobile communication networks work in the same way. The user connects to a base station that transmits and receives the radio frequency signal (as in the GSM, CDMA, WCDMA, TDMA, PDC, or PHS radio standards), which is connected to a fixed network. The user connects to the network and, through this, to a global network like the telephone system or the Internet, either through setting up as an Integrated Services Digital Network (ISDN) or an Internet Protocol (IP) connection. The primary purpose of the mobile network is to set up communication channels between mobile parties and to help with the setup of mobile-terminating calls. Originally developed for voice applications, most mobile communications networks have been enhanced to handle packet data communications. This is true for all second- and third-generation networks; analog networks (which are still deployed in the U.S. and some parts of Europe) cannot handle packet data, only data over a modem connection.

Mobile networks are part of a system, which includes functions to authenticate the terminal, security functions for communications and services, and functions for charging and management of supplementary services, such as location. Location information is handled internally in the network via the signaling system (typically signaling system no. 7, or SS7, or IP). SS7 functions exist to handle the positioning of a terminal, encoded in the internal system. Developers can access those functions using APIs, e.g., Parlay or the LIF MLP API. In indoor location systems that are not based on a mobile system, you have to add these functions.

When connecting, GSM will automatically position the terminal, although the resolution in meters is not very good (and not sufficient for the E911 requirements). When a user connects to the network, his terminal is automatically allocated to the base station transmitting with the strongest field strength. The system knows the identity of the base station that an active GSM user is currently located at, with an accuracy that varies from a few hundred meters in urban areas to a few kilometers in rural areas. This method is known as the cell of origin (COO).

COO requires no modification to the terminal or networks, thus permitting its use as the positioning for existing subscribers, but is less accurate than the other methods employed. It is also possible (if the operator allows transmission of cell identity to terminals) to use this method for terminal-based positioning, e.g., as a reference source in combination with methods that have a more fine-grained local position.

Moreover, while using this existing communication infrastructure has been found to be promising in outdoor environments for coarse-grained location services (such as some location-based content services), its immediate applicability in indoor environments for more precise indoor LBS is limited. However, in urban areas the cell size of the GSM network might be small enough to allow for indoor LBS applications, especially in public areas like shopping malls, where indoor GSM networks are often deployed.

While GSM inherently allows for a coarse location, this method has several other drawbacks, e.g., the limited data capabilities of GSM networks (a function of the TDMA technology and the frequency management). The frequencies used also create limitations for the cell size. Current wide-area positioning systems, mostly tied to GSM (the alternative on the market is to use GPS), use a mobile positioning center (MPC), which is tied to the operator's network. This creates a potential bottleneck, although the volume of requests has to be quite high to be a problem to current systems. The MPC interfaces to the network; applications work with a location server, which receives queries formulated in either MLP or Parlay requests, and forward queries to the MPC for processing. After receiving the result, the response is packaged in the appropriate protocol and returned to the requestor. This implies that location services are request–response based in the systems envisioned by the standard. An alternative is to use the Internet Multimedia Subsystem (IMS), which allows for an alternative method (publish–subscribe), but since this is not yet deployed to any large extent, we will not cover it in this book.

In a wireless cellular network like GSM, there is an identity built into the mobile device (or base station) that is unique. Both the device itself and the user (through the subscriber identity module (SIM) card) have a unique identity, which is mapped to the MSISDN (i.e., the telephone number). This unique ID is used when locating the mobile device. Mobile phone numbers are allocated by the International Telecommunications Union (ITU) and delegated to national standardized bodies. As of 2004, it is possible for users in the U.S. to retain their number when moving from one operator to another. This means that the number that follows the subscription is no longer tied to an operator or an area (much in the same way as IP addresses are disconnected from the location of an Internet Service Provider when using large IP networks).

The radio network is just one part of the system, however. When the signal has been captured, it requires a method to calculate the position. There are a number of these, the positioning algorithms.

Cellular phone carriers provide user location services using a number of systems, including A-GPS, TDOA, AOA/TDOA, and E-OTD. A-GPS uses the GPS infrastructure to determine user location. A-GPS is assisted by base stations, which provide information to the GPS measurement processing of a cellular phone. The cellular phone can search GPS signals quickly and sensitively using the aiding information. The cellular phone sends the captured GPS signal to the base station. Then the base station calculates the location of the cellular phone. TDOA, AOA/TDOA, and E-OTD are also used by a few cellular phone carriers. In general, these technologies cannot achieve better positioning accuracy than A-GPS.

POSITIONING ALGORITHMS

Different algorithms are used to translate recorded signal properties (metrics such as RSS, AOA, and TOA/TDOA) into distances and angles, where trigonometric functions can come into play. The positioning algorithm is independent of the network used to communicate and derive position information. The algorithm (or, as the case may be, algorithms) is used first to derive the position from the sensor information, then to relate the position derived to a known position (either relative or absolute). The algorithms used today are the same irrespective of whether the system is used in outdoor or indoor positioning. However, the higher resolution in the indoor system may require different algorithms.

Algorithms can be (and most of the time are) used in combination. How to determine which algorithm to use, and which combination of algorithms to use, depends on several things: the resources available for the computation, the frequency of updates required, and the type of data available. In a later chapter, we will go through how to determine which algorithms (or combination of algorithms) can be used, depending on the constraints on the system.

There are, in the main, four different types of algorithms (based on how they process gathered location metric parameters) that can be used to derive the position information from the sensor information:

- Cell ID based
- Proximity (containment)
- Triangulation
- Trilateration

All of the following Wi-Fi-based positioning algorithms depend on an initial *training* phase:

- Centroid
- Particle filters
- Location fingerprinting

Once the training phase is completed, the training data are processed to build a "radio map" of the area. The nature of this map depends on the positioning algorithm used. Once the map is built, a user's device can simply perform a Wi-Fi scan and position itself by comparing the set of access points heard to the radio map. This process is referred to as the positioning phase.

Overall, different algorithms work best with different densities and ranges of access points. Since both of these quantities can be measured beforehand, the best algorithm could be automatically switched in, depending on the current situation.

CELL ID-BASED SYSTEMS

Currently the most widely deployed solution for network-based positioning uses existing data from the network to identify which cell site and sector a user is in. The cell site is derived from the base station identity in the network, and the sector can be determined by checking which antenna is receiving the signal. As a result, location accuracy is dependent on cell size (actually on lobe size, since not all antennas may have equal coverage). This method works for networks such as GSM as well as Wi-Fi, as explained earlier.

The cell ID-based systems can work in two ways: either the terminal derives its position by receiving cell information from the network and calculating the position using a database of cell positions, or the terminal identity is received by the network and the network identifies which cell the terminal is in. In the latter case, the accuracy can be higher, since the identity of the cell is communicated to a system that is used to map the cell identity to the extent of the cell. In case the network identifies the cell, it can also determine which sector of the cell the terminal is in, if the antennas of the base station can be identified. The cell sector can also be determined using trilateration (about which more later).

Future systems may be deployed with smaller cells than in today's GSM networks (although in the present IMT-2000 deployments in Europe, the cells are actually far larger than in GSM). It is argued that this will lead to query and update volumes several magnitudes higher than in today's GSM systems. This may be the case in Wi-Fi systems as well, if the location information is to be handled at a central site.

There is also an effect on the protocol underlying the system. Such systems contain two basic operations: move (location update) and find (location query). Typically, proposed solutions use a multilevel hierarchy of location servers. Each location server node has a well-defined network coverage area. Location updates are triggered by the leaf nodes and are propagated through the directory following a well-defined algorithm. In contrast, queries can generally be directed to nodes at any level in the hierarchy. This often leads to recursive query patterns.

The basic mechanisms employed to improve response time and reduce network traffic are data replication and forwarding pointers. Data replication reduces query latency and query traffic but increases update traffic. Also, consistency control is normally achieved by simple time-outs. Since efficient queries and updates need to make assumptions about parameters of user mobility (such as the call-to-mobility ratio), sometimes systems can adapt dynamically to changing user characteristics. The subscriber profile would be used to store such mobility parameters.

VoIP

Voice-over Internet Protocol (VoIP) refers to the delivery of audio (voice) information over IP in discrete packets, rather than through the traditional circuit-committed protocols used in the public switched telephone network (PSTN). Pinpointing the exact origin of a VoIP call can be difficult to impossible using current technologies, since customers can choose distant area codes and use the service when they are away from home. There are currently a number of techniques to position VoIP calls, each with their own complications, including IP address-based location, hybrid VoIP calling (users connect to the PSTN via gateways provided by the service provider), and Dynamic Host Control Protocol (DHCP)-based positioning. Another option proposed is a DHCP-based solution (http://www.ietf.org/internet-drafts/draft-ietf-geopriv-dhcp-lci-option-03.txt). Location configuration information of varying precision is included in the protocol. The question then is one of how that location information is derived. One solution proposes smart jacks, in which the jack itself can be queried about the MAC address of devices attached (Schulzrinne and Arabshian, 2002). This would entail upgrading countless Ethernet ports with that capability. Barring that, the location of the port would have to be known and mapped. This presents a large burden to the IT manager. In contrast with network-based solutions, Rosum's TV-GPS solution provides the absolute location of the calling device without any need for knowledge of the local wiring system and without a need to keep wiring diagrams up to date (Rosum, 2004). With Rosum TV-GPS, the device itself provides its location as a simple latitude, longitude set. In addition to map-based pinpointing like the Skyhook positioning system, Rosum TV-GPS is a mobile, handset-based solution to the VoIP geolocation problem. VoIP device locations can be kept "warm" through regular position updates, or positioned on an as-required basis. However, unlike map-based pinpointing approaches, by providing caller coordinates, Rosum's solution eliminates the need for time-consuming updates to a centralized in-house database associating VoIP numbers with locations for the purpose of PSAP call routing. 802.11 access points would need to be mapped precisely, or the location of the Ethernet port must be known. Assigning latitude and longitude coordinates to all Wi-Fi nodes is an enormous task, as would be keeping such a database current. This approach eliminates the need to map devices to locations and enables more precise location than that offered by using the nearest router, which could be hundreds of meters away.

MOBILE IP

A variation to cell-based systems is using the Mobile IP and network domains. Mobile IP is the primary IETF standard for supporting mobility on the Internet. It provides transparent support for host mobility by inserting a level of indirection into the routing architecture. By considering the mobile host's home address as an endpoint identifier and not merely as an interface identifier, Mobile IP ensures that the delivery of packets to the host's home address is independent of the host's physical point of attachment. This is achieved by creating an IP tunnel between a mobile host's home network and its care-of address.

Using the Mobile IP, when the mobile host enters a new zone, it must discover the foreign agent (FA) to be assigned a temporary IP address. (Note that the location discovery is dependent on the size of the zone. Typically, it is an operator network, which may mean that the size is equivalent to a country.) By installing a context manager service on the same host of FA, the mobile host imports the context of the current zone from the context server just discovered during registration with FA.

This means that Mobile IP systems can use location-tracking algorithms based on network domains. (Note: This technology is known not to work. Also, it assumes you are using Transmission Control Protocol (TCP)/IP, which may not be true.) These are typically not specifically targeted for indoor or outdoor use (for example, the GUIDE system). Such a connectivity-based approach to track mobile users can also be realized in other communications technologies, e.g., using Bluetooth.

PROXIMITY/CLOSEST AP/CONTAINMENT

The simplest method for locating a mobile device is estimating its position to be the same as that of the AP that it is communicating with. Closest AP gives users granularity equal to the total coverage area of a single access point. It is the simplest and coarsest way to locate a device or user. Since the information of the serving AP is easily available in a Wi-Fi network, it does not require any parameter estimation or any complex location algorithms. The drawback is that its accuracy depends on the coverage radius of the serving AP. In addition to that, an MS may be communicating with an AP that is not the closest due to propagation effects. With closest AP, which is the coarsest approach, a query is submitted to a network management system to find a client based on the MAC address table. But since an 802.11b and g AP has roughly a 100×100 ft coverage area, then locating the user by the closest AP method only tracks Amy to within a 10,000 ft^2 area, or the space of 100 cubicles in a building. With closest AP, an IT administrator submits a query to a WLAN management system to find a client based on its media access control address. The WLAN management system checks all access points to see where the device is associated. Because an 802.11b and g access point has roughly a 100×100 ft coverage area, locating the client by the closest AP method tracks it to within a 10,000 ft^2 area, or the space of about 100 cubicles in a building.

Several indoor positioning systems take advantage of the fact that many Wi-Fi-enabled mobile devices can measure the RSS of incoming packets, which is used for positioning. Mobile devices primarily associate with the AP that provides them with the strongest RSS. As a mobile device roams, it periodically does a site survey of signal quality measures to determine the best AP with which to associate. When the query is submitted, the AP that Bob is associated with responds, and thus the user is found. The question is: What granularity did the closest AP arrive at? If an AP has (roughly) a 50×50 ft coverage area, then locating the user by the closest AP pins the user down to a 2500 ft^2 area. Access points with greater coverage areas have an inverse impact on closest AP location granularity.

TRIANGULATION

Triangulation is where a call goes out to all APs on the network and each AP that hears the user's signal responds to the network management system with the strength of the signal. The system then draws coverage circles around each access point that hears the user, with each circle reflecting the border of the signal strength at which the AP received the signal from the user/device. A triangulation algorithm estimates the location of the MS to be at the intersection of circles with their centers at each AP location and radii determined by a particular metric's values. At least three distinct estimates of the distance of the MS from known fixed locations are required. With triangulation, IT administrators use a WLAN management system to query all access points on a wireless network to see which ones hear the target user's 802.11 signal. Access points that hear the desired device respond to the WLAN management system with their received signal strength indicator (RSSI) information. The WLAN management system then draws coverage circles around each of these access points, with each circle reflecting the border of the signal strength at which the access point received the signal from the user/device. The point where these circles converge is assumed to be the location of the desired device. Specifically, position information (e.g., distance estimates) can be obtained by using any generic wireless infrastructure by using various communication metrics to triangulate from known positions. Typical metrics include the use of RSS, BER, AOA, or time delay (TOA/TDOA), which are used to obtain the distance estimates.

Domain-specific knowledge may reduce the number of required distance measurements. For example, one indoor location positioning system measures distance from indoor mobile tags to a grid of ceiling-mounted ultrasound sensors. The mobile tags' three-dimensional position can be determined using only three distance measurements because the sensors in the ceiling are always above the receiver. The geometric ambiguity of only three distance measurements can be resolved because the mobile tag is known to be below the sensors and not in the alternate possible position on the next floor or roof above the sensor grid.

With triangulation, the system draws coverage circles around each access point that hears the user. Each coverage circle reflects the border of the signal strength at which the AP received the signal strength from the user/device. If an access point heard the device at –65 dBm, then the network management system would draw a circle at where it believes the –65 dBm demarcation is located on a coverage map. If another access point heard the device at –45 dBm, then the coverage circle around that AP might be considerably smaller (meaning the signal strength was higher), and it too is factored into locating the user. When the network management finishes, there will be a number of line intersections. Algorithms are then used to determine the most likely location within the intersections. There is a high probability that the area with the highest density of intersecting lines is the correct location. Triangulation has a granularity of about 10 m (~30 ft). Triangulation will therefore yield a location of 30 × 30 (900 ft²), or about the size of nine cubicles. In free space (i.e., an area with no walls or objects that might block a signal) triangulation will yield fairly accurate results.

Also, a triangulation-based geopositioning system may use a database to store values about distortions in RF signal propagation throughout a building. The positions of walls and other interference can be modeled in the database in order for the geopositioning system to predict propagation effects. This assists geopositioning algorithms by filtering out or correcting known anomalies in signals, so DLOS triangulation can be more accurately performed. In essence, determining location relative to known points is a fairly simple trigonometric function; compensating for errors and anomalies, however, often requires many additional computations.

Advantages include:

- COTS approach.
- Little additional overhead.
- Only determining the position of the AP is required.

Disadvantages include:

- Existence of physical barriers reducing signal strength can skew results.
- Not sufficient precision with standard WLAN installation.
- High deviation in signal strength up to 5 dB/sec.
- Signal strength measurements up to a radius of 5 m can be almost identical.
- Triangulation is divisible into subcategories of *lateration* (time of flight or attenuation), which uses distance measurements, and *angulation* (AOA), using primarily angle or bearing measurements.

Type 1: Lateration

Lateration computes the position of an object by measuring its distance from multiple reference positions. Calculating an object's position in two dimensions requires distance measurements from three noncollinear points. (In three dimensions, distance measurements from four noncoplanar points are required.)

Two approaches to measuring the distances required by lateration are (1) time of flight and (2) attenuation. Time of flight refers to measuring the time difference in transmission and arrival time of an emitted signal (with a known velocity).

Attenuation can be explained as follows: The intensity of an emitted signal decreases as the distance from the emission source increases. The decrease relative to the original intensity is the attenuation. Given a function correlating attenuation and distance for a type of emission and the original strength of the emission, it is possible to estimate the distance from an object to some point P by measuring the strength of the emission when it reaches P.

In environments with many obstructions, such as an indoor office space, measuring distance using attenuation is usually less accurate than time of flight. Signal propagation issues such as reflection, refraction, and multipath cause the attenuation to correlate poorly with distance, resulting in inaccurate and imprecise distance estimates.

Signals travel through the air at the speed of light, or 3×108 m/sec, and radiate outward in all directions in the form of a sphere. If we measure the precise time a signal leaves a transmitter and the precise time the signal arrives at a receiver, we can determine the time of arrival (TOA), the time it takes for the signal to arrive at the receiver. Since we know how fast the signal was traveling and for how long the signal was traveling, we can determine the distance traveled according to the following relationship:

$$distance = rate \times time$$

In two dimensions, one such calculation tells us that the object is located anywhere on a circle with a radius d (distance), centered at the transmitter. In three dimensions, the object is located anywhere on a sphere with a dimension of d, centered at the transmitter. Whether a circle or a sphere is used for the location calculation depends on whether we are trying to locate the object in two or three dimensions. Distance can be calculated with reference to more receivers, which results in additional circles on which the object is located. The only way that the object can be located somewhere on two or more circles is if the circles intersect. The point of intersection of the circles is where the object is actually located. Because circles can intersect at more than one point, three distance measurements are needed, and similarly four different measurements are needed for a three-dimensional system.

The problem with triangulation is that it does not take into account the effects (e.g., reflection, attenuation) that a building will have on signal.

One indoor LPS implements attenuation measurement using low-cost tags. These tags use radio signal attenuation to estimate intertag distance and exploit the density of tag clusters and correlation of multiple measurements to mitigate some of the signal propagation difficulties.

Type 2: Angulation

Angulation is similar to lateration, except instead of distances, angles are used for determining the position of an object. In general, two-dimensional angulation requires two angle measurements and one length measurement, such as the distance between the reference points. In three dimensions, one length measurement, one azimuth measurement, and two angle measurements are needed to specify a precise position. Angulation implementations sometimes choose to designate a constant reference vector as $0\times$.

Phased antenna arrays are an excellent enabling technology for the angulation method. Multiple antennas with known separation measure the time of arrival of a signal. Given the differences in arrival times and the geometry of the receiving array, it is then possible to compute the angle from which the emission originated. If there are enough elements in the array and large enough separations, the angulation calculation can be performed.

TRILATERATION

Trilateration is a method of determining the position of the wireless device as a function of the lengths between the wireless device and each of the APs. Trilateration calculations are performed by the wireless device using the RSSI data, which must be configured with appropriate software (e.g., a client-side module) to accomplish such tasks. As a result, the demands on the wireless device are increased.

Trilateration relies on determining the direction of signals, and then computing the intersections of them. The idea behind trilateration is that a network of sensors (either APs or passive radios) detect signals from a wireless device and report the signal strength of the wireless device. Trilateration derives the position of a terminal by determining which direction it is in and intersecting the directions. This requires at least three signal strength measurements (e.g., from Wi-Fi access points), which means they can be mapped to an approximate distance.

A distance from each sensor is determined by making an assumption that a device with a given signal strength is a certain distance from a sensor based on circular coverage maps. By specifying the location of the sensors, a set of distance measurements from each sensor is estimated. These estimates are combined to identify points of intersection around each sensor to estimate the location of the wireless device. The area with the most intersection points is presumed to be the area of the wireless device.

Angle of arrival, the simplest of the trilateration techniques, requires at least two distinct signals from known locations of Wi-Fi access points. By overlaying the transmission areas of the two access points, an area can be determined, where it is likely that the receiver is located. This does not require much computation, but it does require a database of the known locations of the access points.

With RF signals that radiate in a uniform circle from all devices, the trilateration algorithm has an error rate of ±20 ft in any direction, so now you can narrow location down to within two floors. This is certainly better than nearest access point (proximity method), but still not sufficient to determine if devices are inside or outside a building.

Further, trilateration suffers from significant problems that dramatically reduce the accuracy of its location algorithm. Trilateration assumes that RF traffic is very uniform. Because of the short wavelength of 802.11 signals, trilateration is seriously error-prone due to a wide range of environmental issues from buildings and device interference.

These environmental factors cause four problems with the accuracy of wireless device tracking via trilateration:

1. **Attenuation**: When objects affect the RF signal by reducing the signal strength. For example, walls, glass, and other common building structures reduce the signal strength of an RF signal as it passes through that object. This prevents any RF signal from emanating from a device in a uniform radial pattern. The actual coverage map for any wireless device is rarely a perfect sphere. For instance, a wall might take away –4 dBm of signal strength from a wall; the coverage circles on the triangulation map are usually perfect circles and do not reflect the imperfect effect that walls, glass, and other building materials have on RF.

2. **Occlusion**: When objects completely block an RF signal, a given sensor may not detect a wireless device at all even though it is within range and may be very close to the sensor.

3. **Reflection**: When an RF signal reflects off of objects (walls, projector screens, desks), its path to a sensor is longer than expected and the resulting signal strength will be lower.

4. **Multipath**: When an RF signal is sent, it can follow multiple paths before arriving at a sensor. If the signal does not take the most direct path to the sensor, the sensor will indicate a lower signal strength, making the wireless device appear much farther from the sensor than it really is. Therefore, a signal that was heard from a device at –65 dBm may actually be well within the coverage circle, but it was heard at that signal strength because it did not take the most direct path to the AP path; any given spot on a coverage map can be reached via many different paths.

Trilateration can be used in combination with cell ID technologies. A central server then aggregates the values to triangulate the precise position of the object. If the time can be measured as well (i.e., when the signal is transmitted from the access point, and when it is received at the terminal), more accuracy can be gained — but at the price of more computation.

Once the training phase is completed, the training data are processed to build a radio map of the area. The nature of this map depends on the positioning algorithm used. Once the map is built, a user's device can simply perform a Wi-Fi scan and position itself by comparing the set of access points heard to the radio map. This process is referred to as the positioning phase.

CENTROID

This is the simplest positioning algorithm. During the training phase, all of the readings are combined for a single access point and estimate a geographic location for the access point by computing the arithmetic mean of the positions reported in all of the readings. Thus, the radio map for this algorithm has one record per access point containing the estimated position of that AP.

Using this map, the centroid algorithm positions the user at the center of all of the APs heard during a scan by computing an average of the estimated positions of each of the heard APs. In addition to the simple arithmetic mean, we also experimented with a weighted version of this mechanism, where the position of each AP was weighted by the received signal strength during the scan.

PARTICLE FILTERS

Particle filters have been used in the past, primarily in robotics, to fuse a stream of sensor data into location estimates. A particle filter is a probabilistic approximation algorithm that implements a Bayes' filter. It represents the location estimate of a user at time t using a collection of weighted *particles*.

Particle filter-based location techniques require two input models: a *sensor model* and a *motion model*. The sensor model is responsible for computing how likely an individual particle's position is, given the observed sensor data. For Intel's Place Lab, the sensor model estimates how likely it is that a given set of APs would be observed at a given location. The motion model's job is to move the particles' locations in a manner that approximates the motion of the user.

Place Lab uses a simple motion model that moves particles random distances in random directions. Intel's future work includes incorporation of more sophisticated motion models that model direction, velocity, and even mode of transportation.

LOCATION FINGERPRINTING (PATTERN MATCHING)

RF pattern matching, popularly referred to as location fingerprinting, achieves positioning accuracy within 10 ft (Ekahau system achieves 3-ft accuracy). RF fingerprinting approaches the problem of location by determining how a signal will be received at every grid point within a building. A grid point can be as granular as a half foot. RF fingerprinting builds upon the earlier methods (closest AP and triangulation) to provide granular, higher-resolution location to within a few meters. RF fingerprinting has granularity of about 3 m (or 10 ft). Therefore, it yields a result of 10×10 ft, or 100 ft^2 — about the size of a cubicle. RF fingerprinting builds on the closest AP and triangulation methods to also take into account the effects (e.g., reflection, attenuation, multipath) that a building, objects, or people will have on an RF signal. Access points use RF fingerprinting technology to collect information pertaining to RF topology via a site survey, which creates a grid identifying how every part of a building floor plan looks to all access points. A grid point can be as small as a half foot. To determine the radio frequency at each grid point, a WLAN management system must first predict how the RF will interact with the building.

Due to the harsh multipath environment in indoor areas, other techniques that use triangulation or direction are not very attractive and often can yield highly erroneous results. Instead of exploiting signal timing or signal strength, location fingerprinting relies on signal structure characteristics. It turns the multipath phenomenon to good use — by combining the multipath pattern with other signal characteristics, it creates a signal signature database of a location grid for a specific human-defined physical area (this can be a floor plan layout or a GIS map). The WLAN management system then creates an RF fingerprint with detailed topology information and RF data by tracing rays from every access point in the network and accounting for reflection and multipath to any given destination. The RF prediction involved in RF fingerprinting takes into account the attenuation and reflection off walls and other objects into a building. The WLAN management system then creates a database of coordinates, recording how each access point views that area from a signal strength perspective — creating an RF fingerprint for each coordinate. Each RF signature (or fingerprint) is unique to a given location. (Some systems have used the multipath characteristics of a signal as its fingerprint.) Next, the WLAN management system cross-references real-time client information collected by access points (for example, RSSI) with RF topology information stored in the system for accurate location tracking. This information is constantly updated to account for changes in the RF environment. Specifically, the received signals in a location (*fingerprint* of some characteristic of the signal) are matched with another reference set (that is location dependent) contained in a database. The sensors receive data that have followed multiple possible paths and also have been attenuated, occluded, and reflected by objects in the area. Location fingerprinting algorithms gained importance and are the dominant location determination approach today. Types of fingerprint information include RSS, multipath angular power, or multipath power delay.

Figure 4.14 portrays these various types of fingerprinting information that can be used in estimating a user's location.

Then, to determine the position of a mobile device, the system matches the transmitter's signal signature to an entry in the database. This is called the online phase and it is explained in more detail below.

Figure 4.15 shows how the EPE pinpoints the locations of mobile devices.

Many of the commercial indoor positioning systems using existing Wi-Fi network infrastructure make use of location fingerprinting rather than TOA or AOA techniques for determining the location of mobile stations. The fingerprinting technique is rela-

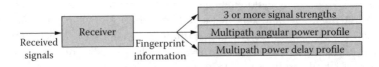

FIGURE 4.14 Location fingerprinting information types.

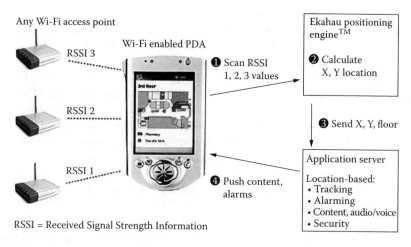

FIGURE 4.15 Location fingerprinting. (From http://www.pocketpcmag.com/_archives/Sep03/images/Sept03_p74_1.jpg.)

tively simple to deploy compared to the other techniques, such as AOA and TOA. Moreover, there is no specialized hardware required at the mobile device. Any existing wireless LAN infrastructure can be reused for this positioning technique.

In comparison, the cell ID location determination method depends on a database of location information, mapping the cells to locations. The fingerprints of different locations are stored in a database and matched to measured fingerprints at the current location of a mobile device. An alternative to passively receiving and identifying the signals is to actively send out a fingerprinting signal. This requires that the base stations be programmed to do so.

In comparison to proximity and trilateration, location fingerprinting enables highly accurate determination of locations inside vs. outside a building. Rather than being negatively impacted by the unpredictable behaviors of RF traffic, RF pattern matching considers the real-life behavior of RF signals and factors in attenuation, occlusion, reflection, and multipath effects when determining device location. The clear benefit of adding RF fingerprinting to location is adding a much more granular, precise, and reliable method to this capability, being able to find clients within a few meters. RF fingerprinting improves on the closest AP and triangulation location-tracking approaches by taking into account the effects that a building or people will have on an RF signal — characteristics such as reflection, attenuation, and multipath. This makes wireless device location tracking more detailed, precise, and reliable.

In general, some of the advantages include the following: Due to the physical and technological limitations associated with other techniques, location fingerprinting schemes remain the most feasible solution for indoor position location. The location of objects can be inferred using passive observation and features that do not correspond to geometric angles or distances. Measuring geometric quantities often requires motion or the emission of signals, both of which can compromise privacy and can require more power. It does not require any hardware modifications to the mobile device (however, it does require specialized hardware in every access point to correlate the multipath characteristics). Also the location fingerprinting algorithm accurately maps RF space to the physical space, like Room ABC, and is not described in terms of (x, y, z) coordinates.

Disadvantages include the following: The observer needs to have access to the features of the environment against which it will compare its observed scenes. Changes to the environment in a way that alters the perceived features of the scenes may necessitate reconstruction of the predefined data set or retrieval of an entirely new data set. For example, using the Ekahau system in a dynamic environment requires recalibration. There is a significant cost in the comparison of measured data with the stored information. This may not be very costly for small areas, but it becomes an increasingly important component as the area to be covered and the number of users becomes large.

A system for location positioning using WLANs can run on a stand-alone server and give the (x, y) position and the floor of the mobile unit. The positioning is based on an *a priori* accomplished calibration map created by collecting sample points of the area. Each sample point received contains the signal intensity and related map coordinates (Zlatanova and Verbree, 2003). The positioning accuracy achieved by Wi-Fi systems is up to 1 m. Despite providing accurate positioning for indoors, however, the WLANs have problems with implementation because they require a reference database for average signal measurements at fixed points throughout a building (Pahlavan and Li, 2002).

An advantage is the high accuracy. Disadvantages include high overhead due to need for existing measurements for every set of coordinates that needs to be uniquely identifiable, and measurements have to be repeated every time the location or even the furnishings of the facility change.

The Offline and Online Phases of Location Fingerprinting

Pattern matching consists of two phases, the offline phase (consisting of a site survey and resulting in a radio map) and the online phase (sometimes referred to as the real-time or positioning phase).

The offline phase is about generating a fingerprint database, specifically collecting location fingerprints through the coverage area by performing a site survey of the desired metric from multiple APs, which are then averaged over time and recorded in a database. In order to determine how the RF will sound at each location in an enterprise, a user must first predict how the RF will interact with the building. Therefore, an RF prediction is required. In this case, the RF prediction is not required prior to deployment, but rather to determine what happens after deployment. When an RF prediction is run, the attenuation of the walls is taken into account. Reflections off of walls and multipath are calculated. At each training location, a model for the joint probability distribution of the visible access points at that location is stored. The training phase results in a collection of probability distribution models for the signals at each point. Maximum likelihood estimation was used to calculate these joint probabilities. Also, the tuples recorded at the training locations were clustered according to APs that were visible at that location. Since the order of the APs seen could vary, the tuples recorded at these locations were not ordered tuples. Usually RSS is used as the desired metric, but AOA or TOA may also be used. Sniffer software is used to analyze the incoming signals and compile a unique signature for each square in the location grid, storing it in the database. The mobile device will report a sample measured vector of RSSs from different APs to a central server (or a group of APs will collect the RSS measurements from a mobile device and send it to the server).

A common signal modeling approach is to record samples of wireless signals from points in a large location grid drawn to cover either the entire floor or occupied areas of a building. The smaller the grid cell size, the more samples that are stored in the database, resulting in higher positioning precision of the system. (The vector of RSS values at a point on the grid is called the location fingerprint of that point.) Signals from the APs used in the calibration are sampled at each location to train a position determination model. The RSS is measured with enough statistics to create a database or a table of predetermined RSS values on the points of the location grid, forming a set of fingerprints resulting in a so-called radio map.

The obtained samples are smoothed using a moving average filter. A database stores entries where each entry is in the form of [x, y, RSS1, RSS2, RSS3 Ö]. (x, y) is the location of the trained point, and RSSi are the signal strength values received from access point i. (Multipoint signal reception is not required, although it is highly desirable. The system can use data from only a single point to determine location. Moving people and objects do not affect the system's capabilities.)

The signal strength method works similarly to the TOA method, with the difference being in how the distance is calculated. Instead of measuring the departure and arrival times, the signal power is measured at the transmitter and the receiver. This information can be used to determine the distance that a signal has traveled. Multiple distance measurements are used similarly to TOA to calculate the location. Both TOA and signal strength are promising approaches for WLAN location sensing. The online phase consists of collecting samples from some access points at an unknown location. The strongest access points are used to determine the cluster to search within for the most probable location. Then Bayes' theorem can be used to estimate the probability of each location within the cluster using the observed tuples and the collected data during the offline phase.

Online Phase

The *online (real-time) phase* is when the positioning system estimates the location of the mobile to be one of the locations previously recorded in the database. Each beacon transmits its data to a pattern-matching system that matches the collection of beacon data to a database of RF signatures. A match (fingerprint) is found and the system identifies the associated physical location for the wireless device. Or, in the case of handset-based location determination, the mobile device performs a scan of its environment and observes the WLAN signals and applies the position determination model to calculate a position in real time. The workload is determined by both the number of marking positions and the number of scanning operations for each position. However, reducing the training workload and improving system accuracy are trade-off problems. The computation is repeated for every coordinate on a map, and for each AP on the map. For a single point on the grid, there may be many different access points that can reach it and hear it. However, each access point will probably hear it at different signal strengths. At the conclusion of prediction, a database is populated inside of the management system. That database contains each coordinate, and how each access point views that coordinate from a signal strength perspective.

With the defined space having been trained, positioning is accomplished within and among the locales by processing actual signal strength data of a mobile device as it moves about or resides in the defined space, and comparing the actual data against the known statistical signal strength model. Specifically, a scan is compared with all of the fingerprints in the radio map to find the fingerprint that is the closest match to the positioning scan in terms of APs seen and their corresponding signal strengths. Same as in triangulation, when a network administrator tries to find a wireless device, each access point replies with the signal strength with which it can hear the device. The management system tries to match the information it gathered from the access points against its database of location fingerprints.

At any one time, a mobile device transmitting in the trained space may be detected by a plurality of detectors, which may be in the same or different locales. A comparison of the actual signal strength data at each access point receiving the mobile device's signal with the signal strength patterns of those access points allows for a determination of the real-time location of the mobile device within the defined space. Such analysis, when performed over time, allows tracking of the mobile device within and among the locales.

The resulting signals from the offline phase are used to estimate the mobile's location. Figure 4.16 shows that each fingerprint corresponds to fingerprint information associated to a known user's location.

The positioning server uses an algorithm to estimate the location of the mobile device and reports the estimate back to the mobile device (or the application requesting the position information). The algorithms can vary that are used to compare a received signal strength vector to the stored radio map in the online (location determination) phase. The most common algorithm to estimate the location computes the Euclidean distance between the measured RSS vector and each fingerprint in the database. Other algorithms use nearest neighbors, neural networks, or Bayesian modeling to relate the sample RSS vector to the fingerprint in the database.

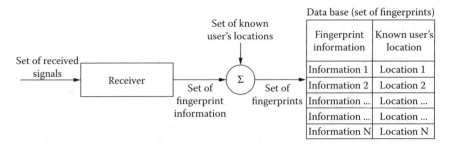

FIGURE 4.16 Matching fingerprints.

Specifically, in addition to triangulation, these include:

- K-nearest-neighbor averaging (KNN): The algorithm searches for K-location entries from the location database having the smallest root mean square error in signal space with the given run-time measurement at the unknown location. Averaging the coordinates of the K-locations gives the final location estimate. The process includes creating a radio map for the area interested by doing random or uniform sampling in that area. After that, the location information is computed by searching the nearest neighbor of the measured signal strength within the radio map. Suppose the positioning scan discovered three APs, A, B, and C, with signal strengths SSA, SSB, and SSC. With the set of recorded fingerprints in the radio map that has the same set of APs, it is possible to compute the distance in signal space between the observed signal strengths and the recorded ones in the fingerprints.
- Smallest M-vertex polygon (SMP): M candidate locations from each access point whose distances in the signal space with the given run-time measurements are searched from the location database. M-vertex polygons are formed by including at least one candidate location from each access point. The smallest polygon is the one having the shortest perimeter. Averaging the coordinates of vertices of the smallest polygon gives the final location estimate.
- Neural networks.
- Baysian modeling: Bayesian filter techniques provide a powerful tool to help manage measurement uncertainty and perform multisensor fusion. Their statistical nature makes Bayes' filters applicable to arbitrary sensor types and representations of environments. For example, Bayes' filters provide a sound approach to location estimation using positioning data along with street maps or signal strength information along with topological representations of indoor environments. Furthermore, they have been applied with great success to a variety of state estimation problems, including speech recognition, target tracking, vision, and robotics (Figure 4.17).

The accuracy and precision reported by most of these algorithms are quite similar. The coordinates associated with the fingerprint that provides the smallest Euclidean distance is returned as the estimate of the position. The mobile's location is estimated to be at the corresponding database entry that minimizes the Euclidian distance Dk in signal space. Averages or weighted averages of multiple nearest neighbors can be used to decrease the variance of the errors.

In general, performing location fingerprinting can be classified into two types of techniques:

- **Deterministic techniques** represent the signal strength of an access point at a location by a scalar value, for example, the mean value, and use nonprob-

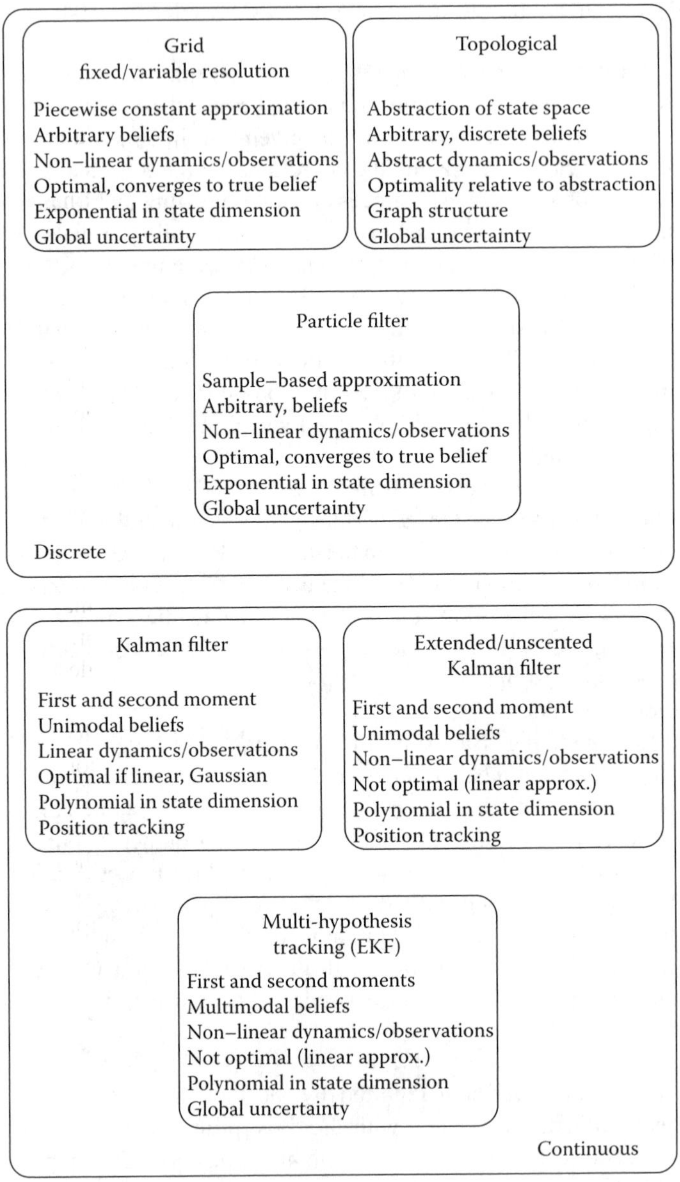

FIGURE 4.17 Bayes' filters for location estimation.

abilistic approaches to estimate the user location. For example, the radar system uses nearest-neighborhood techniques to infer the user location.

- **Probabilistic techniques** store information about the signal strength distributions from the access points in the radio map and use probabilistic techniques to estimate the user location. For example, the Nibble system uses a Bayesian network approach to estimate the user location. Research shows that the probabilistic technique outperforms the deterministic technique.

Figure 4.18 illustrates that for the online phase, specific fingerprint information is obtained from measured received signals and compared with a recorded set of fingerprints (database). A pattern-matching algorithm is then used to identify the closest recorded information of the database to the measured one, defining the corresponding user's location as a result.

Map Locales and Adjacency

Because of the variable nature of signal propagation in indoor environments, it is hard to eliminate the position determination error if only the signal distribution probability is used in the determination procedure. To improve the accuracy of the positioning system, area topology or locale adjacency can be used in a model along with a tracking-assistant positioning algorithm. Databases are used to store static information such as coordinates of access points and building layouts.

Typically, the defined space is comprised of a set of defined regions, areas, or locations (collectively referred to as locales). Each locale is defined within the system in relationship to the digital form of the physical space. A locale may be defined as an interior or exterior space or location, or a combination thereof. For instance, a conference room, office, or waiting area may be defined as a single locale within a defined space.

The digital form or map of the physical space preferably includes the identification of permanent obstructions that will affect the transmission and reception capabilities of the access points, e.g., walls, columns, and so on. The defined space is comprised of a set of defined regions, areas, or locations.

Locales may be defined either prior to or after generation of the signal strength model. Typically, once the digital form of the space is formed, the locales are defined and the statistical signal strength model is then defined. In other forms, an iterative process of defining locales, generating the signal strength model, and (optionally) positioning the access points may be used.

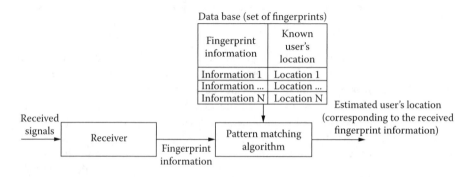

FIGURE 4.18 Online phase of location fingerprinting.

Area topology can be conceptualized as the following. If a locale A is only adjacent to a locale B and a locale C and, according to signal strength data, the mobile device could be in locale B or a locale E, knowing that the previous locale of the mobile device was locale A allows the system to accurately determine that the mobile device is currently in locale B, and not locale E.

The concept of adjacency may be implemented in a state-based approach. In such a case, each locale may be uniquely modeled as a state-within-a-state diagram. Since only a finite number of known next states and previous states can exist for each state, a current state can be determined with greater reliability given knowledge of the previous state and its subset of allowable next states.

Locales may be defined either prior to or after generation of the signal strength model. However, typically, once the digital map of the space is formed, the locales are defined and the statistical signal strength model is then defined. An iterative process of defining locales, generating the signal strength model, and (optionally) positioning the access points may be used.

Figure 4.19 provides a representative top view of a distorted signal strength field pattern combined or superimposed over a digital floor plan for a single access point in a defined space. The figure demonstrates how the traversable locales within the defined space can have different signal strength profiles and how different adjacent locales will also have differing signal strength characteristics.

A combination of approaches may be implemented to locate and track a mobile device through the defined space and from locale to locale:

- Clustering statistics: Using clustering statistics of received signal strength indicator (RSSI) data from one or more access points, a determination of the location of the mobile device can be made with relatively high accuracy.
- Trilateration: A trilateration analysis of RSSI data received from three different detectors can be performed, wherein the location of the mobile device can be determined as a function of the length of the sides of a triangle formed by the three access points.
- Hybrid solution: The results of the clustering statistics and the trilateration are combined to increase the accuracy of the overall determination of the location of the mobile device. This approach can also be performed over time for improved tracking.

Because the defined space may impose inherent restrictions on where the mobile device can and cannot travel (e.g., corridors, walls, rooms, and so on), there is the potential for increasing the number of locales having distinct signal strength profiles. In addition, since it is more likely that the mobile device will travel by passing through adjacent or connected locales, i.e., the device will not be in one area at one moment and then instantaneously appear two or more locales away, this further increases the ability to accurately identify the location and motion of the mobile device within the defined space, since only certain signal strength profile transitions will likely be observed.

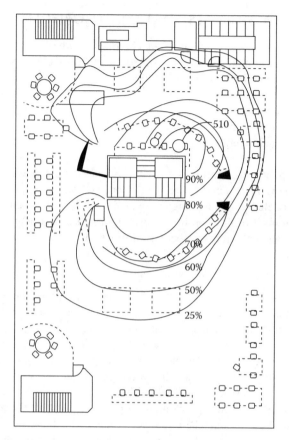

FIGURE 4.19 View of a floor plan and signal strength pattern.

In summary, specifics of the online phase include the following: The system compares the measured signal strength tuple (RSS1, RSS2, RSS3, Ö) with the database. A pattern-matching algorithm is used to search and identify the closest recorded information of the database to the measured one, thus defining the corresponding user's location. For matching, a linear search algorithm can be used (especially if the database size is rather small). However, for a larger database, a search mechanim is needed that utilizes efficient access methods.

There are only a certain number of available locations, though multiple location estimates can be averaged to obtain better estimates. The system must be trained with more locations in order to increase the location accuracy.

Also, a user may have different privileges, access, or rights with respect to functionality or data, depending on the current locale of the user. Transitioning from one locale to another locale may cause loss of privileges, rights, and access, and in some cases, selective loss or delivery of data (see more on security and access privileges in Chapter 9).

POSITION ESTIMATION ALGORITHMS

Besides geolocation metrics and ranging techniques, another factor of a localization system is an algorithm for position estimation. In this section, we discuss three classes of algorithms for position estimation:

- Monte Carlo localization
- Convex optimization
- Iterative multilateration

The most powerful algorithms to date are based on Bayesian inference, in particular Markov models and Monte Carlo localization. Alternately, the environment can be modeled with a topological map, e.g., as a generalized Voronoi graph, as explained above. Localization in this paradigm is based on identifying nodes in the graph from geometric environmental information.

MONTE CARLO LOCALIZATION

Multilateration is the problem of estimating a node's position from ranges to three or more known nodes (sensors). If not all nodes have ranges to at least three sensors, their positions must be estimated through an iterative process.

Iterative techniques have been explored for robust position estimation in sensor networks where nodes in a wireless network can improve the accuracy of their RSSI-based location estimates by dynamically deriving (learning) the surrounding wireless channel properties (Savvides et al., 2001). The algorithm starts with an initial guess of channel properties and tries to obtain node position estimates through a sequence of successive multilateration. The initial set of position estimates can now be used to obtain an initial estimate of the channel properties by providing two crucial components: (1) a large set of inputs for the estimation of the channel parameters and (2) a corresponding error variance that is used as a weight for each input in the channel model estimator.

Iterative techniques incur additional energy costs in communication and are not guaranteed to be completely fault tolerant.

CONVEX OPTIMIZATION

One way to formalize the problem of estimating node (e.g., in *ad hoc* networks) positions in a sensor network is to express relations (angular, range, etc.) between different pairs of nodes (known or unknown) as a set of convex constraints.

Convex optimization techniques have been proposed for solving the position estimation problem in sensor networks in an offline, centralized manner (Doherty et al., 2001). The advantage of this approach is that it requires very few sensors as references since all system constraints are solved globally. However, this algorithm is not very robust to failures when there are ambiguities in measurements.

ITERATIVE MULTILATERATION

Multilateration is the problem of estimating a node's position from ranges to three or more known nodes (sensors). If not all nodes have ranges to at least three sensors, their positions must be estimated through an iterative process.

Iterative techniques have been explored for robust position estimation in sensor networks (Savvides et al., 2001). Iterative techniques incur additional energy costs in communication and are not guaranteed to be completely fault tolerant.

Mapping Infrastructure

Pure positioning provides just mobile device discovery. Still, information about the space that a user or device is in is important. Hence, providing spatial information requires good boundary detection. Moreover, developers want to build navigation applications, which require knowledge of position coordinates and orientation in some coordinate system, in addition to spatial information.

These are discussed in more detail in the next chapter.

REFERENCES

Bluetooth Special Interest Group, *Specification of the Bluetooth System, Core, Version 1.1*, Vols. 1 and 2, February 22, 2001.

Bulusu, N., Heidemann, J., and Estrin, D., GPS-less Low-Cost Outdoor Localization for Very Small Devices, *IEEE Personal Communications*, October 2000, pp. 28–34.

Doherty, L., Pister, K., and El Ghaoui, L., Convex Position Estimation in Wireless Sensor Networks, paper presented at the Proceedings of IEEE Infocom 2001, Anchorage, AK, April 2001.

Fontana, R.J., Experimental Results from an Ultra Wideband Precision Geolocation System, paper presented at the EuroEM 2000, Edinburgh, Scotland, May 30, 2000.

Fontana, R.J. and Gunderson, S.J., Ultra wideband precision asset location system, in *Proceedings of the IEEE Ultra Wideband Systems and Technologies Conference*, Baltimore, MD, May 2002, pp. 147–150.

Kantor, G. and Singh, S., Preliminary results in range-only localization and mapping, in *Proceedings of the IEEE Conference on Robotics and Automation*, Washington, DC, May 2002.

Krumm, J., Cermak, G., and Horvitz, E., RightSPOT: a novel sense of location for a smart personal object, in *Proceedings of ACM Ubicomp'03*, October 2003.

Ladetto, Q. and Bertrand, M., Digital Magnetic Compass and Gyroscope Integration for Pedestrian Navigation, paper presented at the 9th St. Petersburg International Conference on Integrated Navigation Systems, 2002.

Pahlavan, K. and Li, X., Indoor Geolocation Science and Technology: Next-Generation Broadband Wireless Networks and Navigation Services, *IEEE Communications Magazine*, February 2002, pp. 112–118.

Rosum, VoIP Telephony and 9-1-1: A Device-Based Solution to the Problem of Geolocating 9-1-1 Calls, Rosum Position Paper, June 2004 (updated January 2005).

Savvides, A., Han, C.-C., and Srivastava, M.B., *Location Discovery in Ad-Hoc Wireless Networks*, ACM, 2001.

Schulzrinne, H. and Arabshian, K., *Providing Emergency Services in Internet Telephony*, Department of Computer Science, Columbia University, May 2002.

Tsukiyama, T., Navigation system for the mobile robots using RFID tags, in *Proceedings of the IEEE Conference on Advanced Robotics*, Coimbra, Portugal, June/July 2003.

Uchida, F. and Kinoshita, T., New Packet Management Technology on Wireless LAN for Enterprise and Public Networks Enabling a Fivefold Increase in Data Throughput, *Hitachi New Release Digest*, October 2002, http://koigakubo.hitachi.co.jp/nre/02nrde1022.pdf.

Zlatanova, S. and Verbree, E., Technological Developments within 3D Location-Based Services, paper presented at the International Symposium and Exhibition on Geoinformation, Shah Alam, Malaysia, October 13–14, 2003, pp. 153–160.

5 Location Awareness and Navigation in Location-Based Systems

The previous chapter talked about several potential solutions to the problem of indoor positioning, ranging in complexity from placing an infrared (IR) light-emitting beacon equipped with a unique ID in each room, to the use of ultrasonic beacons, to giving the user (human or a robot) a computer vision system to identify his place in the indoor world.

For a host of applications, what is really desired is location within a building or an area, or location relative to other people or objects, whether moving or stationary. Most positioning systems allow autonomous determination of position, yet this information must be communicated by a seperate mechanism in order to be shared. Positioning data are only one aspect of providing location-based services (LBS). Having the location coordinates (x, y, z) obviously is essential; the location would be related to other pertinent information to understand the environment and to provide users with the location of an item or place they are interested in finding. Knowing one's latitude and longitude is useless without additional information such as a map. In other words, positioning is just one part of an LBS, but in addition to just x, y, z information, a navigation location-based service needs some kind of knowledge or representation about its environment. The positioning receivers do not display a map and include only the function to display positioning measurements (e.g, a latitude and a longitude) of the present position of the user and a distance and a direction to a registered landmark. Furthermore, mobile devices have the capacity to organize information dynamically according to the relative physical locations of the user, points of interest (POIs) to the user, and available navigable pathways of various kinds. However, these devices do not determine a route between specified physical locations and do not present navigational instructions for traversing the route.

If necessary, the position information should be transformed into a location. Position refers to a numeric or symbolic description within a spatial reference system, whereas location refers to information about surrounding objects and their interrelationships (topology).

The normalized presentation of location information is calculated from the data delivered by the chosen scheme, such as coordinates of base stations, distance from the mobile device, location area shape, and error margins. Normalized location data should take the form of a standardized presentation of geographical coordinates, e.g., World Geodetic System 1984 (WGS84). WGS84 is a set of U.S. Defense

Mapping Agency parameters for determining global geometric and physical geodetic relationships. Parameters include a geocentric reference ellipsoid, a coordinate system, and a gravity field model. (Global positioning system (GPS) satellite orbital information in the navigation message is referenced to WGS84.) At a later stage applications can transform the location information to their preferred presentation, e.g., from longitude/latitude to a well-known human-readable one. Normalizers take the computed location information from their respective positioning algorithm and transform it into the system's reference presentation. This component is only required if a positioning algorithm's output is based on nonstandard spatial reference systems and reference points.

Moreover, location can be distinguished between noncontained and contained locations. Contained locations contain one or more other locations; in noncontained locations no other locations are contained. This definition of location in the context of this work enables information logistics to regard any type of information that answers the question of where an entity is as a location. Thus, there is no restriction to the type of location information that is processable by LBS applications. Examples of positions are "Boston," "MIT campus," "Room 101," "subway," "way to shopping mall," and "x, y, z coordinate." Among these, "Room 101," for example, is a contained location if it contains other locations that can be detected by the available sensors and that are relevant to the application, such as the furniture in the room. It is noncontained if there are no other entities that can or need to be located within it. Noncontainment therefore is not an unchangeable property of a location, but it varies according to the available sensors and the presence of mobile users or objects.

Representations about the environment are usually in the form of maps, which have geometrical and topological properties, capturing the properties of the environment. Furthermore, they can also be symbolic, in the form of labeled entities, which are typically used for task specification.

Maps relate positions of mobile users and the surroundings (e.g., objects) within their space. Spatial relations (i.e., containment and intersection) constitute a fundamental concept in location-based applications, which is *location awareness*, or localization/topology. Topology is a mechanism to establish spatial relationships between the mobile user and the surroundings (the physical world). Topology is essential to build the spatial relationship (connectivity, proximity, intersection, and membership) between the objects and the features of a network. It is rather a question of knowing a neighborhood and the relative positions between objects. In order to enable indoor navigation, a deeper knowledge of the map objects is required in addition to more information about their topological relationships.

So what's really different about indoors from the outdoors when it comes to maps and navigation? The following outlines the main differences.

INDOOR DATA

The data underlying traditional geographic information systems (GIS) usually contain neither the detail nor the coordinate systems (discussed next) to satisfy the needs of users in an indoor setting. Indoor LBS require microdetailed georeferencing to satisfy the needs of indoor location-enabled applications. It is not enough to georeference a building if the positions of users and other objects inside the building are also relevant. Static objects are used as landmarks (spatial references), and relationships among these objects are crucial for symbolic representation of the whole system. Consider a shopping mall as an example. The level of mapping detail of the underlying location model determines what functionality can be provided to the application. For example, if the application is used outdoors, to navigate between doorways, it is sufficient to return places at the granularity of buildings, whereas indoors the room of the building the user is currently in (or possibly an even finer level of granularity, such as the shelf spaces) is appropriate. As a result, traditional geographic information is not detailed enough to satisfy user needs in an indoor setting.

In order to develop more accurate outdoors maps/networks, enormous surveying efforts are continuously happening (Shibata, 1994; Schiff, 1993; Deretsky and Rodny, 1993). Some of these efforts are being undertaken by government agencies and others are being undertaken by private companies. However, no large-scale efforts of a similar nature are taking place for the indoors.

An indoor spatial information database can be created from the digital maps from such sources as computer-aided design (CAD) drawings used for the construction and the maintenance of buildings. The content of such a CAD database is composed of many objects, such as buildings, rooms, corridors, doors, and points of interest (POIs). As an example, a navigation product can provide users with information on points of interest (restaurants, shops, etc.) in the vicinity of their current location. In terms of CAD attributes, in general, each room in a building is uniquely named or labeled. The room names or labels provide a useful value in a database table as well as an intuitive indicator of floor. For example, Room 101 implies a specific place on the first floor of a building.

In addition to visualization purposes dealing with map content display, these objects are also useful for positioning purposes; this is tied to map matching of a user's position to corridors (paths in a topological map) in order to achieve more accurate positioning. We discuss map matching in more detail later in the chapter. Regarding just map visualization, all the elements of the database may have a label with a list of attributes (for example, a room number). Such a database is useful to find the position of specific points of interest (POIs). The space database can be used as the source of attribute information about buildings in maps, as well as the detail for rooms in a database. This information is useful for monitoring the use of space down to periods within a year (date ranges), and capturing department names, people responsible for space, and functional use coding, as well as many other attributes about buildings and rooms. Additional attributes about these objects can be imported from other databases like staff management. CAD design drawings are a two-dimensional graphical representation of space. All the CAD objects are designed with a geometric view and are composed of shapes, arcs, and circles. The graphic standards used to draw two-dimensional (or three-dimensional) models are based on the description of objects using polylines and shapes.

A set of tools is required to automate the process of data conversion to create indoor maps from CAD files that are useful for indoor LBS and navigation. NearSpace Workshop, for example, contains various tools for importing appropriate data. They can be used to create NearSpace Frames (a combination of raster and vector information) based on other representations; they do not need to be explicitly spatial in nature. By separating the raster and vector data, the visible representation can be made as detailed or as symbolic as required for the purpose at hand. Specifically, these tools are necessary to identify specific objects required for indoor applications. After detection, the objects are clearly identified as door, room, corridor, building, etc., and all unused elements of the CAD files are removed. In addition, the content of diverse databases is difficult to mix because of the various sources of data. As a result, special algorithms must be developed to combine different sources of data. Refer to Chapter 8, where we show how to create indoor maps from CAD files.

Most of this graphical information is not georeferenced in the real world due to CAD systems being based on local two- or three-dimensional coordinates linked to the building. This means that these data are not compatible with georeferenced information of the surroundings or of other buildings; most road networks are digitized in geographical coordinates. For instance, the Geographical Data File (GDF) standard provides a geometric model and a topological model with lots of features. This geographical format is used in many applications like car navigation and outdoor LBS. This data set is limited to the road and its immediate surroundings. Indoor navigation, on the other hand, presents a different case for such location-based applications, especially when entering buildings or when using underground paths. This situation of noncompatible data results in a lack of a seamless mapping/content handoff when the user travels from indoors to outdoors or vice versa. Specifically, georeferenced information can be useful for a local orientation, for example, to find a room from the entrance of the building. In short, without multizonal (i.e., indoor–outdoor) integration in terms of spatial references (and mappings between them), seamless handoff is not possible.

Also, the resolutions of location and orientation measurements in buildings are beneficial. Associated with position are three additional coordinates that provide the orientation of the device. These are roll (rotation about longitudinal axis), pitch (oscillation of the longitudinal axis in a vertical motion about the center of gravity), and yaw (oscillation of the longitudinal axis in the horizontal plane). Each is defined as an angle of deviation, in a predefined coordinate system. These are typically used as parameters to a pointing function to select a particular object of interest (as opposed to inertial navigation functions). Additional sensors, generally within the device, such as a compass to provide a roll angle or tilt meters for pitch and yaw, need to be incorporated in the equipment. The orientation parameters are also defined as signed 32-bit parameters representing decimal degrees with corresponding error parameters. In order to generate appropriate graphical presentations or interaction schemata to describe a navigation task, we need a metric for the quality of location and orientation information.

The *resolution of location* can be measured in meters and indicates the maximum deviation from the assumed position. Accurate radio bearing for mobile devices is technically difficult and costly, and even small deviations can cause wrong implications about the position of the user (i.e., whether he is in front or behind a wall). The quality of position and location measurements is information coverage. Especially in buildings, certain areas will provide no information at all about position and location. Fortunately, this is not critical if the general coverage at decision points is acceptable. Nevertheless, larger untracked areas during the navigation have to be compensated for by appropriate graphical presentations. This problem can be avoided by using optical media, such as IR light, to track the user's position — of course, at the cost of equipping the environment with special hardware. Other approaches try to determine the location with the help of computer vision, but they currently still lack a certain degree of accuracy and reliability. Later in the chapter we present map matching, which improves the accuracy of localization.

The *resolution of orientation* can simply be measured in degrees. For solutions integrated into the environment, e.g., wall-mounted infrared senders or beacons, the resolution depends on the sender's coverage area. Realistic resolutions for an IR-based system in buildings range from 20 to 180°, depending on the form of transmitter and receiver. A more detailed discussion of this approach can be found in Baus et al. (1999) and Butz et al. (2000).

One important observation is that for the task of way finding, a high resolution of orientation and location information is not always necessary. Under the assumption that a user moves on a segment from one decision point to the next without changing direction, it is sufficient for the system to distinguish between these two decision points. The same holds for the resolution of orientation. It can be considered sufficient, if the system can distinguish between all the choices at a given decision point. For example, at a T-junction the resolution of ±45° is good enough to reorient the user correctly to the new direction.

SOURCES FOR LOCATION DATA

Digitized maps and GIS databases for the road network are generally available from various commercial (e.g., TeleAtlas, Navteq) and public sources (e.g., U.S. Census TIGER files), whereas indoor maps of buildings and campuses are not. Moreover, vast amounts of outdoor GI data are publicly available, including geographical features, governmental boundaries, public roads and other transportation, etc.

Indoors, even if data are available online, they are most often created, maintained, and controlled by a private enterprise of some kind. Since large buildings frequently sublet space to tenants, it may be necessary to span several location authorities. Thus, federation issues are central to indoor location authorities. In addition, outdoors, names of entities are typically matters of public record, whereas indoors, names are assigned by the operator of the space in question and may not be publicly available.

Outdoor data authorities and indoor data authorities tend to differ in some subtle yet important ways. Outdoor data are frequently available from a global location authority. Indoor data require connection to some kind of private location authority. Just because a data authority is able to compute the answer to a question does not mean that it is willing to do so. For example, an organization may keep its building floor plans secret from nonemployees. Even if an employee can see the floor plans, there may be some sensitive areas omitted from the published floor plans.

TWO-DIMENSIONAL VS. THREE-DIMENSIONAL

Whereas outdoors humans mostly navigate on the curved two-dimensional surface Earth, in indoor environments humans will navigate in three-dimensional structures. Thus, a different set of navigation tasks become necessary. For three-dimensional structures the problem arises that signal strength penetrates floors (direct vertical metric distance), while humans must navigate along corridors and use stairs (thus on a constrained network) to overcome vertical distance. Moreover, this distance constraint (a derivative of location) must be negotiated through communication or the interface with the system for location finding. We discuss modeling vertical movement in greater detail later in this chapter. For outdoor areas, the level of resolution used to create three-dimensional city models is expected to be sufficient. This theoretically means that there is no need to create a specialized three-dimensional model. Three-dimensional indoor models can be obtained from construction companies. Topological structuring is not of particular interest since small discrepancies are not of importance, which means CAD models have to provide sufficient detail, though it should not be forgotten that three-dimensional topology is desirable for three-dimensional routing. Similar to the node-edge graphs, which use duality to represent space–activity interactions, the node relation structure (NRS) (Lee, 2004) was developed to represent more than just adjacency and connectivity relationships among three-dimensional spatial objects in built environments, based upon three-dimensional Poincaré duality transformation.

The NRS is defined as a set of nodes (three-dimensional entities in primal space) with a set of edges (spatial relationships between three-dimensional entities in primal space) that represent the topological relationships among entities in built environments. The topological relationships between three-dimensional entities can be represented in dual space with sets of nodes and edges. By not dealing with three-dimensional entities, therefore, Poincaré duality helps to resolve the problems associated with data storage and computational efficiency to keep topological consistencies of current three-dimensional entity-based data models.

We show how to create three-dimensional indoor maps in Chapter 8.

SPATIAL REFERENCE SYSTEMS

A coordinate system is a vector space spanned over a domain. It has an origin that is the subordinate-contained location or the system boundary. A position's coordinates are an element of a coordinate system. Each singular coordinate is a component of this element. To give an example, the coordinates specifying the position of the location "Room 101" consist of the room number only. There may be additional coordinates, such as a room description, a room type, the measurements of the room, and so on. These coordinates identify the room uniquely within the associate coordinate system for rooms. The domain a coordinate system is spanned over consists of continuous or discrete values from various dimensions. These dimensions correspond to coordinate types; the values are valid coordinate values. They may be specified by an enumeration in case of discrete values or by setting a valid value range with suitable constraints. The coordinate system, the measurement precision, and the data structures supported by the indoor location infrastructure often differ from outdoor navigation. In general, a coordinate-based spatial reference system is defined by a representation (e.g., cartesian, polar, latitude–longitude coordinates) and a reference frame (e.g., the point of origin). In the outdoor case, there is one primary space to represent. An example of a coordinate system is a room coordinate system with the dimensions number, type, description, and measurements and with valid values for these dimensions. Its origin may be the contained location "Building 1," containing the rooms of the coordinate system. It may also be the system boundary — that is, void — if subordinate locations are irrelevant to the application and there can be no ambiguities. Various coordinate systems may be used, but they typically have well-defined (standardized) transformations between them. There may also be several instances of any type of coordinate system.

Indoors, however, each floor of a building typically acts as a separate universe — two points on different floors may have the same coordinates on their respective floors, but have an unknown relationship in the real three-dimensional world. This presents a challenge for achieving a seamless handoff from one coordinate system to another. And, for dynamic environments where the system is *ad hoc*/self-coordinated (where the assumption that positions of beacons or reference nodes are preconfigured is no longer valid) an important system capability would be for beacons to cooperatively determine ranges to each other and independently form a coordinate system.

Hence, traditional latitude–longitude coordinate systems are no longer sufficient to provide effective and seamless interaction.

One way to achieve the seamless handoff from one coordinate system to another is to attach coordinate systems to indoor entities such as building floors. This approach also works when buildings are more complex, for example, if they have wings or towers, or if two buildings share a parking lot or structure.

Also, the location model should provide a way of representing the coordinate system frame in which the location is expressed, and a transformation function to transform that location to another coordinate system, if necessary.

Plus, LBS applications often need to integrate location information obtained from different positioning systems having different coordinate systems, implying a federation of multiple spatial coordination systems. This requires a framework for representation of position information from various sources (sensors) and a metadata standard (i.e., SensorML) providing a degree of accuracy.

Because no single positioning technology works everywhere, one expects the use of several positioning technologies for LBS applications, each with their own frames of reference, location granularity, and error models. Algorithms are needed for establishing federated coordinate systems that can combine location information from several frames of reference and will allow a new class of seamlessly integrated applications. In terms of implementing multiple spatial reference systems, consider that locations are grouped into location sets. The coordinates of the available locations refer to three different coordinate systems. The first one is a geographical coordinate system used to represent addresses. It might define the dimensions of name, city, and state. Another coordinate system serves to represent a building's floors and defines a single dimension, the floor number. The third coordinate system might be a room coordinate system with the associated dimensions name, type, and number. Additionally, there might be a floor coordinate system.

The origin of the spatial coordinate system is the system boundary, while both the floor and the room coordinate systems' origins are contained locations. As regards the floor coordinate system, its origin might be an address. A room coordinate system's origin may be either an address or a floor. The origins of the coordinate systems described above indicate that the locations to be employed in an application may be part of a hierarchical structure. Rooms are spatially contained in floors or addresses, the latter of which may also contain floors. This way, the root node of any location's structure might be an address, while floors might be contained in a location graph as either nodes or leaves. On the other hand, rooms might always be leaves; consequently, this would make rooms noncontained locations.

In the case of proximity-based localization, usually the relations of users to locations other than their present location are taken into account; users are always considered to be at an address, on a floor, and in a room. Accordingly, the following is true. The application does not need to make use of the location model's ability to explicitly represent spatial prepositions; application does not need to determine the proximity of locations and does not have to calculate whether two locations overlap; and transformation operations are not necessary either, because all possible results of a location's transformation are already contained in the structure of this location type.

The operations of locations that are made use of in the case of proximity-based localization might be those that serve to determine equality and containment of locations. *Containment* is the determination of positions of objects by identifying spatial regions that contain those objects. In other words, location encloses subordinate objects (model entities — places, people, devices, etc.) like a container. The subordinate objects can be but are not necessarily hierarchically related to each other. In this case, the size of a container governs the resolution. Examples are geographical location, symbolic location, and geometric model with location defined as n-tuples, $n > 2$.

Overall, for supporting human-based indoor navigation applications, it can be argued that the detection of space boundaries (symbolic spaces) by an indoor positioning system is more important than accurate coordinate estimates. Robot-based navigation, on the other hand, would require real coordinates.

For dynamic environments (where systems are *ad hoc*/self-coordinated), an important system capability would be for beacons to cooperatively determine ranges to each other and independently form a coordinate system. Here, the assumption that positions of beacons or reference nodes are preconfigured is no longer valid.

However, this would require complex capabilities in beacons. Because the quality of location information can significantly vary with dynamics, the location model should provide a way to extract both statistical distributions and raw location information.

NETWORK PATH PLANNING

A navigation system needs to provide support when defining the travel as well as guidance to the selected destination. Consider an emergency use case. In order to approach the disaster site quickly, emergency responders need to know where the room is within the building and to identify the optimal route from a source node to the destination node. Such procedures are well implemented in car navigation systems based on road databases.

Typically, a GPS receiver or navigation device comprises a digital road map, or GIS database, of the area of travel. A user, such as the driver of a vehicle, inputs a destination into the GPS navigation device and the GPS navigation device calculates a route from the current position of the vehicle, through the network of roads as represented in the GIS database, to the destination. Turn-by-turn directions are relayed to the driver through visual or audible prompts. The current position of the vehicle is determined by receiving GPS signals from GPS satellites. If the GPS signals are obstructed or interfered with, the current position of the vehicle may be very difficult to ascertain and the usefulness of the GPS navigation device is severely degraded. GPS signals may be obstructed, for example, in urban areas with many tall buildings, on roads with many trees or mountains, and in areas with high levels of background radio noise or interference. In order to provide accurate driving direction, the GIS database must be kept current and up to date. Databases, with varying levels of accuracy are commercially available and are released for purchase and installation into GPS receivers at regular intervals. Various dead-reckoning techniques may be used to ascertain the position of the participating vehicle in the absence of a complete GPS signal. For example, an adjacent position dead-reckoning method uses a gyroscope or compass in order to estimate the vehicle's position based on its current velocity and direction. Many dead-reckoning methods use estimation methods such as least minimum square error (LMSE), autoregressive equations, and Kalman filtering.

On the other hand, for indoor navigation, the concept of route guidance must be reconsidered for several reasons: specific design of map database, style of human displacement, and particular needs for users. Depending on the mode of the goal definition, different way-finding strategies can be applied. For example, a metric criterion such as the shortest path can only be applied if — besides topologic information about the structure of the environment — the metric information of the goal is available. Metric information, on the other side, is not required to make decisions in an airport environment, because topologic and semantic information are sufficient for navigation computation.

In order to define the optimal route in the three-dimensional geometric network model (GNM), a well-known algorithm — Dijkstra's priority-first search for finding the shortest path in graphs — is used. This network defines a directed graph with sources and sinks and is then used to solve network flow problems (e.g., emergency evacuation) by determining an optimal plan for minimizing the time required to evacuate the building. Also, there might be a need for recalculating the path on the fly. For example, when attempting to reach the emergency site without knowing which stairways or hallways are feasible and safe, rescuers may be blocked in the middle and have to find an alternative.

A significant problem that arises in *ad hoc* positioning systems is that there may be too few distance constraints to obtain a consistent coordinate assignment. Obtaining a coordinate assignment that is unique up to translation, rotation, and reflection requires that the graph formed by available distances be *globally rigid* in a technical sense. An arbitrary deployment of location infrastructure nodes (either beacons or passive receivers) or sensors will not generally produce a globally rigid structure. Priyantha et al. (2003) shows that *mobility* can help solve these problems. In *mobile-assisted localization*, a mobile object (e.g., mobile device or a robot) wanders through an area, collecting distance information between the nodes and itself.

The challenge is to design movement strategies that produce a globally rigid structure of known distances among the static nodes. The pairwise node distances can be fed into a localization algorithm such as the anchor-free localization (AFL) algorithm (Priyantha et al., 2003), which does not require any anchor nodes that already know their positions. AFL computes an initial coordinate assignment to all the nodes, using the radio connectivity information alone. This initial assignment results in a node layout that resembles a scaled version of the actual node layout, roughly preserving the topological ordering of nodes. AFL then uses an iterative optimization procedure to reduce the sum of squared distance errors between the nodes' true distances and the distances inferred from their current coordinates.

GEOCODING

Of primary interest is that one point in space in the outdoor world (e.g., one address or set of coordinates) potentially represents entire sets of points in the indoor world. For example, a rookie first responder arriving at an emergency site (a specified address or position) in an unfamiliar part of the city might be surprised on arrival to find quite a large building or even a campus of interconnected buildings sharing the same address; he may have been prepared to search a single, small building or home. Realistically, a single building may be expanded into a collection of floors, and then onward to halls, rooms, etc. (logical divisions of indoor space).

Within a typical urban area, E911-allowed margins of error of up to 100 m usually include a range of possible addresses (e.g., "The 400 block of Main Street," which is a collection of indoor spaces). To achieve maximum precision for emergency positioning service in urban areas, the location of a phone may be interpolated as set of geographic coordinates (as in GPS-assisted determination of latitude/longitude). In rural settings, often lacking well-named and sufficiently dense street address networks, representing the position of an emergency caller using coordinates is a highly useful level of accuracy and detail.

A major reason that geographic coordinates alone are not optimal in denser urban settings is that emergency response teams usually travel over the road network to reach an emergency. Typical street address structures of a city can be navigated intuitively, and it is a common practice to geocode or cross-reference geographic coordinates within a reference model of the land-based road network. There are common data standards for performing geocoding or address-matching operations in the outdoor world. Network (vector) data models such as TIGER have achieved high levels of efficiency, descriptiveness, and functionality, such as containing street-line-level attributes for much of the U.S. Census Bureau.

Outdoor-oriented geocoding models extend to the indoor world, where they facilitate location-based service authentication and delivery. Specifically, it is proposed that the process of using any type of position or translating absolute position (represented by coordinates) into relative or contextually symbolic ones (such as "in Room 101") is a critical enabling step for delivering seamless indoor LBS, which by definition makes information available in all forms, whether being specified by an address, (x, y) coordinate, or "in Room 101". The terms *microgeography* and *micro-LBS* can be used to describe the spatial nature and geolocation methodologies of indoor environments for a variety of applications. Note: There is no such thing as a single absolute coordinate reference system; for the sake of our argument, it could be either geocentric cartesian coordinates, polar geographic coordinates (latitude, longitude, elevation), or planar projection coordinates, as defined under a well-known geodetic system. *Relative* is used to describe locations as they relate to their contextual surroundings, such as "Room 201 in Building A" relates to a certain floor (probably the second) and building (A). Relative or contextual locations may include topology, enabling the expression of relationships with other entities through connectedness, adjacency, containment, etc. Relative positioning systems like Active Badge usually sense an object or person being or not being in a room.

Consider also the fact that indoors, even a 3-m potential for error could mean a difference of two or three floors in vertical space and a number of rooms in any horizontal direction. Plus, public and private facilities can be intermingled in wireless space, which does not obey the barriers of walls and floors within buildings. Part of the indoor LBS challenge is to design solutions for a world of heterogeneous, overlapping, and interrelated spatial zones, reflecting the various ways that people and property tend to move and congregate indoors.

The need for a seamless LBS (and the end user's experience) is evident from this example.

The fundamental principle of a three-dimensional indoor geocoding method employs a traditional address-matching technology developed for location positioning of outdoor phenomena. A TIGER-type reference database model can be comprised of the three-dimensional geometric network model (GNM). An edge in the three-dimensional GNM, a hallway line segment, contains indoor address (room number) range attribute data, which are the low (F_RA_l) and high (T_RA_l) room numbers on the edge's left side and the low (F_RA_r) and high (T_RA_r) room numbers on its right side. Unlike a street address, the indoor address for a room or suite is not always assigned as a number in sequence, or is named as the "Liberty Room." In order to standardize the indoor address (using a sequential numbering), translation tables can be used.

ONTOLOGY

Lastly, ontology of pedestrian navigation differs indoors when compared with the outdoors and includes route selection criteria, appropriate route optimization strategies, and route structures.

In terms of system feedback and user interfaces, valuable information must be included to guide users in their interaction adequate to scale, physical, and social contexts.

POSITION TYPES AND SPATIAL REFERENCE SYSTEMS

Navigation depends on the type of application. It can be relative or absolute. Commonly navigation data are given as latitude, longitude, and height. Speed, acceleration, and bearing can account for the dynamics in the position of moving objects. But the navigation data are only useful in the context of a map system, i.e., a reference. It is thus important to differentiate further between absolute and relative navigation data as well as position, orientation, and distance in the stationary state. Moreover, the relative position of a user depends on his relation to objects in the vicinity; the absolute position is the position relative to a reference position.

ABSOLUTE POSITION

Absolute positioning is when objects have specific (x, y, z) coordinates (globally expressed in latitude, longitude, and altitude) or are positioned as a metric offset from a fixed reference. There is a spatial reference unique to each specific position. Absolute information provides three-dimensional position coordinates in standard format relative to the Earth, using WGS84 as the reference system (Department of Defense, 2000). Latitude and longitude are defined with respect to the WGS84 ellipsoid representation of the Earth. Altitude or elevation is defined as the distance above the ellipsoid (representing the surface of the Earth). Furthermore, an absolute position allows the determination of positioning information of disjunct systems independently, in reference to the same point in the inertial system. Also, dynamic navigation systems can surpass this problem partially by adding motion in form of a history of past positions or velocities to the problem's solution.

Additionally, the attribute of direction complements the absolute position. Some local positioning systems, such as MIT Cricket and Active Bat, provide absolute positions where a user is situated at (x, y, z) coordinates. On the other hand, this means a trade-off since the beacon has to be provisioned with its own position.

To enable its beacons (sensors) to gather the distance information (ratio of height to distance) for terminals, MIT Cricket implemented a local coordinate system using four active beacons instrumented with known positions within the space. The beacons are configured with their specific (x, y, z) coordinates and broadcast this information on the radio frequency channel, which is sensed by the receiver in the terminal. A Euclidean (geometric) model makes it possible to define a terminal's position plus orientation information by a rotation with translation matrix in an orthogonal coordinate referential. Absolute positioning using the Wi-Fi cell ID positioning method works as follows. Wi-Fi access points (APs) transmit the Wi-Fi signal, augmented with their physical (fixed) coordinates. This can then be used to estimate the location of the mobile host.

In general, a coordinate system is defined by a representation (e.g., cartesian, polar, latitude–longitude coordinates) and a reference frame (e.g., the point of origin). For example, MIT Cricket uses latitude–longitude as its coordinate system for Cricket beacons. The benefit of using latitude–longitude has to do with indoor–outdoor seamless position handoff. This way, mobile applications can keep the same position representation and reference point as the user roams from indoor

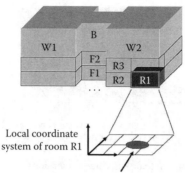

/de/stuttgart/keplerstr/9/floor1/wing2/room1: circle (X, Y)

FIGURE 5.1 Local coordinate system.

to outdoor. The disadvantage of using latitude–longitude, however, is that it is not practical to deploy Cricket beacons using latitude–longitude, with just CAD drawings, that is.

It might be easier to define a set of local reference points throughout a building (e.g., corner of a floor or a room) and configure beacon coordinates with respect to the local reference frames. Within each local reference frame, it is possible to express coordinates in terms of latitude–longitude. However, for indoor applications, cartesian coordinates provide more meaningful representation than latitude–longitude coordinates (e.g., (1 m, 1 m) and (1 m, 2 m) vs. 80.234 latitude and 142.433 longitude).

MIT Cricket's floor map is divided into its own reference frame so that all Cricket beacon coordinates are configured using the same coordinate system as defined in the corresponding floor plan drawing.

The benefits of using separate reference frames include the following:

1. Beacons on the same floor are already assumed to have the same z value.
2. Each floor plan is a two-dimensional representation of the spatial boundaries.
3. No need to find arbitrary reference points during deployment (just simply follow the convention as set in the CAD drawing).
4. Coordinate system is switched at the same time as a new floor map is downloaded (since the reference frame is subdivided in the same way as the map).
5. Consistency between the beacon coordinate and floor plan coordinate is maintained if the original floor plan adopts a different coordinate system between different floors.

In addition, the strength of the radio signals arriving from more than one AP can be used by the terminal to relate its position to the access points. Either the distance from the AP can be estimated by the time of arrival (if there is a timing sequence in the signal from the AP), or the direction of the signal can be used to

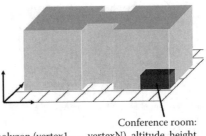

Conference room:
polygon (vertex1, ... , vertexN), altitude, height

FIGURE 5.2 Local coordinate system with three dimensions.

trilaterate the terminal, if the antennas used can determine direction (something that is not the case on most APs today). Some rooms may have one or more Wi-Fi APs, but the same trilateration model can be used. The accuracy of the system might be limited by the cell size, intervening walls, and the nondirectionality of antennas. Handling of location to a room can be done with a tolerance level (snapping). This is applied in the location fingerprinting positioning method.

The main advantage of the absolute positioning systems is the high accuracy that they support when estimating the position of an object. A main drawback is the deployment costs and the operation maintenance of location-specific infrastructure, which has to be added to the Wi-Fi infrastructure.

Using absolute coordinates gives more flexibility in application development, since it is possible to reference a fixed point in space, and — this point being globally unique — to use it as a database key. Other systems can reuse the database for other applications, e.g., by defining different areas of interest. However, if the objective is to keep track of the relation of objects in space, relative positioning may be more appropriate.

RELATIVE POSITIONING

Relative positioning can be defined as the process of evaluating the location and orientation (direction of movement) by integrating information provided by diverse sensors. The integration starts in a certain moment at an initial position and is continuously updated; i.e., the movement of the user is continuously tracked. An advantage of such a relative localization is that tracking can be done relative to some object the application is interested in, for instance, the location where GPS positioning fails. Relative location can be defined as the position with respect to an arbitrary location mark defined as the origin. A relative location coordinate system is characterized by the origin coordinates and a reference system. Several alternative reference systems can be supported as required. Coordinate transformation functions are used to perform conversions between differing representations of location.

The Open GIS Consortium (OGC) Coordinate Transformation Services Implementation Specification (OGC) provides interfaces for general positioning, coordinate systems, and coordinate transformations. Coordinates can have any number of dimensions. So this specification can handle two- and three-dimensional coordinates, as well as four-dimensional, five-dimensional, etc.

Relative tracking has been initiated indoors with the goal to provide high-accuracy tracking of the human body (e.g., head, hand, full body) basically for augmented reality (AR) systems. Plenty of research employing a variety of sensing technologies deals with motion tracking and registration, as each technology has strengths and weaknesses.

Vectors linking two points give information on their positions relative to each other. When no reference points are given, the solution can be rotated and mirrored through an arbitrary axis. Each position added to a system solution reduces the problem one degree of freedom at a time. Common to *ad hoc* networks, there are no fixed reference positions. Here, triangulation can only resolve the position of nodes relative to each other. They can then still float and be rotated arbitrarily in free space. The geometric navigation solution derived here is always the relative distance between points.

A system of many known positions that has no reference location or orientation is only fixed in itself, not in its position in space: if only the position of one point and no orientation is given, the system can be mirrored and rotated through any axis leading through this point. The translatory movement is prohibited by this first known position. When a second position is introduced into the system, the rotation of the system is further restricted to the axis through these two points. A two-dimensional problem would thus still have two possible solutions. A third known position finally fixes the system in three- and two-dimensional space.

Usually, indoor LBS applications do not require metrics like distance and time due to the small scale of the indoor space. For example, walking inside a shopping mall does not require distance measurements, but a more meaningful representation of location, that is, where things are in relation to fixed objects. Instead, a mere value of proximity ("next to") or containment ("in the room") to some object is sufficient. Each object can have its own frame of reference. For example, the Active Badge LPS provides relative position (symbolic location), which encompasses abstract ideas of where something is. For example, the person is "in the room" and therefore not in the hallway.

Consider the following as an example: There are two possibilities with the Wi-Fi positioning approach: absolute positioning (the Wi-Fi APs have absolute (fixed) coordinates) and relative positioning (the Wi-Fi APs have symbolic locations, i.e., "room 2, floor 201," "sector 1, hallway Z"). Relative positioning expresses location of objects or users in relation to one another (i.e., "in Room 1," "near," "next to").

Similar to the absolute positioning with Wi-Fi, relative positioning works by referencing the Wi-Fi AP that the user is accessing to its location. Instead of relating this to a fixed position, it is related to the access point ("near" or "far from"). Wi-Fi access points (APs) transmitting a signal augmented with their identity, e.g., "AP1: L_R101; R_R104," can be used to estimate the location of the mobile host in relation to themselves.

If there were access points or beacons in each room, and the rules for binding the user to the access point were strict, then the user's position could be related to the access point he was bound to. This would require cells to be small enough to fit into a room (and not overlap to other rooms). Alternatively, a database could be created that used the fingerprints of the radio network (field strengths and identities) to describe the rooms in terms of the radio network.

Moreover, each AP has an access range, which may or may not be defined to a specific metric distance (i.e., range of 10 m). The point is that it does not need to be for such applications as the Buddy Finder, because knowing that AP1 is "next to" AP2 will allow referencing (matching) that the user accessing AP1 is "next to" the user accessing AP2. Also, if a mobile device can detect signals from two APs that are known to be relatively far apart, then the device's location can be described as somewhere between these two APs. Decreasing precision in favor of logical statements such as "between object A and B" is a cell ID-base interpolation method that works with both relative and absolute positioning.

Relative positioning can also be used in location fingerprinting systems. The positioning database, which stores IDs of each AP and the radio characteristics of each room, can potentially define more regions and be more specific (decreasing calculations to determine the user's position).

An example of a relative local positioning system is Active Badge. Also, point-of-sale logs, bar code scanners, and systems that monitor computer login activity are symbolic location technologies mostly based on proximity to known objects (i.e., the network segments in an Ethernet network, or the interfaces it connects to in a router; this is then mapped to a database of the physical layout in the system). Purely symbolic location systems typically provide only very coarse grained physical positions. Using them often requires multiple readings or sensors to increase accuracy, such as using multiple overlapping proximity sensors to detect someone's position within a room.

The Buddy Finder application can function based on relative positioning where a location-based trigger would alert the user when his friend is in the proximity. For this purpose, the Active Badge LPS is sufficient where positioning is based on being sensed in a location (i.e., a room). However, it is essential to be able to tell with certainty which room a person is in, or which side of a partition he is on.

REFERENCING ALGORITHMS

For any location-based service to give meaning to the mobile terminal's location, it must fit that location into a world model. The relative and absolute models described previously tie into world models, with very different characteristics. This world model might be a geographic map based on Euclidian geometry (such as might be produced by a GIS), or it might be a more symbolic representation of space such as a network graph describing the connectivity relationships among physical objects of interest.

The world model expresses the location by describing how to reference it to other objects. World models implicitly underlie the mapping infrastructure, yet they are rarely set in a proper theoretical framework by going back to the basics of what a location (space) can be in pure mathematics and, more importantly, symbolically (semantics). Though these models are purely abstract, they are frequently used in association with a particular location-sensing technology, from which they retain only relevant characteristics that can be mapped to a corresponding notion of location. Systems can be built that are independent of the location-sensing techniques, however.

Based on these world models, applications become possible that can access information of the real world originating from sensors and additional, aggregated information. This way, information systems that build upon this information space can take advantage of the user's location, activity, and environment to select and provide information that is relevant to the user according to his profile (preferences, access rights, identity, etc.) and his current activity, like shopping. The choice of model depends upon what you are trying to do. For instance, will a typical requirement be navigation or merely proximity or presence?

Different types of location models provide different abstractions between terminals and the data provided by various location-sensing technologies.

In general, location models can be classified into four types:

1. **Symbolic**: Describes location and space in terms of names and abstractions. Unlike the geometric model, humans and computational devices can understand this model better. However, they lack the precision of geometric models in terms of metrics for location and distance. Nevertheless, predominantly symbolic models work because location awareness is most intuitive when using named locations.
2. **Topological (or structural)**: Location entities as subsets or neighborhoods of space, but also their structural relationships (i.e., adjacency, connectivity).
3. **Geometric**: Allows points, areas (two-dimensional), and volumes (three-dimensional) to be modeled; however, a point in geometric space has no relationship to what it points to. The resolution of this model is as fine as the units of measurement used.
4. **Hybrid**: Represents a logical step forward in combining the advantages of the geometric, topological, and symbolic model types in order to overcome their respective disadvantages. As a consequence, the hybrid model is more complex, requiring greater amounts of data.

Numerous world models have been defined in different application domains. Once again, one size does not fit all applications, and we will describe the predominant approaches that have been adopted to date.

In the wide-area location-based systems, there is a world model built in (based on the characteristics of the network). This makes certain assumptions about the underlying system, which may not be correct in indoor location systems. One advantage of indoor location systems over outdoor location-based systems is that they are able to determine position with a higher accuracy. This depends partly on the fact that the position sensor information is more accurate, but it is also a function of the possibility to relate the world model more accurately to the position information. One way to do this is to assign coordinates to positions (absolute positioning). Since the measurement accuracy is higher, it is possible to assign coordinates in the indoor system with higher accuracy than in the case of outdoor systems.

It is, however, not enough to geo-reference a building if the positions of users and other objects inside the building are also relevant. To begin with, buildings are three-dimensional. Information about buildings contains at least one dimension (the height information), which is not relevant in traditional location-based systems (since they may contain height information, but there is no assumption that the objects are stacked on top of each other, like they are in a building).

As a result, the world model underlying traditional geographic information systems usually contains neither the detail nor the coordinate systems to satisfy the needs of users in an indoor setting.

Frequently, however, the absolute position is not needed; all you need is to relate the position of an object to yourself — relative positioning vs. absolute positioning. This implies that the information you need is not the coordinates, but the proximity of the object.

Whichever world model you use, information related to position must somehow be stored, modeled, and mapped. The storage is typically handled in a spatial database; the modeling depends on the algorithms that can be associated with the world model. The mapping is the role of the mapping infrastructure, and to create those, location (or space or world) models are essential. It is also about the content or context data that can be associated with that location data for specific location-based services.

Part of the mapping infrastructure is a modeling language for the exchange of this information. Local positioning systems structure information according to the position of the objects, but since their different properties (i.e., sensor type and position determination algorithms) lead to different ways to express and model them, there is a technical challenge with respect to data integration and sensor fusion. Hence, a common data modeling language is needed in order to be able to integrate sensor data from different systems. Building a new indoor location-based application is relatively easy, and gathering and organizing the data is not so hard — as long as it is done in isolation. However, if data from one application need to be integrated with data from another application, a common data modeling language is essential.

One example of such a modeling language has been developed by the RELATE project. This project investigates relative positioning in the specific context of tangible interfaces that involve spatial arrangement of physical interaction objects on two-dimensional surfaces, such as white board or tables. RELATE is an approach that uses dedicated positioning technology to obtain finer-grained relative position, targeted at close-range operation.

The research in the RELATE project is driven by the positioning requirements of tangible interface systems composed of physical interaction objects. Tangible interfaces have recently attracted considerable research interest, as part of the paradigm shift toward ubiquitous computing, aiming to provide interaction in ways that are intuitive and seamlessly integrated with people's activities in a physical world.

Thus, sensors that are also integrated in an infrastructure can be used directly by devices nearby that do not have access to the infrastructure or do not want to use it, because an uplink to the infrastructure may be too costly, either financially or in terms of energy consumption.

For scalability and abstraction, locations are typically organized hierarchically in both the geometric and symbolic models. For example, the ActiveMap service from ArialView Awareness System uses a symbolic model with a location containment hierarchy. In their geometric models, Nelson and Ward use an R-tree index and a quad-tree index, respectively, to facilitate location searching and updating. The EasyLiving project from Microsoft Research uses a geometric model with flat layout that works well for small scales (i.e., a room). Trying to take advantage of both symbolic and geometric models, Leonhardt and Magee (1996) propose a hybrid model, in which a location contains both a symbolic name and geometric coordinates. The symbolic name and geometric coordinates can convert to each other via predefined predicates. Such a combined model shields the details of underlying sensors and can support applications that need or could use both symbolic and geometric location information. Containment and intersection are the most frequently seen and probably most powerful relationships in a location model.

Symbolic information and attributes might also be applied to an intentional naming system (MIT's INS by Balakrishnan) or diffusion (Estrin). Both positioning and name diffusion may be applied to the same end of resource and service discovery.

Also, mapping infrastructure providers (i.e., content providers) have to figure out how to seamlessly provide location content utilizing local location model types and data formats.

Also, different indoor location-based applications deal with different scales. For example, both the Buddy Finder and the Product Finder can operate on a building scale (i.e., "Notify me when Bob is in the shopping mall" and "Show me the location of the store that has the product that I am looking for"). In addition, both applications could require a room-scale location model (i.e., "Notify me when Bob is in the same room" and "Show me the aisle where I can find the product"). As a result, it is important to make sure that the underlying location model is suitable for the desired functions. In addition, having the appropriate scale and federation of data would enable faster querying and data processing.

EXPRESSING WORLD MODELS

The location model should provide a way of representing the coordinate system frame in which the location is expressed, and a transformation function to transform that location to another coordinate system, if necessary.

This knowledge is often referred to as symbolic reasoning, allowing decision making about the world. Symbols with geometrical properties redirect the immediate surrounding of the object or person being navigated in order to allow safe navigation. In indoor navigation, these types of symbols include walls, doorways, or obstacles such as tables, chairs, and people. On the other hand, a model of the large-scale structure of the area is required to enable planning of routes to fulfill an entire path planning.

The mapping infrastructure is about storing and managing position information data as well as content data associated with the location. While the most basic location-based applications can answer the standard question "Where am I?" to varying degrees of accuracy, there exists a need to frame "where" in the context of a modeled environment (or world) in order to move beyond simple inferences of position to a better understanding of what "where" relates to contextually. Hence, a world data model is about contextual information, which includes location.

To explain the significance of the mapping infrastructure, let us look at how these are used in the Buddy Finder and Product Finder applications. The Buddy Finder application will function based on relative positioning where the proximity of one user in relation to another user is enough to cause a location-based trigger that will notify both users that they are in the same area (i.e., hallway). The Product Finder application, on the other hand, would need more precise positioning that is not based on proximity or containment, but on the exact (x, y, z) coordinates (absolute positioning) of a product in a store isle.

The argument can go both ways regarding the need for absolute positioning (and navigation) for such applications as the Product Finder that will help the user find the product. Relative positioning could be good enough to locate a user in relation to the aisles in the stores and notify her that this is the aisle for the product that she is looking for, adding a symbolic description for the location of the product and notifying the user with "post 3, 2nd shelf from the bottom." In any case, absolute positioning is seen to be valuable to disabled customers (those that are either blind or in wheelchairs) and for use in robotics, and deserves further exploration in terms of mapping and modeling location and navigation.

An example of a world model can be taken from the Sentient Computing Project by AT&T Cambridge, which has developed the Active Badge and the Active Bat LPS. This project explored what users could do if computer programs could see a model of the world. By acting within the world, users would be interacting with programs via the world model as through a user interface. While humans can observe and act on the environment directly, application programs observe and act on the environment via the world model, which is kept up to date using sensors and provides an interface to various actuators. If the expressions used by the model are natural enough, then humans can interpret their perceptions of the world in terms of the model, and it appears to them as though they and the computer programs are sharing a perception of the real world.

Another example of a world model is the Augmented World Model (AWM) from Nexus, which provides the whole location context for context-aware applications, for both the indoor and outdoor worlds. Two main components responsible for the main aspects of the world model are the spatial data and the position information of mobile objects. This includes the representations of geometric objects (static and mobile) using geographic coordinates, and also virtual objects (i.e., virtual bill-boards, virtual Post-its, or virtual kiosks) with which the real world is augmented. Virtual objects provide, among other things, links to external information spaces like the web. Information is organized in a database, called the ContextCube. The intent is that in an *ad hoc* environment, mobile devices can communicate directly with any ContextCube in their transmission range using a wireless communication interface. In such an environment you typically have a very heterogeneous set of devices. Hence, interoperability plays an important role and can be achieved by using an open protocol as provided by the Augmented World Markup Language (AWML) and the Augmented World Query Language (AWQL). Using the same protocol for both, the infrastructure and the *ad hoc* mode offer the advantage of only having to implement a single interoperability protocol for the ContextCube.

SYMBOLIC

Some types of symbolic models that represent symbolic location information include the cell model, zone model, and domain model. These models are not mutually exclusive. A processing pipeline for location data could utilize all of them in different processing stages. These concepts were first proposed in Leonhardt and Magee (1996). The cell model assumes that each location-sensing system represents an object's location in terms of a well-defined geographical area (e.g., a room, a shape on the map, an IR or RF cell, etc.) or cells. This way, cells are the symbolic locations in the model. Cells might potentially overlap due to having more than one sensor system. The inclusion relationship among cells might not be represented in a cell model, making it simplistic. The cell model preserves the accuracy of incoming information because it defines symbols for all locations referenced in sightings. However, it is awkward to use for spatial queries because of overlaps that it cannot represent. For example, it is impossible to enumerate all objects within a given cell in this model.

With the zone model, in a cell space, the intersection of the cells generates a set of mutually exclusive (i.e., nonoverlapping) locations or zones. Each zone is part of one or more cells. The cells are divided up into nonoverlapping zones. Zones do not overlap. As a result, the zone model can be classified as an exclusive symbolic model. A single-zone space can accommodate an arbitrary number of cells, the only prerequisite being knowledge of their respective overlaps. Thus, different and overlapping cell spaces generated by different location-sensing systems can be integrated. Moreover, since zones do not overlap, a located object can be in at most one zone at a time. Therefore, within a zone space, the movements of one object can be modeled by a single finite-state machine. Hence, a zone space is a natural framework for persistent object tracking and movement prediction. However, the zone space has some shortcomings. First, there is no notion of abstraction or multiresolution processing. Second, the zone space for one located object may be entirely different from another's if both are visible to different location sensor systems.

With the location domain model, a location domain is a symbolic location that can be ordered with respect to other location domains. Any set of location domains is partially ordered by the "contains" relationship. This ordering rejects the spatial inclusion of the underlying geographical areas. As in the cell model, domains are allowed to overlap.

When employing this model, a location service typically uses a predefined set of domains to represent the locations of the located objects. Located objects join and leave domains as they move in geographical space. If a located object is a member of a particular location domain, it must also be a member of all "parents" of this domain. To preserve strong consistency, changes in the domain membership should instantly propagate up the domain hierarchy. However, implementations may choose a weaker level of consistency. An arbitrary set of location domains does not necessarily include a "root" location, that is, a location domain spatially containing all other domains. However, it is desirable to include such a domain in the domain space used by a location service. This root domain can be called "anywhere." The partial ordering of location domains is a very powerful mechanism, which can be exploited to gain scalability and manageability. Because they are sensor independent and support multiresolution processing, location domains provide a flexible framework for both client interaction and management operations. A key aspect of developing LBS infrastructure is the organization of spatial reference information, which can take a variety of forms. For indoor spaces, spatial modeling can take the form of abstracting indoor spaces into symbols (such as rooms, hallways), and labels for these features are associated to users for positioning. This type of spatial approach eliminates some of the complexity required by geometric models. However, without distance attributes, symbolic models are more suited for filtering information by information spaces and intuitive mapping and navigation than detailed spatial and topological analysis.

In short, the advantages and disadvantages of mapping out the indoor world using symbolic models are:

Advantages:
- Implicit representation of spatial relationships (containment, closeness). For example, Room 200 implies a specific place in a building (second floor), intuitively distinct from Room 100 (first floor).
- Supports algorithms for handling some location queries.

Disadvantages:
- Lack of position precision.
- Sometimes an inefficient method to compute distance.

HIERARCHIES AND LOCATION DATA MODELS

The following is a discussion of mapping out or modeling the indoor world using symbolic location modeling. In this type of model, spatial location may be defined implicitly rather than explicitly, by reference to more or less abstract concepts relevant to a given world model, i.e., a semantic frame of reference.

A location entity, such as a shopping mall or university campus, is decomposed into several intersected subspaces: building 1, building 2, building 3, etc. Each of these buildings is divided into smaller composing subspaces (floor 1, floor 2, floor 3, etc.), until enough level of detail is reached to reference or map objects (i.e., a printer, Wi-Fi access point, product, etc.) in a space, which in this case is a room, a store, or a shelf in a store.

The relationships between the levels may be seen as a hierarchy, or they may be regarded as an object system, with inheritance between the branch nodes and the leaf nodes. The parent–child link in the graph implies super- or subspace relationships between two spaces. These divisions could be used as location entity providing a spatial reference for users (or other objects), which will themselves be defined by some supposedly well-known characterization rather than their physical properties. It is up to the location service designer to decide how to decompose the physical environment. The location service needs to maintain a hierarchical-style data structure for the space tree and handles queries of spatial relationship (i.e., containment) based on this data structure.

A fundamental principle behind any location model is that each significant object, such as a room doorway in a building, is uniquely named. It would be unusual, for instance, to have two Room 101's in a single building. Therefore, the label of the doorway combined with the name of the building provides a useful key value (unique ID) in a database table, as well as an intuitive indicator of location for users. Note that in traditional information models for location systems (based on maps), this is usually accomplished by using a coordinate system.

With such a constructed space tree, it is easy to tell whether the containment relationship exists between two physical spaces. The model is also capable of answering connectedness queries (i.e., "near," "next to") that exist between two physical spaces. Simple queries include "Where am I?" "Who/what is there?" and neighborhood discovery. This implies that it can also be used as a basis for service discovery.

The challenge is to extract information intuitively from the room number. Users know that Room 201 might very well be located on the second floor, but there is no absolute rule saying this is so. A simple TYPE can be added to a building record in a database and all associated room numbers can be topologically arranged (see the section on topological models, below).

A symbolic map, sometimes referred to as a spatial model graph or a spatial tree, represents these different levels of spaces by nodes. Information is considered to be affiliated with a location, hence linked to a node in the spatial model. Edges (or arcs) between the nodes define how these places are connected.

For example, a shopping mall can be represented as a set of spatial model graphs, each graph representing a floor. Each node is a place of interest (i.e., store, restaurant, ATM), which can be expanded into a more detailed scale or hierarchy, as explained above. Moreover, these nodes can have symbolic addresses or location IDs. The edges (arcs) between the nodes, in this case the stores, represent connections, which in this case are the shopping mall's hallways. Each store represented as a node may have associated secondary nodes representing shelves or products. Links between floors can be represented by special arcs (i.e., represented as dotted lines), which have specific properties indicating the type of connection (stair, elevator, escalator, ramp) and the floors they link to. In the parking lot floors, each node could represent one parking spot and the edges (arcs) are the paths to elevators and stairs. This model can also represent access privileges. For example, thin dotted lines between rooms and doors (both represented as nodes) can represent that the door is open. Of course, the binding of information to nodes may be more abstract (i.e., not human readable) if this is what is required by the system.

Using these properties, a location-based service is able to generate paths according to specific user needs. If, for example, a disabled person wants to visit the center using a wheelchair, the navigation service will only use floor-linking arcs with an elevator or ramp connection, thus providing the user with a personalized path, a process that can be automated so that the user receives a map that is appropriate to his context. This can equally well be used to derive personalized shopping maps (provided the nodes have information about the content of the shops) or to control cleaning robots.

The properties of users could also be represented symbolically, associated with profiling a particular user (e.g., authorizations, security constraints, etc.).

In terms of mapping out the positioning, infrastructure (beacons, Wi-Fi APs, or sensor points) can be represented by a node (which can be associated with specific physical properties in various ways, e.g., through visualization). The nodes are physical objects in space, but unlike the geometric model, they are represented in terms of relative positions, such that AP1 is "next to" or "adjacent to" Store ABC (as opposed to absolute positions of (x, y) coordinates in the geometric model).

Computation of the graph-like structure of rooms and hallways from maps to improve the performance of location estimation and enable path prediction can be accomplished. In addition, being more than just a hierarchical arrangement, a spatial model graph permits intuitive traversal among node relationships, while still allowing hierarchies to be modeled.

In addition, containment relationships between two-dimensional geometric shapes are a good way of formalizing vague spatial relationships. Simpler abstractions fail to capture complexities in the environment that are obvious to the user, while more sophisticated ones risk being too complex for the user to understand. In addition, people are very well-suited to reasoning about and remembering two-dimensional geometric shapes.

The main disadvantage or limitation of symbolic location models is their inherent lack of geometric attributes and precision. A symbolic model is unable to compute distance accurately and represent location precisely. For example, it may be difficult for the system to calculate the remaining distance between "the user" and "the nearest printer" if it is operating with them just as symbolic names.

One way to compensate for the lack of geometrical information is defining some hierarchy of information in the environment. Another way is to divide a grid map into distinctive parts and define their topological relation.

Hierarchical location domains can be specified in the Z notation (Potter et al., 1991). The Z notation is well suited for specifying state-based systems. Automated checking tools are available. An ordering relation for locations reflects spatial inclusion, which may include asymmetry and transitivity. Asymmetry means, for example, that a building containing a room cannot at the same time be contained by that room. The location ordering must also be transitive. For example, a desk in a room in a building must also be in the building. Overall, hierarchical structuring is beneficial, because multiresolution processing is the key to reducing complexity and ensuring scalability. Data naturally lend themselves to multiresolution processing, which can be used to address scalability and performance problems.

MASTER WORLD AND SECONDARY WORLDS: HIERARCHICAL TREE STRUCTURES

The concept of the master world and a secondary world can provide a standardized solution for a uniform definition of the world. The uniform definition can be defined in terms of a hierarchical tree of nodes (the master world), where each node represents some aspect of the world.

One or more hierarchical tree structures are defined that uniquely identify geographical divisions of the space or physical or logical entities. Each tree has multiple nodes, and at least one node from each tree is linked. Figure 5.3 illustrates this.

Objects and services can be associated with individual nodes on the tree, the nodes providing a universal reference when attempting to locate the objects or services. A mobile device has access to one or more of the tree structures and can utilize the tree structures to ascertain its current location (and context). The device determines its location by traversing one or more of the tree structures to ascertain information that is associated with the individual nodes of the tree structure. The device is capable of autonomously determining its location within a master world and one or more secondary worlds. Once the computing device has determined its location, it can then take part in serving location-based services.

A master world is defined and is a hierarchical tree structure that represents a universally acceptable description of the world. One or more secondary worlds can be defined and constitute entity-specific (e.g., organization) views of the world that link with the master world. A secondary world can describe the location tree of an organization.

With a secondary world, individual entities (such as businesses or organizations) can define their own particular worlds, which need not necessarily conform to the master world view of the world. That is, while the master world is essentially a physical hierarchical representation of the world, the secondary worlds can be physical or logical representations of each individual entity's worldview.

Based upon a user's calculated position, these various services that are associated with secondary world nodes can be offered to the user. In addition, because the user's context is determined relative to the master world, other services that may not be associated with a particular secondary world can be offered.

Both the master (or at least a portion of it) and the secondary worlds can be either locally maintained on the computing device or accessed, e.g., via the web or

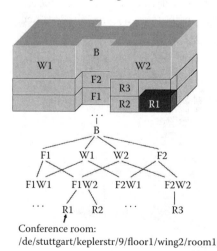

Conference room:
/de/stuttgart/keplerstr/9/floor1/wing2/room1

FIGURE 5.3 Local coordinate system and hierarchical tree structure.

some other mechanism, so that a user can derive his location and context. For example, the secondary world can be downloaded onto the computing device so that a user can derive his context within the secondary world.

Each node is connected to at least one other node by a branch and has various attributes associated with it that assist in LBS computing. Exemplary attributes include a unique ID, name, geographic entity class, latitude/longitude, relative importance, and contextual parents, to name just a few.

This hierarchy approach is useful because it can be used to determine the relative location of a place anywhere in the world and at any definable granularity. The master world's accurate standardized geographic dimension attribution can be easily accessed by both providers and consumers. Services and product providers (or third parties such as search engines, networks, and Yellow Page database directories) can use the nodes of the master world by assigning a standardized persistent geographic reference to all commerce locations or points of interest (POIs). These commerce locations or points of interest can be considered as leaves on the tree structure.

Defining the Master World

The master world can be defined as a politically correct and publicly accepted hierarchical tree structure that catalogs physical location or geographic divisions of the geographical space. The master world is defined in such a way that many different classes of administrative and geographic entities across space are included. Figure 5.3 shows an exemplary hierarchical tree structure that represents a portion of the master world. The master world contains multiple nodes, with each node representing some type of geographic division of the geographical space.

Each node is uniquely identified by an ID (e.g., entity ID). In addition to the unique IDs, a URL is associated with the tree structure and provides a context for the tree structure.

The Bluetooth LP group is developing a syntax that can express location information for a place, which includes an address (outdoor and indoor info):

```
<? xml version="1.0" ?>
<locpos>
<addr>
   <CountryCode>PT</CountryCode>
   <State>Alentejo</State>
   <City>Evora</City>
   <Street>R. Romao Ramalho, 59</Street>
   <PostalCode>7000 Evora</PostalCode>
   <Building>
        <Name>Colegio Luis Verney</Name>
        <Floor>2</Floor>
        <Room>222</Room>
        <Tel type="work,voice">+351 266745371</Tel>
```

```
    </Building>
  </addr>
  <env xmlns:"www.uevora.pt/bluetooth/lp/dtd/university.dtd">
    <Name>University of Evora</Name>
    <Department>Computer Science</Department>
    <President>Prof. Salvador Abreu</President>
  </env>
  <reference url="..." Label="...">
    (Contains description of web presence)
  </reference>
  <date>
```

The date, time, and time zone offset in ISO 8601 format:

```
  </date>
</locpos>
```

As an example, consider the following: an airport will be included in the master world, but references to individual airline business locations might be leaves on the tree that are tagged by the airport's entity unique ID (the secondary world).

Another example is traveling from the airport to the hotel, you decide that you want to find the nearest Starbuck's. Consider that your handheld computing device is a cellular phone and that Carrier ABC is the carrier. A cellular provider/operator might define its own secondary world, which might, for example, be designated in terms of cell nets. By virtue of having Carrier ABC's secondary world on your computing device, you are able to ascertain your location in Carrier ABC's secondary world and, accordingly, your location in the master world. Consider that Starbuck's also has a secondary world that links with the master world. By executing a search application on your device, you are able to ascertain the location of the nearest Starbuck's as well as travel directions thereto. All of this is possible because your device has access to the master world and one or more secondary worlds. In this example, the master world provides a mechanism to daisy-chain two or more secondary worlds together. This is possible because the secondary worlds have at least one reference or link into the master world.

Defining the Secondary World

Individual organizations can define their own secondary worlds. This gives organizations a great deal of flexibility in providing goods and services and, more broadly, increasing the efficiency of their organization. In one embodiment, a software tool is provided that enables individual organizations to define and maintain their own secondary worlds. Each secondary world can be uniquely identified as a namespace (e.g., an eXtensible Markup Language (XML) namespace). This ensures that any overlap in names between the secondary world and the master world will not result in a collision.

It is necessary to name the objects of interest. See Needham (1993) for a discussion of naming. A namespace for locations can address the following issues:

Location names must be hierarchical. This is necessary for scalability. Also, fixed locations are intuitively hierarchic. The design of the namespace should intuitively reflect the physical location space. However, this must be a loose coupling since the namespace should be less dynamic than the location space. In particular, movement of a mobile location should not necessitate a change in the namespace. Individual locations have characteristics that are important for location awareness. Locations can be either fixed or mobile, and their extent can be specified by either area geometry or an abstract concept like a room number. While these distinctions should be recognizable in the notation, it is undesirable to reflect this in the semantic model. Additionally, there are a number of preexisting naming schemes for locations. Path names for fixed symbolic locations can assume a static acyclic graph with a root location. There may be more than one name for a location. Such a namespace is very similar to the Internet's Domain Name System (DNS). For example, "Cambridge/MITCampus/Building1/101" identifies a particular room at MIT.

Also, it is important to know the reference coordinate system. Coordinate tuples can be used to identify a geometric position. For example, \WGS84:(0.??W, ??N, 0)" identifies a point in the Boston area. The reference coordinate system is WGS84 (World Geodetic System 1984). And natural language used in abstract locations contains a number of instantly recognizable and fairly unambiguous location abstractions, e.g., "room", "building", or "city". The names for symbolic positions are hierarchical path names. Hence, they allow the construction of hierarchical location spaces. We assume that there is set of well-known symbolic reference points, such as "Boston" or "MIT", which can serve as root locations. Note that a symbolic position usually implies a certain coverage area. By requiring the explicit specification of the area, we force such hidden knowledge to be made explicit. With this naming scheme we can concisely refer to locations of interest. Additionally, a resolution scheme is required in order to map names to locations. The resolution of location names depends on the nature and capabilities of the underlying location model. The resolution process requires that the location corresponding to the resolved name be identified.

In terms of building a context-aware data structure, the software might receive input from a source that specifies information that pertains to physical or logical entities. Or a system administrator might physically enter information about the structure of the secondary world that he desires to define. This information can include information about buildings, divisions, conference rooms, and the like. The software then processes the information to define a hierarchical tree structure that has a context. Once the tree structures have been built and linked, they are ready for traversal in a manner that enables location and context to be derived from one or more of the nodes.

CONTENT INFORMATION

Content information can be used to determine a tree structure and a node's entity ID.

The device receives position information from a beacon. This information is collected and mapped by the location service into a node in the master world or secondary world. The hierarchical trees can then be traversed to determine the device's accurate location in both the secondary world and the master world.

Transmitted information can include an entity ID–universal resource locater (URL) pair and location unique ID–URL pair. The entity ID gives the identification of a node in the master world, and the associated URL gives a protocol to communicate with the master world. The URL might, for instance, link to a server that can provide additional context information that uses the entity ID. This indicates a node on a secondary world that corresponds to a current location, and the URL gives a protocol to communicate with the secondary world. For example, the URL can link with a server that is hosting the secondary world. This server can then be queried to discover more information about the secondary world (i.e., secondary world tree structure, location of associated resources, etc.). With these two types of unique IDs (along with the URLs), a device can now traverse the master world or secondary world to determine its location.

GeoZones: The Master World Index

Furthermore, one additional concept adds value to the hierarchical approach of structuring the world: a geozone. A geozone is essentially a spatial indexing mechanism by which the master world is subdivided into individual zones. Specifically, the use of an index scheme can identify peer-level nodes by virtue of geographical proximity. This indexing scheme makes use of a quad-tree algorithm to define the geozones. The zones are subdivided through the use of a quad-tree algorithm that is dependent on a density function (although many other spatial index approaches can also be used). The spatial index breaks the geographical area (map) into homogeneous cells of regularly decreasing size. The quad-tree segmentation process can continue until the entire map is partitioned based on many different end result criteria, including the density of the number of items (e.g., points of interest) in each quad. This approach provides a form of spatial index that accelerates spatial selection and content identification. To complete the spatial indexing scheme to provide each node with a defined geozone, a quad-tree algorithm is applied to the nodes and can be based upon a desired density of, for example, points of interest that are to occur in any zone. Once all of the zones have been defined, each zone is given a unique ID (e.g., top/left and bottom/right latitude and longitude pairs). Each of the nodes of the master world is then assigned a zone in which it is located. Once a desired density level is achieved (density might be defined in terms of points of interest per zone), each node on the master world is assigned a particular geozone. Geozones enable proximity calculations to be computed in a fast and straightforward manner. By definition, the master world provides a hierarchical structure of entities (nodes) that cover the entire globe. Upward navigation within the hierarchy is quite natural. Efficient navigation downward requires geographic proximity awareness. Additionally, there are possible scenarios that will require jumping from branch to branch in order to successfully return values in a query, or for more accurate calculations of distances to close leaves attached to nodes other than the original source node.

Topological Maps and Location Data Models

Topological maps reflect the large-scale structure of an environment containing information of low complexity. The space is segmented into topological units (places), and the connectivity of these units is given. Such a map is mostly represented by a graph structure with nodes standing for important places in the environment and edges defining how the places are connected. These types of maps are very compact representations and are usually easy to construct due to their low complexity. Another advantage of these maps is that they only contain information that hardly changes over time (rooms or corridors). Hence, they are still valid after, for example, refurnishing an office space. However, they are harder to use for navigation purposes than metric maps, because only limited knowledge about the user's surroundings is available.

Since these types of maps lack a lot of geometrical information, they are often combined with metric maps. One way to do this is defining some hierarchy of information in the environment. Another way is to divide a grid map into distinctive parts and define their topological relation. Other times, only the interesting parts of the environment are mapped geometrically (e.g., ATM locations inside a shopping mall), while others (e.g., corridors) have a pure topological representation.

In order to get the interconnections between places, it is necessary to have relations between the objects involved. Some relations are modeled implicitly when the geometry is modeled, e.g., which rooms are next to each other. However, this may not be sufficient if the geometric model lacks information about doors between rooms. In this case, this relation has to be explicitly modeled.

Hence, in a topological representation, there has to be a node in front of each door, at each corridor crossing, and at other places of interest. Each node has a location in a fixed coordinate system.

Topologic relations structure space. Several topologic categorizations of environments are suggested in the literature. Lynch (1960) mentions five urban design elements to describe the setting of a city. Arthur and Passini (1992) give a typology based on the structuring features of built environments, called circulations. The four-intersection theory of topologic spatial relations between sets (Egenhofer and Franzosa, 1991) defines relations in terms of the intersections of the boundaries and interiors of two sets. Evidence for cognitive hierarchical organization of space was deduced from distance and direction judgments (Hirtle and Jonides, 1985).

Spatial model graphs may be seen as enrichments of topological models, modeling not only location entities as subsets or neighborhoods of space, but also their structural relationships. This is implicitly the kind of model underlying the cell layout of cellular networks, where adjacency relationships between cells are used for the communication (and positioning) handover of a locatable entity from one cell to another.

Graph theory defines an arc as a set of curves in the Euclidian plan. An arc can be completely characterized by a finite sequence of points. A node is a point at which an arc terminates or begins, or a point at which it is possible to move from one arc to another. A link is a linear element between two adjacent points.

For the campus example, this model is not completely appropriate since it represents the network in two dimensions only. The main axes of the network are defined by the centerline of the corridors and the street axes between buildings. The doors are represented with their projection points on the centerline of the corridors. The projection point of each door is regarded as a node.

The link/node model in Figure 5.4 allows the construction of a two-dimensional network with the topological relationships between the main objects of one building's floor. The connection between floors is possible by modeling staircases and elevators as vertical links. In order to realistically represent the buildings, the altitude can be used as a third dimension. The third coordinate of each node is the altitude of the floor on which it is found.

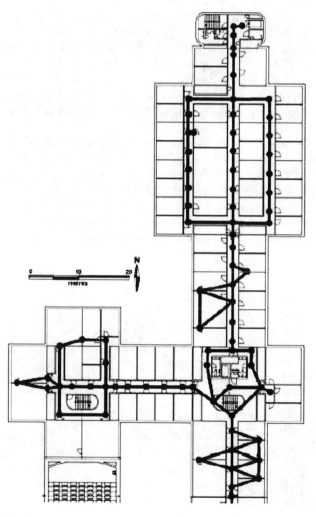

FIGURE 5.4 Spatial model graph: links and nodes.

Connectivity relationships between corridors and elevators and stairways can be represented using the spatial graph to represent adjacency relationships between area objects in two-dimensional GIS. The spatial graph has a corresponding spatial partition, where the nodes in dual space are the units of the partition (polygons in two-dimensional GIS), and two nodes are linked by an edge (in dual space) when the corresponding spatial objects are adjacent in primal space. Figure 5.5 shows the three-dimensional view of the links between floors.

Adjacency is but one particular case of a spatial relationship. A complementary hierarchical model loosely underlies most of the semantic models used in directories, but it is also an implicit model for the space within a building, as decomposed in floors, rooms, cabinets, etc.

Research on mobile robot navigation has produced two major paradigms for mapping indoor environments: grid based and topological. The map can be associated with more information using cognitive maps on several levels of abstraction, where one of these levels was topological. This can be further extended by learning a spatial semantic hierarchy of an area, and combined with the grid-based approach using a topological map specifically for office environments, assuming angles of 90° between all corridor parts. Corridors can also be divided into large cells, where each cell defines a topological unit, as done for the indoor navigation system Dervish.

Topological maps are very compact representations of the environment, based on a symbolic description of the environment, where the objects are related to each other and laid out in a two-dimensional space. Such maps are usually easy to construct due to their low complexity. Another advantage of these maps is that they only contain information that hardly changes over time (rooms or corridors). Hence, they are still valid after, for example, refurnishing an office space or moving people around. They work less well in offices with cubicles, if the cubicle positions can be moved.

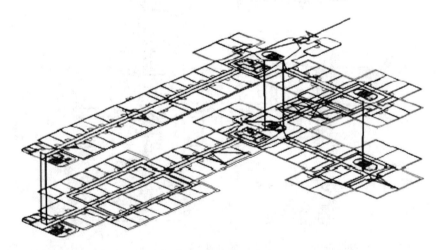

FIGURE 5.5 Spatial model graph: three-dimensional view.

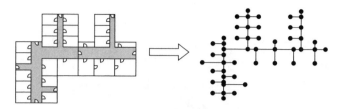

FIGURE 5.6 Transforming of CAD model into a node/link model.

Also, while grid-based methods produce accurate metric maps (with absolute measurements), their complexity often prohibits efficient path planning and problem solving in large-scale indoor environments. Topological maps, on the other hand, can be used much more efficiently, yet accurate and consistent topological maps are often difficult to learn and maintain in large-scale environments, particularly if momentary sensor data are highly ambiguous.

The topological model will mostly be a common implementation for indoor spaces. A large number of buildings, rooms, and corridors can be transformed in a network representation of a venue (e.g., campus). Figure 5.6 shows the process of data modeling for a part of a building.

A node/link model is composed of the following elements: corridor, way, road, path between buildings, room, and hall, as well as stairs, lift, door, and POIs.

Many other objects could be added to this list. However, these elements are essential for the procedure of route guidance and map matching (described below). For the needs of navigation, all objects are represented as points or links, which are connected. This kind of data structure is well adapted for the computation of routes.

Figure 5.7 shows an abstract model for route guidance.

The node is the basic element of this abstract model. A node can be connected to a door, a point of interest, an access point, or a link. Each link is composed of a starting node and an ending node. All elements of the network have attributes necessary for the computation of routes:

- Type of link (straight, stairs, lift), length of link
- Restricted/special access privileges
- Type of node (crossing, connector)

This model has been developed in a project for the implementation of route guidance algorithms (Büchel, 2003).

GEOMETRIC MAPS AND LOCATION DATA MODELS

Location awareness in many applications means processing of geographical coordinates. Geometric location data are provided as a set of coordinates with respect to some reference coordinate system. For GPS, this is often longitude, latitude, and altitude in the WGS84 reference system (Drane and Rizos, 1998). Location areas (such as cities, buildings, or rooms) are represented by the coordinates of two-

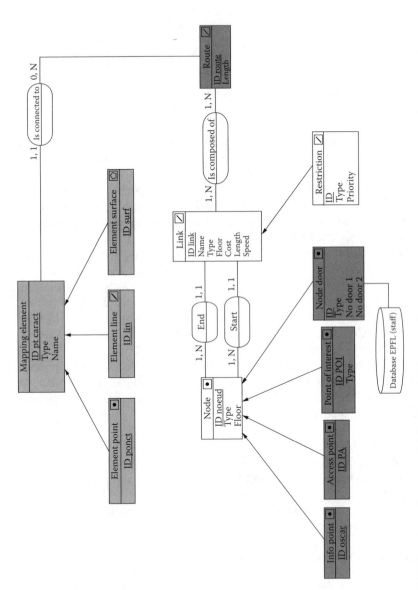

FIGURE 5.7 An abstract model for route guidance.

dimensional areas or three-dimensional volumes, in contrast to the purely name-based symbolic models.

Geometrical maps are a rather detailed representation of an area. They are rich in information, which makes them convenient to use for navigation purposes given a good estimate of the robot's position. However, the large amount of information poses constraints on the storage capacity of a system. Further, these maps are rather hard to construct due to their detailed description of the environment. There are two main types of these large-scale representations: *grid maps* and *feature maps*.

Geometric location maps are made up of geometric features, such as lines, points, and polygons, that have metric (and nonmetric) attributes. Examples of points are objects with a finite extension in space, such as water fountains and electric outlets. Examples of lines are objects that have one-dimensional extensions when drawn in a two-dimensional representation, such as doors and hallways, and examples of polygons are objects that have a two-dimensional extension, such as rooms and buildings.

Geometric points represent physical objects in space (i.e., beacons, or emitters such as Wi-Fi APs) with an absolute position ((x, y, z) coordinates). If the emitters represent the location of a store, they can be presented together with reference information, e.g., a URI, a binary code, or some other method of representation. Points can be connected with lines, which will then share the same properties (of having a set of coordinates that are absolute); hence, distance can be computed by connecting the points.

Grid Maps

A grid map divides the environment in many equally large cells of square shape. Each of these cells holds a number reflecting the probability that it is occupied. *Occupied*, here, means that anything could be there: a wall, a smaller obstacle, or even a person. Since there is no distinction of different objects, range sensors are the optimal choice for creating and using grid maps.

Feature Maps

A feature map not only reflects the occupancy of cells, but also contains information about the actual objects occupying the locations. This representation consists of features and their coordinates in the environment. These features are often lines or general geometric beacons like points, corners, and walls.

The complexity of a feature map can vary from minimalistic models containing only walls to a complete CAD model of the environment.

Spatial Relations in Geometric Maps

In order to compute spatial relations (i.e., containment and intersection), geometric attributes must be present in a well-defined spatial reference system, such as shapes, extensions, point coordinates, etc. Geometric models define an n-dimensional space, and the locations are points in this space that can be uniquely specified and accurately represented by a tuple of numbers (x, y, z). However, there are sometimes mismatches in the meaningful precision of the coordinates in various locations. An example of a spatial (coordinate) reference system is the one utilized by MIT Cricket.

Position and layouts are key to geometrical models. Position is ideally modeled by a point (i.e., a set of coordinates with respect to a reference coordinate system). However, no positioning or tracking system is error-free. So the located object's position is best represented by an area of positioning uncertainty. The located object's layout is modeled in the same way as are "normal" location areas.

Various coordinate systems may be used, but they must have well-defined transformations between them (or else they become meaningless to derive measurements from). For example, each floor of a building typically acts as a separate spatial reference space — two points on different floors may have the same coordinates on their respective floors, but have an unknown relationship in the real three-dimensional world. One way to solve this problem is to allow each space to have its own local coordinate system by specifying the origin point and three axes of x, y, and z. This is done within the CMU Aura Project's hybrid location model. This approach also works when buildings are more complex, for example, if they have wings or towers, or if two buildings share a parking lot or structure. In principle, there is no limit to how spatial structures can be mapped onto a coordinate system.

While coordinates are easy for computers to manipulate, and for humans to visualize graphically, they may not convey intrinsic meaning to humans, and listing them in text may be rather tedious. Within a geometric model, the relevant objects have to be identified, which requires symbolic information that is provided in the form of attributes. This information is used to restrict a query to return only relevant objects, such as rooms or buildings. The complexity of a feature map can vary from basic models containing only walls to a complete CAD/three-dimensional GIS model of the environment. Coordinates can also be used as database keys, but if this is done, an intermediate step may have to be taken (since the point of information may be different in its spatial extent from the point that is represented in coordinates). Coordinates may represent a very small point, but the information may represent an entire room; hence, if a query is directed at the point, a computation is required to determine that the point is within the room, and derive the information to be returned.

In short, the advantages and disadvantages of mapping out the indoor world using geometric location models are:

Advantages:
- Precise location and distance computation (due to built-in geometric attributes). Computations are typically Euclidean, and many rich relations can be computed without prestoring them.
- Indoor location positioning systems that rely on the geometric model to generate streams of position fixes that are independent in the sense of describing a location without external reference. There is a shared reference grid for all located objects, which can be transformed into a relative location, if needed.
- Allows to define points or areas for which there is no name in the hierarchical name system.

Disadvantages:
- Hides hierarchical relationships (so it needs extra specification to enable deduction of spatial relationships).

In short, there are several ways of mapping the structure of an object (i.e., a building) to an information space. A terminal will also be able to map the characteristics of the signals it receives from positioning systems to an information system, which describes its coordinates in a way that can be mapped to coordinates in the information system.

INDOOR NAVIGATION AND NAVIGATION ALGORITHMS

An LBS needs to guide a user from a present location to a specified destination. Indoor navigation might seem different from car navigation (on the street network). Whereas car navigation is a relatively even process, which follows the road geometry, mobile users walking inside a building, on the other hand, are less predictable.

Indoors, mobile users do not have a set of well-defined pathways because they can roam about in practically any free space within the environment, whereas cars are mostly restricted to travel only on the road network.

Nonetheless, indoor venues do constrain mobile users to some level according to the layout of the venue, which can be modeled (explicitly with topological models) within the indoor navigation system. Mobile users can walk forward, backward, and can change the direction of their walk at any time, or they can move upstairs or downstairs.

Vertical displacements must be taken into account due to users traveling upstairs and downstairs.

Explicitly to topological models, the techniques used for computing a path between any indoor origin and destination are not really different from those used in computing paths outdoors on the road network.

A navigation system for pedestrians includes the following:

- An input module for input of a departure place and a destination of the user
- A database with route data representing position and connection of each route consisting of a route network (a map)
- A route calculation module to calculate a route from the departure place to the destination by referring to the route database
- Optionally, a landmark database to store landmark data for the pedestrian's confirmation
- Optionally, a landmark representing a signpost of the departure place, the destination, and the route
- Optionally, a landmark selection module to select the landmark data corresponding to the route calculated by the route calculation unit from the landmark database
- A presentation unit to present a route guidance for the user by using the calculated route and the landmark data.

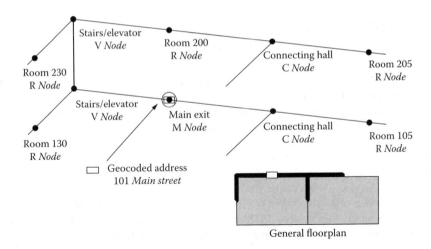

FIGURE 5.8 Semantic descriptors of indoor three-dimensional hall and room network.

SPATIAL DATABASE SYSTEMS

In order to enable navigation, a digital map database is needed to store guidance procedures.

Many aspects related to the representation and management of spatial information have already been studied in the area of spatial databases. A spatial database management system (SDBMS) is a database system that offers spatial data types in its data model and query language and supports these types in its implementation by providing at least spatial indexing and efficient algorithms for spatial join. SDBMS are characterized by the ability to manage large collections of geographic objects. Traditionally, SDBMS deal with geographic objects, each of which consists of two components, a spatial object describing the location, shape, etc., of the object in space and nonspatial or descriptive attributes. For example, in the AT&T Sentient Computing Project, for computing applications each real-world entity is represented by a Common Object Request Broker Architecture (CORBA) object, which possesses a type, a name, a location, a set of properties, and an interface. Thus, in the Sentient Computing Project, a quite formal approach to context modeling complying with the object-oriented modeling paradigm is pursued. The CORBA objects are persistently stored in a relational database. Sentient computing applications contain a spatial monitoring service that is responsible for the transformation of location data. The spatial monitor generates information about the containment and overlapping of two-dimensional locations. Furthermore, it detects location-related events and provides applications with notifications about relevant location changes. Thus, in sentient computing an event-driven programming style is used. In addition, a method for automatically storing generated data, along with the points in time at which these data were created, has been developed. As a result, a timeline is constructed that contains both the generated data and some information from the world model. This timeline can be queried by means of a temporal query language.

The main approaches to geographic space modeling can be divided into entity-based and field-based models. In entity-based models, geographic objects or entities are the primary objects. They are composed of an identity, a spatial object, and a common description. The main abstractions used to represent spatial objects are points, polylines, and polygons. Field-based models, in contrast, view space as a continuous domain. Each point in space is assigned one or more attribute values by means of continuous functions. Concerning the modeling of collections of spatial objects, various models such as the spaghetti model, the network model, and the topological model have been developed. The main problem in the implementation of geographic space models is the transformation of the infinite point sets in Euclidean space into finite representations that are processable by computers. For this purpose, different representation modes such as tessellation, vector mode, and half-plane representation are made use of.

Spatial relationships can be classified as topological, direction, and metric relationships. To achieve a general extensibility of databases in terms of user-defined spatial data types, the concept of abstract data types (ADTs) is used. Querying in spatial databases first of all involves the connecting of the operations defined in spatial algebra to the features of a database system's query language. This also includes extensions made to existing query languages to enable them to handle spatial data. In addition, since an interactive use of SDBMS requires the presentation of input and output graphically, a graphical representation of spatial data types is needed. Furthermore, data structures for spatial data types as well as algorithms for the operations of spatial algebra are prerequisite to implement SDBMS. In addition to noncontained operations, spatial indexing and the support of spatial join are of particular importance. At the system architecture level, an integrated architecture using extensible database systems has recently been gaining the most acceptance and is made use of in several SDBMS, such as Postgres, Monet, or GéoSabrina.

NAVIGATION ALGORITHMS

Due to the specific design of indoor maps (discussed at the beginning of this chapter), special algorithms must be developed to combine different sources of data within the navigation process. An application domain-specific data model for indoor applications is needed. Navigation of physically challenged people and of robots are two examples where the database must fit with specific algorithms in order to deliver reliable navigation information to the user (Gilliéron, 2003).

Navigation systems include a computation functionality to execute an algorithm for determining a route between a source location and a destination location using a map (network) database. A path of motion can be divided into certain segments, each segment consisting at least of four parts belonging to different categories: *starting point*, *reorientation* (orientation), *path/progression*, and *ending point*. Since the paths we describe are continuous, the ending point of one segment serves as the starting point for the next one. Segments are mostly separated by changes in direction, and the resulting presentation has to communicate a change in direction clearly.

The algorithm manipulates a graph representation having vertices and edges, wherein each of the vertices corresponds to a respective one of the paths, each of the paths corresponds to exactly one of the vertices, each node corresponds to at least one of the edges, and each edge corresponds to exactly one node. Furthermore, each edge represents an adjacency between two of the vertices. The algorithm designates a source vertex from among the vertices that represents a beginning path and a destination path in the set of beginning and destination paths, respectively.

The algorithm also designates a destination vertex from among the vertices that represents an ending path that is the other of the source path and the destination path. The algorithm then designates the source vertex as seen, designates each remaining vertex as unseen, and places the source vertex in a head position of a first-in-first-out queue. The algorithm thereafter executes a loop while the destination vertex is designated as unseen and the queue is not empty. In the loop, when all vertices adjacent to the head-positioned vertex are seen, the algorithm removes the head-positioned vertex from the queue and, when a next vertex is next in the queue, places that next vertex in the head position. Alternatively in the loop, when at least one vertex adjacent to the head-positioned vertex is designated as unseen, the algorithm considers in a consideration order each unseen vertex adjacent to the head-positioned vertex and, after all of the unseen vertices adjacent to the head-positioned vertex have been considered, removes the head-positioned vertex from the queue and, when a next vertex is next in the queue, places that next vertex in the head position. Note, each of the vertices cannot be simultaneously designated as both seen and unseen.

Known shortest-path algorithms do not specify the order in which the vertices adjacent to the head-positioned vertex are considered. As stated above, a vertex is adjacent to the head-positioned vertex when the vertex represents a street that intersects the street represented by the head-positioned vertex. The algorithm of the invention can be adapted to consider the adjacent vertices in a consideration order such that when multiple shortest paths exist in G, the algorithm has a tendency to discover that shortest path which traverses the minimum physical distance. In this context, the shortest path will be referred to as the simplest path (e.g., the path with the minimum number of street changes), and a reference to the "physically shortest path" will mean a path traversing a minimum physical distance. Therefore, when at least two of a plurality of possible routes between the source location and the destination location are the simplest routes (i.e., both require an identical minimum number of street changes), the algorithm preferably is biased toward determining the physically shortest route (i.e., the route traversing the minimum physical distance) of the at least two possible routes. This biasing can be accomplished by ordering in the consideration order.

When the algorithm is adapted to manipulate the preferred graph representation, and the computation module is adapted to determine a simplest route of a plurality of possible routes between the source destination and the source location, the algorithm can include any shortest-path algorithm that can manipulate the preferred graph representation to determine a simplest path between some member of the set of source paths and some member of the set of destinations.

For example, such a shortest-path algorithm determines a minimum cost sequence of streets between some member of the set of origins and some member of the set of destinations, and when the graph representation is unweighted, minimum cost means a minimum distance along the sequence of paths, so that the resulting path presents the fewest number of path changes to the traveler. Exemplary shortest-path algorithms that can be used to manipulate a weighted graph representation are Dijkstra's algorithm or the Bellman–Ford algorithm.

Based on this specific model, route guidance and navigation algorithms can be integrated in order to develop applications with particular requirements. This might mean that the content of the database to be used must be upgraded to be compatible with a data model designed for navigation. The best route computed by the system is used as input for the navigation process and for providing navigation information to the user.

Moreover, the determination of the path preferably balances descriptive simplicity of the path with a minimal total distance traversed along the path. Note that distance usually applies to geometric models; however, geometric properties like distance can be added to topological models (associate a path with a distance attribute). Existing systems optimize for distance, time, or simplicity. Optimization is usually done using an individual factor, but not in combination.

Once the path has been determined, navigation guidance can provide a natural language description of the actions the user must take to follow the path. For example, the navigation instructions may instruct the user to pass by or turn at specific apparent landmarks that are likely to be of interest to him. Guidance to the user is provided through user interfaces, which we discuss at the end of this chapter.

MAP-MATCHING ALGORITHMS

In addition, a navigation system must have the capability to provide a position even during bad reception of positioning sensor signals. When both the user's location and the underlying map (network) are very accurate, the reconciliation problem is thought to be straightforward — simply "snap" the location obtained from the positioning receiver to the nearest node or arc in the network (map). However, in some situations it is not possible or desirable to improve the accuracy of the map/network and the user's position enough to make a simple snapping algorithm feasible. Such situations arise for many reasons.

First, it may not be possible to use dead reckoning or other data sources. Second, even if it is possible to develop a network/map that is accurate enough, such a network may not always be available. For example, the mobile device may not have sufficient capacity to store the complete, accurate network at all times, and hence may need to either store inaccurate/incomplete networks or download less detailed networks from either a local or central server. Third, many facilities will probably never be available from map/network vendors and will need to be obtained on the fly from the facility, probably with limited accuracy. For example, vendors may not provide detailed networks/maps of airports, campuses, large parking facilities, and shopping centers.

In this section we explore map-matching algorithms that can be used to reconcile inaccurate locational data with an inaccurate map/network. Point-to-point, point-to-curve, and curve-to-curve matching are discussed, and in all three cases algorithms that only use geometric information and that also use topological information are considered.

A map database can be used to update and verify the position given by the navigation system. This process of verification is called map matching. Two inputs (data sources) are used for map matching. These are:

1. The map database (paths/links, nodes)
2. The list of positions recorded by the positioning system

Georeferenced information can be used to update and verify the position given by the navigation system. This process is well known in car navigation and is called map matching (Shan Hung and Chuan Su, 1998). Map matching is used to improve navigation and guiding through more accurate positioning with the help of a map. Map matching is done in an automated fashion and usually in cases of positioning signal fading or no signal. A raw travel path was stored in the data logger and was available for the so-called postcalibration. The basic idea of this process is based on the comparison of the raw travel path with the link/node view of the topological map's database.

Manually or in disconnect mode, a navigation system could depend on user assistance to determine location. The user would be presented with a series of marks in the vicinity of his last known location. The user could then identify the one closest to his current location to inform the system. While this approach may work well for a handful of applications, it is unlikely to scale to a venue with literally thousands of POIs, since the user's task would become too time-consuming.

Map-matching algorithms require good connectivity between the links and nodes within a topological model. Developing a set of map-matching algorithms for pedestrian navigation is a challenge because the travel path of people is not always similar to the geometry of the mapping data. The development of algorithms is based on the comparison of topological elements from the travel path and the database. This approach will limit mismatching, e.g., to match the current position with a wrong link. The connectivity relationships among three-dimensional spatial units are the combinations of the connectivity relations in the horizontal directions (on a floor) and the connectivity relations in the vertical direction (among floors). The connectivity relationships of three-dimensional units in horizontal directions on a floor can be derived from the connectivity relationships among polygons in two dimensions (e.g., between corridors and stairways or elevators).

Both azimuth and distance must be considered in the process of map matching. In car navigation, a significant change of direction is easily detected by the signal of the gyroscope. The travel path of a pedestrian is subject to major variations, which do not always fit with the mapping data. Therefore, map-matching algorithms have to be robust for pedestrian navigation to filter misleading information.

CHANGE IN DIRECTION

The change in direction involves a crossing or a junction. The navigation system detects that the direction is changing. This information has to be checked with the content of the database. If the turning point is correctly estimated, it shows that the point is corresponding to a junction node of the database.

A map-matching algorithm has the capability to identify specific nodes on the network, such as a point of intersection of corridors where the user is turning. Therefore, a rapid change in orientation could be detected automatically and the updated information could be provided for a recalibration of the navigation system. The positioning errors are not only random, but increase with time because of the accuracy of the sensors. For these reasons, a map-matching algorithm is useful.

The final result of the map-matching algorithm is the coordinates of the point that represents the actual position (the *map-matching point*) of the pedestrian on the link/node model. When a nodal matching is performed, the map-matching point adopts the coordinates of the nearest node. When the algorithm selects the correct link (point-to-curve matching), the map-matching algorithm calculates the coordinates of the projection of the measured point on the link. As a final result, the algorithm creates a file that contains the coordinates of all the map-matching points.

During a vertical movement the algorithm performs a nodal matching considering the vertical nodes only. The vertical nodes are situated relatively far from each other. This fact allows the increase of the horizontal search limit for the matching process. This is very useful when raw points in a staircase have to be matched.

SOME KNOWN MATCHING ALGORITHMS

The following algorithms perform map matching using only *geometric* information. That is, they make use only of the shape of the arcs and not the way in which they are connected.

Geometric Point-to-Point Algorithm

This algorithm takes into account the coordinates of the local mapping system. *The position is matched to the closest node in the network.* The navigation system needs only to determine the distance between the user's position and each point in the network sequentially, storing the closest point found along the way. The most natural way to do this is to calculate the Euclidean distance. (Note: Other metrics can also be used.) The Euclidean distance of two points $x = (x1, ..., xn)$ and $y = (y1, ..., yn)$ in Euclidean n-space is computed. (Note that it is not necessary to determine the distance between t and every node and shape point in the network.) A number of data structures and algorithms exist for identifying all of the points near a given point (often called a *range query*) (Bentley and Maurer, 1980; Fuchs et al., 1980).

One limitation of this algorithm is that the way shape points are used in the network affects this algorithm the most. To minimize this problem, more shapes can be included for every arc. This, unfortunately, increases the size of the network.

Geometric Point-to-Curve (Link) Algorithm

This algorithm identifies the link that is closest to the user's position. The most common approach is to use the minimum distance from the point to the curve. The following equation gives the distance from a point to a line:

$$c = \sqrt{\frac{\left(\left(y_1 - y_2\right)^2 x_p + \left(x_2 - x_1\right)^2 y_p + \left(x_1 y_2 - x_2 y_1\right)\right)^2}{\left(y_1 - y_2\right)^2 + \left(x_2 - x_1\right)^2}}$$

where c is the distance from Pt (xp, yp) to a line determined by two nodes (1 and 2), and Pt and both nodes are presented with their coordinates.

Once the points of the pedestrian's travel path are determined, the navigation system can associate them with the network model. The map-matching algorithm uses a combination of point-to-point and point-to-curve matching. It is also based on the calculation of the weight of each candidate link.

The most delicate part of the algorithm is the choice of the appropriate weighting parameters. They must be carefully estimated in order to provide a correct association between the raw position of the user and the points in the database. On the other hand, the implementation of vertical nodes is very useful for matching the user's position during a vertical movement.

Calculating the minimum distance between a point and a line segment is slightly more complicated than calculating the minimum distance between a point and a line in some cases. Calculating the minimum distance between p and the line segment is straightforward since it is the same as the minimum distance.

Geometric Curve-to-Curve Algorithm

This algorithm considers positions simultaneously by matching to the arc that is closest to the piecewise linear curve, defined by the points. Of course, this requires having some measure of the distance between curves, and there are many ways to define the distance between two curves.

NAVIGATION USING DIFFERENT TYPES OF MAPS

As in the outdoor world, the indoor world is made up of transportation networks, but they are largely pedestrian and foot powered. Instead of streets and addresses (the road network), there are hallways and doorways. Outdoors, however, the need to represent three-dimensional objects is minimal; bridge overpasses and underground tunnels can still be easily depicted graphically because there is typically no other transportation-related feature stacked above or hidden beneath them. The indoor world, however, is a stacked environment, where a position on one floor may be a public space and the very same (x, y) position on the floor above might be a private or secured space. The concept of point-to-point navigation, when vertical

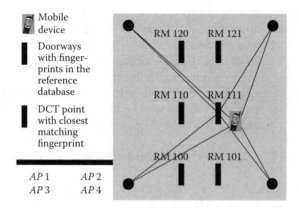

FIGURE 5.9 Interpolating user position using building context.

travel is removed, is similar in the cases of both outdoor and indoor navigation (see Figure 2.3).

Critical to modern outdoor navigation and mapping is street-line spatial modeling. In this technique, such as that used in the U.S. Census Bureau's TIGER/line data model (also MapQuest and other online mapping services), a street is broken into component segments connected at intersections. Plum Street in AnyCity, U.S., for example, could be broken down into segments with address ranges from 100 to 99, 200 to 299, 300 to 399, and so on. This facilitates not only the quick lookup of a specific address on a given street, but it also reflects the broad understanding that people have, whereby addresses are sequenced numerically (and also usually split odd and even, on the left and right of the road).

NAVIGATION USING SYMBOLIC MAPS

There are three main ways in which a representation can be mapped onto the real space: symbolic, topological, and geometric. The symbolic map represents the world in relative terms, describing the relation between objects rather than how they are associated with some larger model of the world. This means that objects only exist in relation to each other, and that location-based services that use symbolic descriptions only have a concept of objects in terms of other objects, and that there is no absolute reference.

Direction sensing based on a symbolic model is based on the trigger effect, such that when User1 left area AP1 and entered area AP2 (a few seconds later), the direction can be deduced that User1 is moving up the corridor (passing AP2 going toward AP3). In giving navigational instructions to the user, they can be given in terms of symbolic reasoning (as opposed to geometric reasoning requiring metrics), for example, "When you pass X (the ABC store), you will see Y (the ATM) on your left."

"Handing off" locations from room to room, e.g., tracking a mobile user, as "in Room 101," then as "in Hallway X," then as "in Room 102," can be performed with a specified tolerance level (snapping).

A parent–child link in the graph implies super- or subspace relationships between two locations. These divisions could be used as a location entity providing a spatial reference for users (or other objects), which will themselves be defined by some supposedly well-known characterization or context rather than their physical properties. It is up to the location reference designer to decide how to decompose the physical environment. For example, if an individual wanted to get walking directions from Room 501 in Building A to Room 222 in Building B, the designer of the application might break the path calculations into three contextual parts:

1. Path from Room 501 to the appropriate exit of Building A (an exit that faces Building B)
2. Path from exit of Building A to entrance of Building B (an entrance with optimal path to Room 222)
3. Path from entrance of Building B to Room 222 (selecting the optimal vertical conveyance, as in 1)

Or, in terms of super- or subclasses within a symbolic hierarchy:

1. Node "Room 501" to exit of zone "Building A"
2. Zone "Building A" exit to zone "Building B" entrance (i.e., outdoor space)
3. Zone "Building B" entrance to node "Room 222"

where a node is a room or information space and a zone is a building or collection of nodes. Note that a zone may simply represent an aggregation of all spaces within a single building, or it may be a collection of information subspaces served by, for example, a single AP within a building.

Labeling System

With people-friendly labels (sometimes addresses), spatial details are intrinsic to the naming convention of the symbolic spaces. In other words, from the label of the location, certain inferences should be possible. Consider the following example:

Boxxer Building
101 Main St., Room 612
Anytown, U.S.

The facts derived from this location label are:

City: Anytown, U.S.
Street: Main St.
Address: 101
Place: Boxxer Building
Room: 612

Spatial inferences from these facts include:

City: I can go to Anytown, U.S., to be closer to this location.
                   (I have the size of an average city in my mind.)
Street: The location is contained on the street line called Main St.
                   (I can travel the length of Main Street to find this address.)
Address: 101 would imply near the beginning of a street line (intersection).
                   (After a short distance, I can notice if street numbers are increasing or decreasing, then go in the direction of decreasing)
Room: 612 should be on the sixth floor, if it is a multistory building.
                   (I would expect 612 to be between 600 and 620 or so; therefore, I could search for 612 by range of numbers, then narrow in, using the numbering sequence as a main guide.)

Navigation and movement are possible based on relative positioning (symbolic models) since direction can be based on the trigger effect (User1 "left area AP1" and "entered area AP2" ⇒ User1 is moving up the corridor). The economy of this method of navigation — particularly on mobile computers and in formats that people can memorize for short periods — is the way a path can be stored in a database and visualized. With symbolic spatial navigation, the labels are selected and stored economically, without intrinsic geometric properties, although it is still useful to store geometric attributes for such problems as travel path calculations. In symbolic space, a person in a room can think, "Keep walking until you reach the printer," and know how far to expect to walk because of his familiarity with the room context. He knows he will not be walking for miles. Similarly, on a path from Room 111 to Room 141, a person should feel intuitively confident about his direction if he sees Room 120.

Important locations in symbolic navigation and travel paths can be described as points of decision and points of clarification. For example, a person may need to walk past 50 rooms in a series of hallways to get from Room A to Room B (each door number could be in a navigation waypoint file, helping make sure the person is on the right path). But if the rooms are in a long hallway (no chance to navigate off the path), then all 50 rooms are not needed to describe the travel path (see Figure 5.10). The points of decision in this case would be the start and end points, and the points of clarification would be one or more rooms along the path to assure the traveler that he is going in the right direction. The required data for symbolic navigation can be reduced to a GPS waypoint-style list, but with landmarks (symbols) instead of mathematical representations. And this list can be intuitively displayed in text format on lesser capable devices.

A rationale for this type of navigation system is found in current research on machine learning and symbolic (cognitive) reasoning vs. geometric reasoning. Machine learning and related algorithms are focused mainly on numeric paradigms. It is, however, widely supposed that intelligent behavior strongly relies on ability to manipulate symbols (symbol strings). For example, concepts of average and similarity can be applied on strings. These concepts are easily defined for numerical data and form building blocks for many learning, adaptation, and clustering algorithms. By defining LBS data structures and navigation models using symbol strings, we can apply current algorithms to symbolic data. Having chosen the distance measure,

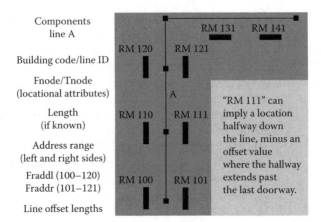

Components
line A

Building code/line ID

Fnode/Tnode
(locational attributes)

Length
(if known)

Address range
(left and right sides)

Fraddl (100–120)
Fraddr (101–121)

Line offset lengths

RM 131   RM 141

RM 120 | RM 121

RM 110 | RM 111

A

RM 100 | RM 101

"RM 111" can
imply a location
halfway down
the line, minus an
offset value
where the hallway
extends past
the last doorway.

FIGURE 5.10 Travel path from Room 111 to Room 141 could be clarified by including Rooms 120, 121, or 131 in a GPS-style waypoint file format.

one can define the average of a set of strings as, say, the string with the smallest distance from all strings in the set. The similarity can be again defined as inverse or, even better, negative distance between the strings. With those two measures and substituting reference vectors by reference strings, one can construct self-organizing maps of symbol strings.

Standard geometric, self-organizing maps are usually used for mapping complex, multidimensional numerical data onto a geometrical structure of lower dimensionality, like a rectangular or hexagonal two-dimensional lattice. The mappings are useful for visualization of data, since they reflect the similarities and vector distribution of the data in the input space. Each node in the map has a reference vector assigned to it. Its value is a weighted average of all the input vectors that are similar either to it or to the reference vectors of the nodes from its topological neighborhood. For numerical data, average and similarity are easily computed. However, for nonnumerical data — like symbol strings — both measures tend to be much more complicated to compute. Still, like their numerical counterparts, they rely on a distance measure. For symbol strings one can use Levenshtein distance or feature distance.

NAVIGATION USING TOPOLOGICAL MAPS

An elaborate topological model with good connectivity is required for navigation purposes. The topological (link/node) view of a street network has a significant advantage in supporting navigation, since a path through a network is readily expressed as a series of decisions at nodes.

Topological maps are suitable for indoor navigation not only for mobile users, but also automatic or robotic vehicles or autonomous robots, which require higher precision and accuracy from the data model than is available in a symbolic map. The main structure of the map containing qualitative information about the large-scale connectivity of the environment is redirected using a spatial model graph, describing the world in terms of nodes and arcs. Nodes stand for important places in the environment and locations where a change in the navigational strategy may occur, and arcs connect these nodes.

Hence, in a topological representation, there has to be a node in front of each door, at each corridor crossing, and at other places of interest. Each node has a location in a fixed coordinate system. The arcs that connect these nodes can be of three different types: room, corridor, or door.

Because the data are derived from CAD files, the implementation of navigation functions is very limited. The structure of the design files is composed of points, arcs, and polylines, which represent objects like buildings, rooms, doors, corridors, and stairs. Moving from this primitive data model to a new topological model requires specific algorithms in order to handle the data. For example, a door composed of lines and arcs in a CAD form can be replaced by a single point in the topological model as part of the conversation process.

To enable navigation, nodes have to be assigned to a physical location, the precision of which depends on the system used to derive position (i.e., the point of interest of a door is assigned to a position midway between the doorposts, threshold, and upper beam). Further, nodes in rooms can be positioned at places that are important for the navigational task. This placement allows the navigation system to effectively keep track of its position and orientation. Nevertheless, these coordinates need not be very accurate, because the nodes in combination with the robot position estimate are only needed for task switching. Context information and a predefined world model enable the coordination scheme to switch between subtasks. These representations constitute symbols, on the basis of which the system makes decisions. These symbols must be anchored in the real world (using the location determination system).

Using a topological approach, the positioning system provides an initial position, which enables the navigation system to select all surrounding links from the database. The initial position is then overlaid on all selected links. Each solution has a probability function that depends on the projected distance. The link with the best probability is selected. Then the exploitation of the topology of the network is envisaged. This process is based on the estimation of the best path compared to the travel path given by the navigation system.

Traveled Distance

In order to determine the traveled distance along a path, it is necessary to evaluate the real travel path. The topological model is imposed to rectify the position given by the navigation system. Figure 5.11b shows an example where the real travel path is not following the straight geometry of the road. Therefore, it is necessary to detect the displacements perpendicular to the selected link. The angle between the origin of the traveled link and the current position is computed, thus allowing deduction of the traveled distance.

The movement (change) from one link to another is possible when reporting the traveled distance along the link.

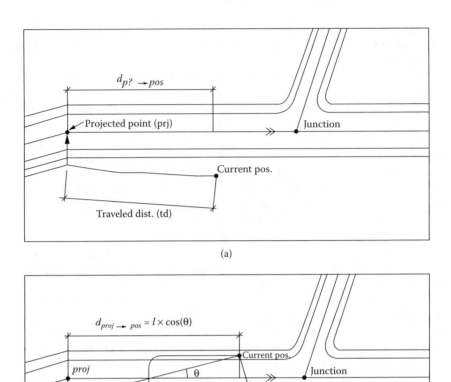

FIGURE 5.11

Note that actual guiding of the robot through a door, for example, is controlled by the behaviors that extract the precise location of the door posts from sensory data. Local geometrical representations parameterize these behaviors.

Topological maps are harder to use for navigation purposes than metric maps, because only limited knowledge about the objects surrounding is available. Nevertheless, it also contains some minimal geometrical information redirected in the properties of the different nodes, which defines their location in the world. Nevertheless, the link/node view of a network has a significant advantage in supporting navigation, since a path through a network is readily expressed as a series of decisions at nodes. The classical methods developed to find optimum paths through networks, such as the shortest-path algorithm (Dijkstra), are based on link/node structures. Car navigation systems are based on a link/node view of the street network, which has significant advantages in supporting navigation. The algorithms developed to find optimal paths through networks, such as the shortest path, are based on link/node structure (Büchel, 2004).

The implementation of a shortest-path algorithm is simple when the cost function is defined for all links (edges) of the database. The length of a path is the sum of the costs of the links. As more than one solution exists, the algorithm has to propose the solution with the minimal cost. Depending on the number of parameters, the cost function could be more complex.

The computation of the best route can be based on a Dijkstra's algorithm, which is specifically designed for a continuous and oriented graph.

Such procedures are well implemented in car navigation systems based on road databases (i.e., TIGER, NAVTEQ, TeleAtlas).

The algorithm has to estimate the shortest path between a start node and an end node given by a user who has a specific profile.

Each link has two nodes (start, end), which are georeferenced by a set of coordinates and altitude in a spatial reference system. One link has attributes such as the type (horizontal, vertical), the length, and information concerning the access privileges. The entrance to some buildings or rooms could be limited to a certain group of people. For this reason, this limitation must be taken into account in the navigation routines. For example, the computation of the best route has to output results that fit with the user's profile (e.g., this person has the right to enter the building xy).

The topological view of the network is designed for an optimal implementation of navigational algorithms. The basic topological structure allows for a simple use of cost functions. Each link has a list of attributes that can be used for the computation of the cost, which is mainly based on the time to travel to a link from start node to end node. The implementation of a shortest-path algorithm is simple when the cost function is defined for all links of the database. The length of a path is the sum of all individual costs of the links.

For example, the cost for a user to travel upstairs or downstairs is given by a mean time of travel. Access restrictions also can be included in the cost function, such as the access doors of the elevators and the staircases are represented by vertically connected nodes. The access restriction is given by a very high cost value. The cost for the vertical links (elevators, staircases) is given by a mean travel time. See equation below:

$cost_i = f_i(l,s,a,t)$
l: length of link i
s: speed (m/s)
a: access rights
t: type of person

NAVIGATION USING GEOMETRIC MAPS

Geometric location modeling maps the objects in the world onto a grid, which means that the objects are positioned absolutely, and that there is a way not only to give their position, but also their extent in space (assuming a fine enough granularity in the positioning system). This type of precise navigation is essential for robots, and for guided vehicles such as wheelchairs.

Navigation is achieved by calculating the distance traveled with respect to the distance remaining in reaching the destination. The ISO TC/211 standard for navigation, "19133 Geographic Information: Location-Based Services Tracking and Navigation," describes the data types and operations associated to those types for the implementation of tracking and navigation services. This international standard is designed to specify web-based services that may be made available to wireless devices through proxy applications, but is not restricted to that environment. Another example is the OpenLS APIs, which focuses on providing the underpinning for navigation (and network analysis).

MAPPING CONTENT TO POSITION

Mapping infrastructure providers (i.e., content providers) have to figure out how to provide location content utilizing various types of location models (which may be different in the local and global contexts, or may indeed vary in the local context as well).

A mapping infrastructure combining both the geometric and symbolic location models makes it possible to support LBS at all relevant scales, in all types of positioning-type environments. Depending on the application, there may also need to be a mapping between the two types of world models, which enable the received position information to be mapped out either geometrically or symbolically, depending on the application.

In order to derive the interconnections between places (or nodes), it is necessary to have relations between the objects involved. Some relations are modeled implicitly when the geometry is modeled, e.g., which rooms are next to each other (topology). However, this may not be sufficient, if the geometric model lacks information about doors between rooms. In this case, this relation has to be explicitly modeled. Depending on the application, different types of relations may be important. For example, in case of a system where a user is expected to walk between rooms, a relation expressing which rooms have connections with other rooms may be important, but in the context where the application is an inventory system for office equipment, the important relation may be in which room an object is situated. It should also be made clear that an object can have multiple relations at the same time, and that the relations may be expressed in different formats, even though they are describing the same type of relation.

Most information can be modeled in a geometric location model, since it expresses the whereabouts of objects in space. The symbolic location model (spatial model graph) can be overlaid on the geometric model, complementing it with the relations between the objects. The computational systems managing the models will vary considerably, depending on whether the models are managed in the terminal or in the network, as will the communications protocols required.

The communications protocol may bind them together or handle them separately (one locally constructed, the other downloaded; both downloaded, but dependent on different positioning sources; etc.). This is an optimization problem that will most likely vary from system to system. The constraints on the communication protocol include not only the various communication-centric aspects, but also the computational aspects (i.e., How often do the models have to be recalculated? What are the requirements on data? etc.).

The other aspect that has to be managed is the computational aspect. The integration of the geometric and symbolic location models can be achieved through bundling nodes in the space tree with geometric attributes. The resulting tree is called a geometric space tree, which has geometric attributes embedded into the nodes that include:

- Shape (indicates the geometric shape of space — cylinder, cube, sphere, etc.)
- Extension (for specifying volume along with the shape attribute)
- Origin
- Rotation matrix

Another possible solution to provide seamless mapping (location) content for LBS is the layered model architecture of location proposed by France Télécom. Fulfilling the requirements of combining several location model types, the researchers have created an architecture template where *locus* is user or object and *loci* is a locatable entity, such as a point of interest that draws a conceptual analogy to the layered abstraction levels of network protocols (International Standards Organization (ISO) network layer stack). Based upon such a general framework, location could become integral to all software (service) infrastructures (i.e., service/device discovery and management), as a basic attribute of all networked entities, supporting location-based queries.

In this location layer model, the bottom layers are closest to the physical properties of space, and as the higher layers get more abstracted away and closer to concepts understandable to users. This is similar to the ISO network layer model, moving up from physical connection to multiple-access control (MAC) addressing, to Internet Protocol (IP), then to DNS and possibly Universal Description, Discovery, and Integration (UDDI) addressing, in network-based identification protocols. This location layer model consists of two vertical domains of representation, orthogonal to layers. The flow of information is vertical through each of these, using horizontal relations at each layer.

By a loose analogy with the physical layer of networking protocols, these layers correspond to the physical location-sensing and identification technologies used, which are much more varied than physical networking technologies. As such, they comprise technologies that identify users (*locants*) and technologies that locate them in space, which may be the same, but not necessarily. It is only by combining the two that tracking a user is made possible. A sonar- or radar-like or vision-based ranging system can be used as a relative positioning technology, but may have to be used with a secondary identification technology for actual tracking.

In order for a location model to be practical, a federation should exist, with more application-oriented models dependent on more fundamental models (but not vice versa). The development of detailed models is a costly task (especially when extending them to detailed world models). Therefore, different applications should be able to share the same model information. Having such a common model may increase interoperability between applications and make new classes of applications possible (due to easier integration). The basic requirement for such an approach is a common language for describing and querying location information. An example of such a query language is Nexus's AWQL and Open GIS Filter Encoding specification.

Such means as the mapping infrastructure combine several complementary models of physical space (topological, metric, Euclidean, symbolic/semantic) and make it possible to support LBS at all relevant scales, in all kinds of environments. A template for this infrastructure may draw a conceptual analogy to the layered abstraction levels of network protocols (the ISO model). Based upon such a general framework, location should become integral to the software (services) infrastructure (i.e., all the service/device discovery and management components) as a basic attribute for the communication (querying) among all components.

Despite the developments presented in this chapter in terms of location models and modeling languages, a number of challenges still exist regarding the modeling of indoor location and navigation. These challenges include the following:

- Managing complexity and scalability: As models increase in complexity, the management and integrity of the information becomes a critical design issue. In addition, the design of a model should not only take into account the potentially large number of entities in a single environment, but also factor for multiple environments linked together.
- Transient environments and aggregation of sensor data: Designing a model that successfully bridges the difference between administrative, social, and home environments is challenging. Focusing the design on a single environment may obscure difficulties when applying it to another environment type. Many environments will support one or more differing LPS. Aggregation of this multiple sensor data would rely on an abstract location model not directly connected to or dependent upon a particular LPS.
- Inference beyond position: While determination of position remains important, there is potential for greater contextual inferences to be made from a model in terms of representing conceptual, logical, and physical connectivity.
- Open and extensible model: The task of providing location information for the model should not rely solely on a single source. The ability for other providers to supply additional information is desirable. In order for a model to evolve along with changes in the environment, it and the sensing technologies employed must be easily extensible and adaptive.
- Ontology for location: The decision of how to describe space is not a trivial matter; however, a common means to represent location across various different models may be useful. Semantic location information can be powerful for many tasks, but it remains an open problem to gather

and represent both semantic position tags and detailed geometric location in a single system.

It is important to realize that a symbolic model does not need to be less accurate than a geometric model, as the real issue is whether the underlying data model and services are based on a symbolic or geometric data model. The accuracy is more dependent on the positioning system than the data model; it is quite possible to calculate a very precise symbolic location, e.g., state that an object is closer than a few millimeters. The symbolic location may, however, also depend on whether any vector information is assumed, e.g., if an object 200 m away is moving toward me with a velocity of 200 m/sec, it will be very close to me in 1 sec, and equally far away in one more second. Time is an additional constraint in the system, which may also be handled either symbolically or geometrically (i.e., What is the scope of time measured in the system?). The time measurement may be synchronized with an absolute clock, or it may be constructed internally in the system (in relation to the internal functions of it), or both — for example, using location to snap to a room or floor and then give services (what is close or navigation) based on being in the room vs. using geometry and Euclidian distance as the basis for judging what is close and far (without caring whether there are walls or buildings in between).

In most cases where a simpler approach (i.e., a symbolic location model of a diagrammatic floor plan layout) might suffice and be significantly cheaper for some basic LBS application, there are additional LBS applications of interest that can only be implemented with a geometric location model, especially those that require arithmetic calculations (i.e., distance measurements). Hence, it is important to consider whether the simpler approach will suffice for all applications that may be needed in the long term. If not, it makes sense to implement a mapping infrastructure (location model) that will handle all applications from the simple to the sophisticated. And depending on what kind of LBS application is to be provided, different semantics need to be considered as well.

Overall, the mapping infrastructure must be extensible enough to federate different ways of modeling and abstracting away physical space. It should be sufficiently accurate to fulfill the needs of all sorts of specific LBS applications while being general enough to be independent of LPS and sensor technologies. There are two goals that should be reached to fulfill these requirements. The first one is to handle and generate location queries (which, in addition to a position filter, may also include events) forwarded to the interested users. These events may also be filtered by an intermediate service to retain only the relevant activities (i.e., moves), depending on the location model and the application concerned. The OGC Filter Encoding specification is an example for a standard way of querying and filtering using XML for spatial operations. The other is to respond to location queries based on location events (i.e., location-based triggers — Buddy Finder and Product Finder). The response has to be provided in real time, preferably in a standard way, for other services to integrate when or if they are needed.

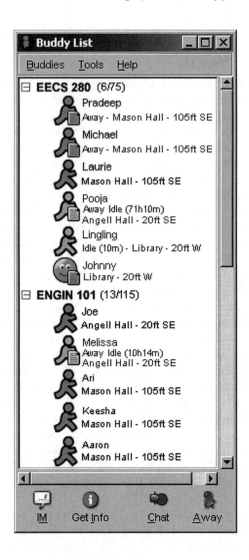

FIGURE 5.12

CONCLUSION

When discussing LBS, two issues in general are discussed interchangeably: (1) the issue of identifying/measuring a location through a system and (2) the issue to support human movement to that location (guided navigation). The former is a technical/measurement issue (e.g., GPS or Wi-Fi) and handled typically through absolute geolocation approaches. The latter is a user–computer interaction issue (where relative space descriptions become relevant).

Indoor navigation is based on diverse technologies, thus involving a high level of integration to combine navigation system and databases. Compared to car navigation, indoor applications are built on hybrid systems and various structures of data. The challenge results in the implementation of data models, which are designed to integrate specific navigation algorithms. At the end, the integrated system has to provide reliable information to the user, especially for very demanding applications (e.g., safety, rescue).

The overall challenge is to build a positioning system that provides accurate position information and navigational capabilities on a global scale — indoors and outdoors. We discuss a hybrid approach in Chapter 10.

Because the CAD floor plans of buildings and campus are not represented in a form that can be readily processed by indoor navigation systems, it is likely that the deployment of indoor navigation infrastructures will be conducted on a custom basis, that is, before indoor building data become more available and widely distributed, similarly to what is going on with outdoor street data.

One of the problems to date with attempting to solve the LBS computing problem is that every proposed solution has its own approach, data structures, processes, and the like. There is little if any standardization between the various approaches. Moreover, there is no one standardized view of the world that would unlock the potential of LBS computing. Standardization can be achieved at the foundational level by defining a universal view of the geographical space. This can be achieved using the hierarchical nodal structure presented in this chapter.

Overall, LBS developers and implementers must determine their own requirements for accuracy, detail, and functionality, keeping in mind that if continuity and consistency of a seamless user experience are goals, interoperability strategies in indoor positioning data structures, models, and methods are desirable.

PERFORMANCE OF NAVIGATION

The evaluation of the positioning accuracy could be based on the comparison of the measured positions after calibration with the content of the map database. In this case, we assume that the user was walking in the middle of corridors, following the main nodes of the network.

The performance of all of the algorithms discussed in this chapter can be improved if topological information is also used. Consider improving the point-to-point matching, for example. Given an initial point, the topology of the network makes it possible to reduce the set of likely arcs dramatically. This kind of information could prevent one from mistakenly matching the start node to the initial point.

Furthermore, mistakes in map matching influence the overall performance. In some situations, directions do not change much as a result of small errors in the map-matched location. In other cases, the directions change dramatically. Scott et al. (1997) provides alternative paths on varying the map-matching algorithm in different situations.

Overall, point-to-point and point-to-curve matching are unlikely to work very well, especially when there are errors in the position or errors in the network representation. Hence, other, more complicated algorithms must often be used.

The main challenge when doing map matching is to control the position of the user at regular time intervals. Also, providing a recalibration of the navigation system can be a challenge.

User Interfaces (GUIs)

Visualization of information is a very important component of any LBS. After determining location, that information should be relayed to the user of the system in an intuitive and concise manner. In an LBS application, a graphical user interface (GUI) can present the user with a detailed map of an area and pinpoint the user location on that map. Previous to GUIs, this information might have been relayed to the user via a textual description. GUIs also provide interactive features such as zoom and multiple viewpoints that are not possible in a text-based system. Overall, a GUI enhances the functionality, usefulness, user-friendliness, and interactivity of an application. Building name, floor name, and (x, y) coordinates will tell the user of the system where a target is located; however, displaying that information on a map will make it easier to analyze. So, once the location of a client is determined, the most obvious and clear way to relay that information to a user is to display it on a map of the area where the client is located.

Symbolic location within GUIs should be the norm. For example, Herecast provides location-based services on a Wi-Fi device. At its simplest level, it can tell the user where he is. More advanced services can use location to enhance information lookups, publish presence information, and create unique games. Herecast uses a symbolic naming system — instead of using coordinates such as 42.9875, –81.2915, it expresses a user's location in terms an ordinary person would use (for example, the name of the building). Every wireless access point broadcasts a unique identifier, which can be used to tell it apart from other access points. That identifier can also be used as a landmark to identify a particular location.

Voice UIs

In contrast, verbal route descriptions consist of a sequential description of route segments, including physical elements and basic motor activities (e.g., walk, turn). Verbal descriptions also differ from maps in that they describe the path of motion from an *egocentric* frame of reference, i.e., as seen by the navigating person.

Some verbal route descriptions mention regions and spatial relations between objects in the current environment. The user undertakes a mental journey, during which elements in the environment are localized in relation to his current position or to each other from an egocentric point of view. This *route perspective* is helpful to convey knowledge about path segments and landmarks, the so-called route knowledge.

Landmarks are elements of the environment external to the observer, serving to define the location of other objects or locations. They are memorable cues selected along a path and enable the encoding of spatial relations between objects and paths, leading to the development of a cognitive map of the environment. Landmarks are generally used in navigation tasks to identify decision or destination points or to convey route progress. They influence expectations, provide orientation cues for homing vectors, and suggest regional differentiating features.

If the user's position and orientation are vague or missing, the system must provide information for the user to locate himself and to determine his actual orientation in space, or design the user dialog in a way that helps the user to fill in the missing information. But in order to align the map to the walking direction, the system has to ensure the user's correct orientation. This task can be accomplished by advising the user to reorient himself toward a landmark (e.g., by prompting a text: "Turn around until the stairs are to your left and the lift is to your right").

REFERENCES

Arthur, P. and Passini, R., *Wayfinding: People, Signs, and Architecture*, McGraw-Hill Ryerson, Toronto, 1992.

Baus, J., Butz, A., and Krüger, A., One way interaction: interactivity over unidirectional links, in *Proceedings of I3 Workshop on Adaptive Design of Interactive Multimedia Presentations for Mobile Users*, March 7, 1999, www.i3net.org.

Bentley, J.L. and Maurer, H.A., Efficient worst-case data structures for range searching, *Acta Informatica*, 13, 155–168. 1980.

Büchel, D., Développement d'une solution de navigation robuste pour l'environnement construit, diploma thesis, EPFL, 2004.

Büchel, D., Méthodes de guidage applicables au plan d'orientation de l'EPFL, internal report, EPFL, 2003.

Butz, A., Baus, J., and Krüger, A., Augmenting buildings with infrared information, in *Proceedings of the International Symposium on Augmented Reality ISAR 2000*, Los Alamitos, CA, 2000.

Department of Defense, *World Geodetic System 1984 (WGS84)*, 3rd ed., NIMA TR8350.2, Department of Defense, January 2000, www.wgs84.com.

Deretsky, Z. and Rodny, U., Automatic conflation of digital maps: how to handle unmatched data, in *Proceedings of the Vehicle Navigation and Information Systems Conference*, 1993, pp. A27–A29.

Drane, C. and Rizos, C., Positioning systems in intelligent transportation systems, in *Intelligent Transportation Systems*, Artech House, 1998.

Egenhofer, M.J. and Franzosa, R.D., Point-set topological spatial relations, *IJGIS*, 5, 161–174, 1991.

Fuchs, H., Kedem, Z.M., and Naylor, B.F., On visible surface generation by *a priori* tree structures, *Comput. Graphics*, 14, 124–133, 1980.

Gilliéron, P.Y. and Merminod, B., *Personal Navigation System for Indoor Applications*, World Congress IAIN, Berlin, 2003.

Hirtle, S.C. and Jonides, J., Evidence of hierarchies in cognitive maps, *Memory Cognition*, 13, 1985.

Lee, J., 3D GIS for Geo-Coding Human Activity in Micro-Scale Urban Environments, paper presented at GIScience 2004, 2004.

Leonhardt, U. and Magee, J., Towards a general location service for mobile environments, in *Proceedings of the Third International Workshop on Services in Distributed and Networked Environments*, Macau, June 1996, pp. 43–50.

Lynch, K., *The Image of the City*, MIT Press, Cambridge, MA, 1960.

Needham, R., Names, in *Distributed Systems*, ACM Press Frontier Series, 2nd ed., Mullender, S., Ed., Addison-Wesley, Reading, MA, 1993, chap. 12, pp. 315–327.

Nissanka, B., Priyantha, H.B., Demaine, E.D., and Teller, S., *Mobile-Assisted Localization in Wireless Sensor Networks*, MIT Computer Science and Artificial Intelligence Laboratory.

OGC, www.opengeospatial.org/docs/01-009.pdf.

Potter, B., Sinclair, J., and Till, D., *An Introduction to Formal Specification and Z*, Prentice Hall, Englewood Cliffs, NJ, 1991.

Priyantha, N., Balakrishnan, H., Demaine, E., and Teller, S., *Anchor-Free Distributed Localization in Sensor Networks*, Technical Report 892, MIT Laboratory for Computer Science, April 2003.

Schiff, T.H., Data sources and consolidation methods for creating, improving and maintaining navigation databases, in *Proceedings of the Vehicle Navigation and Information Systems Conference*, 1993, pp. 3–7.

Scott, K., Pabón-Jiménez, G., and Bernstein, D., Finding Alternatives to the Best Path, paper presented at the Annual Meeting of the Transportation Research Board, 1997.

Shan Hung, P. and Chuan Su, T., *Map Matching Algorithms of GPS Vehicle Navigation System*, Reng Chia University, 1998.

Shibata, M., Updating of digital road map, in *Proceedings of the Vehicle Navigation and Information Systems Conference*, 1994, pp. 547–550.

6 Existing Indoor Location Systems: How They Work

Many indoor positioning systems have been developed to demonstrate that indoor positioning is possible, and that it can be done with high accuracy. This means that most of the systems discussed here are developed as stovepipes, containing all seven components we pointed out in Chapter 2. It also means that they may not have application programming interfaces (APIs), and that the communications protocols are only intended for communication between the components.

FEATURED INDOOR LOCATION POSITIONING SYSTEMS

MIT CRICKET

Cricket is an indoor location system for pervasive and sensor-based computing environments, which is part of MIT's Project Oxygen. Cricket provides the hardware, software-based algorithms, and a software API for location-aware applications running on handhelds, laptops, and sensor nodes. Applications discover their space identifiers/logical location (e.g., which room or portion of a room they are in), position coordinates (e.g., indoor global positioning system (GPS) coordinates), and orientation (the Cricket compass). Features of the Cricket positioning system include:

- Assisted configuration of an *ad hoc* beacon coordinate system
- Accuracy of distance measurements
- Accuracy and latency of real-time tracking of the listener's position within the *ad hoc* beacon coordinate system

Cricket achieves 1 to 3 m accuracy (with portion-of-a-room granularity of 1.3 × 1.3 m) by separating the processes of tracking services and obtaining location information. The system enables service discovery where multiple resource discovery systems can be accommodated due to separating the process of tracking services and obtaining location information. User applications do not advertise themselves unless they want to be discovered by others. Applications learn about services in their vicinity via an active map that is sent from a map server application, and interact with services by constructing queries for services at a required location. As the mobile device moves in a building, the navigation software running on it uses the listener API to update its

	MIT Cricket	Ekahau	Intel Place Lab (open source)	Skyhook Wireless WPS	Microsoft Research RADAR	Rosum TV	AeroScout	BLIP Systems BlipNet	GPS
Deployment Range	Building	Building/local area	Metropolitan area	Metropolitan area	Building/local area	Metropolitan area	Building/local area	Building	Global (not indoors)
Position Calculation	Mobile device	Server (Ekahau Positioning Engine)	Mobile device	Mobile device	Server	Rosum Location Server or Mobile Device	Server (AeroScout Positioning Server/Engine)	Server (BlipServer Positioning Engine)	Device
Position Method	TOF Lateration & proximity (with one beacon).	Location fingerprinting using signal strength	Map-based pinpointing and triangulation	Map-based pinpointing and triangulation	Location fingerprinting & triangulation (lateration)	"Multi-lateration" (a distance measuring technique)	TOF Triangulation (TDOA for absolute location; RSSI for symbolic location)	Inquiries and pagings;	TOF lateration
RF Signal Used	RF (418-MHz) + ultrasound	802.11	802.11	802.11	802.11	Terrestrial Broadcast TV	802.11	Bluetooth	GPS
Transmitter	Beacon	Existing WiFi nodes	Existing WiFi nodes	Existing WiFi nodes	WiFi devices	Existing TV towers	Active RFID tags or WiFi devices	Mobile device	Network of 28 Satellites managed by US DoD
Receiver	Listener	Ekahau Client	Place Lab client (open source)	Skyhook Wireless client	Standard WiFi AP	Rosum RTMM chipset (proprietary)	AeroScout Location Receiver (TDOA), AeroScout Exciter (choke-point); standard WiFi AP (RSSI)	BlipNode (Bluetooth AP)	Commodity receivers
Accuracy	1-3m (3.3-10ft)	1-3m (3.3-10ft)	20+ m (66+ ft)	20+ m (66+ ft)	2-4.3m (6.6-14ft)	30-50m (98-164ft) indoors; 5m (16ft) outdoor	1-5m (3.3-16ft)	10cm-10m (4in-33ft)	3-20m (10-ft) outdoors

MIT Cricket allows applications running on user devices and service nodes to determine their physical location. It does not require a grid of ceiling sensors with fixed locations because its mobile receivers perform the timing and computation functions. It tracks and stores location information for services and users in a centrally maintained database. Cricket sends beacons to disseminate information about a geographic space to listeners (calculates nearest beacon). A beacon is a small device attached to some location within that space, and it sends name/semantics of the space to the listener. Typically, it is obtained by the "owner" of the location (e.g., the occupant of a room in an office or home or a building administrator) and placed somewhere unobtrusive like on a ceiling or wall. Cricket does not attach any semantics to the space information advertised by the beacon: any short string can be disseminated, such as the name of a server to contact to learn more about the space or a name resolver for the space to discover resources. Cricket beacons are inexpensive and more than one of them can be used in any space for fault-tolerance and better coverage. Crickets are now commercially available and more than 15,000 units have been sold in the first year.

Ekahau's proprietary WiFi-based positioning system is a Java-based software that includes Ekahau Client, Ekahau Manager, and a Ekahau Positioning Engine (EPE). The Ekahau Client retrieves the signal strength (RSSI) and other information from the network cards and interrogates the client device Wi-Fi radio chip for RSSI values and simply communicates these data to the EPE, which is a positioning server that provides mobile devices and asset TAG location coordinates (x, y, floor) and tracking features to client applications. EPE includes a stand-alone manager application for site calibration, logical areas, live tracking, and accuracy analysis. Ekahau also offers its Site Survey tool for WiFi network visualization and optimization.

Intel Place Lab consists of a centralized database containing information on landmarks and client side code to perform the position calculation. The client side Place Lab software exists in a downloadable form for multiple platforms. On almost all operating systems it is as simple as extracting files from an archive and executing a single command to get started. The database consists of a single table containing a unique identifier for each landmark (for instance, the MAC address of a WiFi access point), a human readable name (e.g., the SSID of a WiFi access point), and the landmark's latitude and longitude. Before a user can perform a position calculation, some (or all) of the database must be downloaded and stored locally on the user's device. The Place Lab client software contains a MapLoader application used to connect to the database and download the information for a selected region. The database is available online via PHP forms on the Place Lab web site and anyone is able to contribute to the data. Included in the Place Lab software download is a Place Lab stumbler application. This application is run on a mobile device in conjunction with a connected GPS device to first scan for landmarks and then log anything seen with a latitude and longitude (derived from the GPS device). When finished stumbling, the user uploads the log file created using a form on the Place Lab web site to contribute any new information to the database.

Microsoft Research RADAR operates by recording and processing signal strength information at multiple WiFi access points positioned to provide overlapping coverage in the area of interest. It combines empirical measurements with signal propagation modeling to determine user location

Rosum TV-GPS positioning system utilizes unmodified broadcast TV signals for position location. There are 2800 GPS transmitters and 4500 TV transmitters in the U.S. alone, and they are well-correlated with population centers. Rosum designs and manufactures the chipset, software, and infrastructure components of its TV-GPS system. Rosum also has developed a local-area TV-GPS technology that is portable and deployable for supporting first responder police, fire, and rescue personnel.

The AeroScout system provides location information using 802.11 architecture by measuring the time of arrival of packets from a set of active RFID tags to a set of location receivers. The advantage of the WiFi-based RFID is that the tag(s) can be detected by commercially available wireless routers and bridges. The AeroScout system consists of a set of active RFID tags and a number of specialized location receivers (long-range RFID readers.) The tags contain tiny radios that broadcast at a pre-defined rate. The tag uses a standard 3.3V LiH battery (a common camera battery). The tags, without modification, only transmit their Media Access Code (MAC) address, an 8-byte code that indicates who produced the radio and the serial number of the tag. AeroScout recommends that the tags transmit every 10 seconds to maximize battery life. The specialized receivers are called "location receivers" (LRs) and are typically connected to an Internet bridge. This allows the LRs to be in contact with a Java-based "Location Engine" (LE), located on the user's computer. At least three LRs are required to estimate the position of the RFID tag. The LRs work on the principle of time-of-flight (TOF) measurements. That is, the LRs estimate distance to a tag by comparing the TOF of beacon transmissions from other LRs with the TOF of the tags. When setting up the LE, the locations of the LRs are entered. Thus, the distance to an LR is known and a comparison of the flight times will give an approximate distance. Triangulation between three or more LRs is used to find the actual position. Since many beacon packets arrive, the position is often an average of 50 or more measurements at a time. Thus, the system is very precise.

"BLIP" stands for Bluetooth Local Infotainment Point and enables Bluetooth devices to access local information. BLIP Systems BlipNet utilizes Bluetooth to gather data from mobile sensors placed on objects and combines the data with position data. BlipNet connects PDAs and mobile phones to local networks using Bluetooth wireless technology. Paging and inquiry are performed only occasionally when a link is established. Dependent on the use case, BlipNet may perform periodic inquiries and pagings to detect moving devices to ensure handover between APs. The advantage of the system is that it requires considerably less power than WLAN based systems, so there are commensurately longer periods before recharging is needed. Accuracy is flexible; you can get position information with an accuracy of to a few meters from a large area by mounting a grid of BlipNodes, or you can get precise position information at a gate by mounting a BlipNode on the gate and detecting when devices are down to 10 cm from the BlipNode. The distance between BlipNode and "item" to be tracked can be up to 100 meters. The distance between BlipNodes depends on the accuracy that is wanted.

current location. The location information can then be sent to a map server so that the application can obtain updates for the map display, or the location information can then be used to appropriately advertise itself and its location to a resource discovery service.

The services learn their location information by using their own listener devices, avoiding the need for any per-node configuration. Services appear as icons on the map that are a function of the user's current location. When a new static node (e.g., printer) is attached to the network, there is no need to configure it with a location or any other attribute. Within a few seconds, the listener infers its current location from the set of beacons it listens to, and informs the device software about where it is located via the API.

Once location information is obtained, services advertise themselves to a resource discovery service, such as the MIT Intentional Naming System (INS), Internet Engineering Task Force (IETF) Service Location Protocol, Berkeley Service Discovery Service, or Sun's Jini discovery service. These resource discovery services handle service and device mobility within the naming system. Cricket's goal is to develop a system that allows applications running on user devices and service nodes to learn their physical locations. Cricket applications include location-aware applications that enable users to discover resources in their physical proximity, active maps that automatically change as a user moves, and applications whose user interfaces adapt to the user's location.

Most other systems are based on a cellular approach, in which either the mobile device detects its cell or the system determines which mobile devices are in each cell. Being a location-support system (rather than a conventional location-tracking system), Cricket does not track and store location information for services and users in a centrally maintained database. It helps devices learn where they are and lets the applications decide to whom to advertise this information. As a result, user privacy concerns are easier to meet than in systems that always register the user's location. In contrast, systems like the Bat that have the central controller know where each wall- or ceiling-mounted device is located. This has two disadvantages. First, user privacy is compromised because a listener now needs to make active contact to learn where it is (in Cricket, a listener is completely passive). Second, it requires a centrally managed service. Cricket's beacons advertising location information are self-contained and do not need any infrastructure for communication among themselves. Cricket's architecture inverts the architecture of the Badge and Bat systems, which use passive ceiling-mounted receivers that obtain information from active transmitters attached to devices carried by users. Most of the other systems are based on a cellular approach, in which either the mobile device detects its cell or the system determines which mobile devices are in each cell.

MIT CRICKET ARCHITECTURE AND COMPONENTS

MIT Cricket's Spatial Information Service (SIS) enables CricketNav to provide spatial info about accessible paths and space-dividing features. This includes the path network used for navigation (origin to destination) and use of structural information (walls, doorways, elevators, staircases). This can be used in other location-aware applications ("Where is the nearest printer?").

Moreover, Cricket SIS supplies static spatial information, which includes absolute Euclidean distance and relative path distance (more useful metric). Cricket SIS also models cost of a path (shortest path vs. fastest path). For example, the nearest printer may be located across a hallway vs. located on the next floor directly above the user's current position (the latter is physically closer in terms of distance, but it is not faster). The cost of path calculation requires spatial information and its cost model, where a weight factor is assigned to a path. An example is a path that has less doorways and staircases, or is accessible by wheelchairs.

Cricket SIS can be extended with real-time (dynamic) state queries with the purpose of enabling applications to apply their own relative path distance metrics. For example, a real-time cost model could incorporate the state of smoke detectors or a door lock to determine if the path is accessible (useful for emergency crews). However, dynamic state queries and updates should be handled by a separate service (e.g., COUGAR), separate from the SIS.

Cricket SIS serves the purpose of supporting navigation and location-aware applications. Supported operations included the following:

- Shortest path: Use of a spatial model graph where a node represents an intermediate point (waypoint) and an edge represents an unobstructed path between two waypoints. Each edge is annotated with a feature (e.g., door boundary). Each waypoint is annotated with a coordinate and space descriptor (hence, applications can use either coordinates or spaces).
- Find structural features: Used in map rendering to extract features of a specified region. Cricket provides position in symbolic (e.g., current space) or physical (coordinates).
- Find a reference coordinate (center position of a given space): A space is converted to a coordinate so that a query circle can be specified. *Center position* is defined as the position of the corresponding text label in a computer-aided design (CAD) drawing.
- Find a space (that contains a given coordinate): Useful for testing the consistency of reported space and coordinate values.
- Given a symbolic space and a coordinate pair, locate a coordinate for minimum distance: Used for position estimation if the "Find a space" operation does not apply; corrects implausible position estimates that are sometimes reported by Cricket.

Calculations such as shortest path are better off done on the mobile device (by the application), because the communication overhead is greater than local computation. In terms of the SIS server API and fetching floor maps, the floor maps are stored in a centralized SIS server. Applications send a map identifier (consisting of a building ID and floor number) to fetch the desired floor map. The map is transmitted over the Transmission Control Protocol (TCP)/Internet Protocol (IP). The network location of the SIS server is known: SIS service may be discovered via a service discovery mechanism such as INS. Or, the SIS server's port and address may be advertised by Cricket beacons.

Positioning

Cricket uses a combination of radio frequency (RF) and ultrasound technologies to provide a location support service to users and applications. Wall- and ceiling-mounted beacons are spread through the building, publishing information on an RF signal operating in the 418-MHz AM band. With each RF advertisement, the beacon transmits a concurrent ultrasonic pulse. Listeners attached to devices and mobiles listen for RF signals and, upon receipt of the first few bits, listen for the corresponding ultrasonic pulse. When this pulse arrives, they obtain a distance estimate for the corresponding beacon. The listeners run maximum likelihood estimators to correlate RF and ultrasound samples (the latter are simple pulses with no data encoded on them) and to pick the best one.

Cricket uses active beacons and passive listeners, which has two significant benefits. First, it is not a tracking system where a centralized controller or database receives transmissions from users and devices and tracks them. Second, it scales well as the number of devices increases; a system with active transmitters attached to devices would not scale particularly well with the density of instrumented devices. Third, its decentralized architecture makes it easy to deploy.

This does not mean it is hard to manage; a centralized front end allows easy management and control. Cricket can determine which space a device is in by detecting boundaries to within about 2 ft. Beacons are placed in a 5 × 5 ft grid. So, it can give location granularity of 5 × 5 ft, and it can determine angles to within 3 to 5° of the true value. Beacons are placed at the ceiling where they have better line of sight, which gives more accuracy.

Cricket uses *beacons* to disseminate information about a geographic space to listeners. A beacon is a small device attached to some location within the geographic space it advertises. Typically, it is obtained by the owner of the location (e.g., the occupant of a room in an office or home, or a building administrator) and placed at an unobtrusive location like a ceiling or wall. Cricket does not attach any semantics to the space advertised by the beacon. Any short string can be disseminated, such as the name of a server to contact to learn more about the space or a name resolver for the space to discover resources.

A listener is a small device that listens to messages from beacons and uses these messages to infer the space it is currently in. The listener provides an API to programs running on the node that allow them to learn where they are, so that they can use this information to appropriately advertise themselves and their location to a resource discovery service. The listener can be attached to both static and mobile nodes. For example, when a user attaches a new static service to the network (e.g., a printer), he or she does not need to configure it with a location or other any attribute; all the user does is attach a listener to it. Within a few seconds, the listener infers its current location from the set of beacons it hears and informs the device software about this via the API.

This location information can then be used in its own service advertisements. When a mobile computer has a listener attached to it, the listener constantly listens to beacons to infer its location. As the computer (e.g., a handheld computer carried by a person) moves in a building, the navigation software running on it uses the listener API to update its current location. Then, by sending this information securely to a map server (for example), it can obtain updates to the map displayed to the user. Furthermore, services appear as icons on the map that are a function of the user's current location. The services themselves learn their location information using their own listener devices, avoiding the need for any per-node configuration.

Positioning Platform

The system has listener APIs. Cricket version 1 (v1) provides a simple and general API: the listener passes all distance samples from each beacon to the attached host device, which implements all the processing to obtain the host's location. Cricket v1 listeners interface to a host using a RS232 serial interface. This turned out to be inconvenient for mobile users because it required an unwieldy and obtrusive cable and was a significant barrier to wider adoption. Cricket v2 continues to provide raw access to the information collected at the listener to host applications. Cricket v2 provides a more convenient compact flash interface. To enable easy integration with sensor platforms, Cricket v2 also provides a connector to the Berkeley Mote/Crossbow Mica platform. Additionally, Cricket v2 listeners also perform a significant amount of embedded processing, which allows them to be used with a variety of host devices, including sensors that do not perform any Cricket processing. The assumption (and reason for v2) with v1 was that users would not be interested in changing the firmware running in the beacon and listener. However, Cricket designers found that some users wanted to make changes to beacon scheduling, listener filtering, etc. However, the use of a commercial compiler, and software that was tightly coupled to the underlying hardware, made such changes both expensive and time-consuming. To overcome this shortcoming, the designers have rearchitected Cricket v2's embedded software and implemented it in the TinyOS environment.

A part of the software implemented for receiver nodes, called the LocationManager, runs on the host device that has the listener hardware attached to the serial port. The LocationManager listens on the serial port for any data coming from the listener hardware. The listener inference algorithm (MinMode) analyzes distance estimates and is also implemented within the LocationManager (this provides greater flexibility). The listener sends both the location information and the measured distance to the corresponding beacon, to the LocationManager for each valid RF reception.

Asynchronous to the reception of distance estimates and listener computations, applications running on the host device connect to the LocationManager and retrieve current location information using a datagram socket (User Datagram Protocol, or UDP) interface. This allows for the possibility of obtaining this information from a remote node elsewhere on the network, which might be useful for some applications. Cricket has not yet taken advantage of this facility in its applications.

Advantages and Limitations

Cricket uses active beacons and passive listeners, which has two significant benefits. First, it is not a tracking system; it does not rely on any centralized management or control where a centralized controller or database receives transmissions from users and devices and tracks them. Cricket helps devices learn where they are and lets them decide to whom to advertise this information. Also, there is no explicit coordination between beacons; it provides information to devices regardless of their type of network connectivity. The sytem architecture inverts the architecture of the Badge and Bat systems, which use passive ceiling-mounted receivers that obtain information from active transmitters attached to devices carried by users. It helps devices learn where they are and lets them (applications) decide to whom to advertise this information. As a result, user privacy concerns are adequately met.

In contrast, systems like the Bat that have the central controller know where each wall- or ceiling-mounted device is located. This has two disadvantages. First, user privacy is compromised because a listener now needs to make active contact to learn where it is (in Cricket, a listener is completely passive). Second, it requires a centrally managed service, which does not suit Cricket's autonomously managed environment particularly well. Cricket's beacons advertising location information are self-contained and do not need any infrastructure for communication among themselves. Second, it scales well as the number of devices increases; a system with active transmitters attached to devices would not scale particularly well with the density of instrumented devices.

Some of the limitations of the system are the following: lack of centralized management or monitoring, and computational burden (and consequently power burden) that the timing and processing of both the ultrasound pulses and RF data place on the mobile receivers.

How It Works

Cricket uses a combination of RF and ultrasound signals to provide location information to both mobile devices and static nodes in a decentralized, uncoordinated architecture. Wall- and ceiling-mounted beacons are spread throughout the building, publishing location information on an RF signal. With each RF advertisement, the beacon transmits a concurrent ultrasonic pulse. A beacon is a small device attached to some location within the space it advertises. Typically, it is obtained by the owner of the location (e.g., the occupant of a room in an office or home, or a building administrator) and placed at an unobtrusive location like a ceiling or wall. Note: Cricket does not attach any semantics to the space advertised by the beacon; any short string can be disseminated, such as the name of a server to contact to learn more about the space or a name resolver for the space to discover resources.

The most common way to put together a Cricket location system is to deploy actively transmitting beacons on walls or ceilings and attach listeners to host devices (handhelds, laptops, etc.) running Cricket application software, whose location needs to be obtained. Listeners attached to devices and mobile devices listen for RF signals

FIGURE 6.1 Cricket beacon (and listener).

and, upon receipt of the first few bits, listen for the corresponding ultrasonic pulse. Note: Beacons and listeners are identical hardware devices (Figure 6.1). Listeners that have line-of-sight connectivity to the beacon and are within the ultrasonic range will receive this pulse and obtain a distance estimate for the corresponding beacon by taking advantage of the difference in propagation speeds between RF (speed of light) and ultrasound (speed of sound). The listener runs algorithms that correlate RF and ultrasound samples (the latter are simple pulses with no data encoded on them) and to pick the best correlation. Listeners are attached to the host device using an RS232 serial connection.

A Cricket unit can function as either beacon or listener, or can be used in a mixed mode in a symmetric location architecture (which may be appropriate in some sensor-computing scenarios), all under software control. You can attach a variety of sensors to a Cricket device using the 51-pin connector on the Cricket. There are also research prototypes of Crickets with a compact flash (CF) interface, which may be a more convenient form factor to attach to handhelds and laptops than the RS232 interface. These devices may become available in a few months. They will be software- and protocol-compatible with the RS232 version.

Cricket beacons disseminate information about a position to listeners in two forms. Each beacon periodically broadcasts two forms of location information on an RF channel: its space identifier and its position coordinates (accuracy is between 1 and 3 cm). Even in the presence of several competing beacon transmissions, Cricket achieves good precision and accuracy quickly. Listeners within radio range receive the broadcasts with the space identifier and position coordinates. In addition to determining spaces and estimating position coordinates, Cricket provides an indoor orientation capability via the Cricket compass. This facility is not yet commercially available (it is a research prototype). Because RF travels about 10^6 times faster than ultrasound, the listener can use time of arrival (TOA) between the start of the RF message from a beacon and the corresponding ultrasonic pulse to infer its distance from the beacon — measuring the one-way propagation time of the ultrasonic signals emitted by a beacon, taking advantage of the fact that the speed of sound in air (about 1.13 ft/msec at room temperature) is much smaller than the speed of light (RF) in air. Every time a listener receives information from a beacon, it provides that information together with the associated distance to the attached host using the Cricket API. The listener (or software running on the host device) infers its position coordinates based on distances from multiple beacons whose positions are known, and software running on the host device can associate itself with the space corresponding to the nearest beacon. On each transmission, a beacon *concurrently* sends information about the space (space identifer and position coordinates) over the RF channel, together with an ultrasonic pulse. Listeners within radio range receive the broadcasts with the space identifer and position coordinates. When the listener hears the RF signal, it uses the first few bits as training information and then turns on its ultrasonic receiver. It then listens for the ultrasonic pulse, which will usually arrive a short time later. The listener uses the time difference between the receipt of the first bit of RF information and the ultrasonic signal to determine the distance to the beacon. Of course, the value of the estimated distance is not as important as the decision of which is the closest beacon.

Cricket software allows applications running on mobile and static nodes to learn their physical location by using listeners (a location receiver hardware that is attached to the host device or node) that constantly listen and analyze information from beacons spread throughout the building. The listener receives RF and ultrasonic signals, correlates them to each other, and uses the difference in RF and ultrasonic signal propagation times to estimate the distances to the different beacons, and by inferring the distance, can compute the space they are currently in. Listeners estimate distances to individual beacons using the time difference of arrival (TDOA) technique. Listeners use an inference algorithm to determine the space in which they are currently located by listening to beacon announcements (use of concurrent radio and ultrasonic signals to infer distance). This inference algorithm overcomes multipath and interference. Listeners are passive and beacons are active. Beacons use a randomized algorithm to transmit information.

The Cricket beacon and listener hardware are identical; they just run different software. There is only one embedded software image, and a run time configuration switch determines whether a given unit is a beacon or a listener. By default, each Cricket node is configured to run as a listener. The Cricket embedded software runs in the TinyOS environment. At least two Crickets are needed to operate the system: at least one beacon and at least one listener. In the current version of Cricket, the listener is usually attached to a host using a serial cable. In the future it is expected to have a listener with a compact flash interface. The host device (to which the listener is attached) must run software to process the data obtained from the listener. One way to process this data on Linux computers (including handhelds) is to use cricketd, which processes information obtained over the serial interface to obtain various location properties. In particular, cricketd runs on Familiar Linux (on iPAQ handheld computers), as well as on standard laptop/desktop versions of Linux and Windows (under Cygwin). Currently, there is no support for cricketd on handhelds running Windows Pocket PC.

Cricket Hardware and Software

The Cricket hardware consists of the beacon and listener. They both take two AA batteries. NiMH rechargeables are better in terms of their capacity. Using a set of fully charged NiMH batteries, the beacons last about 1 to 2 weeks in continuous operation. The Cricket beacons are the larger modules with the red wire antenna. A typical beacon has a transmission range of about 10 m (depends on the environment; one can also use various simple hardware hacks to increase or decrease this range). Each beacon is preprogrammed to broadcast a unique space identifier, which is used by the listener to identify the user's location. Hence, there should be at least one beacon in each space (e.g., a room). The beacons can also be used to define two or more spaces within the same room. In this case, a pair of beacons can be placed at about 1 to 1.5 m apart. This effectively defines a virtual boundary that divides the spaces represented by each beacon. Information on obtaining the Cricket hardware and software can be obtained from Cricket's web site (http://nms.lcs.mit.edu/ projects/cricket/#download). The software for Cricket v2 (both embedded software and higher-layer software that runs on laptops/handhelds) are available from the MIT Cricket web site. The software is under an open-source license and can be used for education, research, and commercial purposes as long as the requirements in the copyright notice are followed. Commercially available Cricket hardware units (for Cricket v2) are available from Crossbow Technologies (http://www.xbow.com/). Cricket available from Crossbow Technologies may not be preloaded with the embedded software when shipped individually (to program the Crickets you will need a MIB510CA programmer).

Manufacturing and Dissemination

MIT researchers have had more requests for Cricket units than they have been able to handle. The difficulty of providing hardware units to interested users was underestimated. It was found that few users were interested in taking the hardware design available on the web and making their own units. It was hard to justify spending money on hardware support, especially because the design changed on a regular basis. As a result, MIT researchers were unable to satisfy several dozen requests over the past 2 years. For Cricket v2, MIT researchers have partnered with Crossbow, who will make units available for purchase and will provide customer support.

Two programs come with the Cricket sample application:

1. `Beacon Finder`, which displays the distance measurement statistics for each beacon
2. `Beacon Configuration`, which executes the two phases of configuring the *ad hoc* beacon coordinate system and position tracking

Use the `Beacon Finder` to verify that the listener is within range of all beacons. Do this by holding the listener underneath each beacon and verify that (1) it has a row entry for NUMBEACONS different beacons and (2) none of the entries have excessively long last-update times (>5000 msec). Otherwise, some beacons are out of range of the listener and cause the beacon coordinate configuration phase to fail. Please adjust the beacon placement until all the beacons are within range. The BeaconConfigDemo application has two operating phases:

1. The first phase configures an *ad hoc* beacon coordinate system for the set of active beacons.
2. The second phase tracks the listener's position in real time. Beacon-ConfigDemo then uses the position tracking to let the user draw polylines, rectangles, and circles in its window.

There are two versions of Cricket. Cricket v1 used separate RF transmit and receive circuits. A problem with v1 was that when batteries on beacons ran down and voltage dropped, the receiver unit failed before the transmit unit, causing the carrier sense mechanism to fail and leading to poor performance. This problem was corrected in v2. Cricket v1 beacons are not optimized for good power consumption. The two AA batteries powering a beacon needed to be changed every 2 to 3 weeks. To handle deployments where batteries could not be changed this frequently, MIT researchers successfully powered beacons with solar cells placed near indoor lighting. The Cricket v2 design reduces power consumption by using submodules that can be powered down, by using low-power chips, and by implementing better beacon scheduling algorithms. The reduced power consumption of v2 beacons enables them to be effectively powered using small solar cells. Although Cricket v1 works well at moderate beacon densities, high deployment densities (12 or more beacons all within range of each other) cause problems. This is most common in deployment situations where a large number of redundant beacons are used to protect against batteries running out. This problem is mainly due to the poor noise immunity of the Cricket v1 radio (amplitude modulation and surface acoustic wave-based receivers).

Cricket v2 overcomes the noise problems using a better radio based on frequency modulation and a super heterodyne receiver (CC1000 from Chipcon), and appears to perform well at high densities. It is more accurate and energy-efficient than Cricket v1, and has a new software stack that runs on TinyOS; it has better support for continuous object tracking, and it has support for various autoconfiguration algorithms. All Cricket software currently runs on both Linux and Windows platforms that run JDK1.3 or above. The software consists of applications and a daemon, which processes readings from the Cricket listener, attached to the serial port, and exports the processed reading to the applications.

EKAHAU POSITIONING SYSTEM

The Ekahau system uses a wireless local-area network (WLAN) to track tags equipped with WLAN access cards.

It works by measuring signal strength data that is correlated with location. Signal propagation and advanced probabilistic mathematics are used with site calibration for positioning, which takes into account the actual radio environment characteristics by means of recording Wi-Fi network data from various network locations/access points (APs). We discuss more on how this works in Chapter 8.

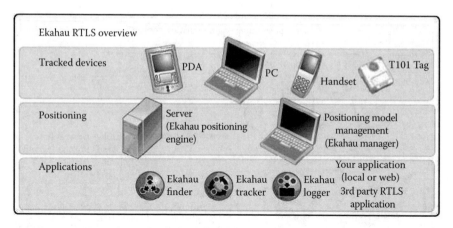

FIGURE 6.2 Ekahau RTLS overview.

Ekahau uses standard Wi-Fi signals to estimate locations, rather than time-based measurements that require proprietary infrastructure and can be expensive. Timing-based methods tend to work poorly in complex environments and are highly susceptible to errors due to common obstructions — such as the human body, walls, and furniture — that may block the line of sight from the receiver to the tracked objects. Deployment issues are discussed in Chapter 10.

Ekahau achieves accuracy of 1 to 3 m, with a site survey and calibration that requires up to 1 h/1200 m^2 (or 10,000 ft^2). Since RF prediction tools can provide only the best estimate of what the final network plan would look like, it is beneficial to have a combination of tools. Ekahau does provide the Ekahau Planner as an off-site planning tool and the Ekahau Site Survey (ESS) as an on-site verification and network optimization tool. We discuss how to deploy Ekahau in Chapter 10.

EKAHAU SYSTEM COMPONENTS

The Ekahau RTLS consists of the components in the following sections, and shown in Figure 6.2. Figure 6.3 shows the system components found in Ekahau.

Ekahau Client

The Ekahau Client is a network agent program that runs on a client device (PC laptop, PDAs, Wi-Fi tag, etc.), which is a device that is tracked by the system. Ekahau Client is a small service that runs in the background of each device to be tracked, negotiating with the Positioning Engine Server.

The Ekahau Client can also be embedded into devices such as VoWi-Fi handsets, Wi-Fi-based telemetry modules, and virtually any device that uses an 802.11 radio. The Ekahau Wi-Fi tags can be attached to a mobile object or asset, or carried by people. The Ekahau Client interrogates the client device Wi-Fi radio chip for received signal strength indicator (RSSI) values and simply communicates these data to the Ekahau Positioning Engine, which in turn calculates the location of the device the client is on. The amount of bandwidth and processing power consumed by the client

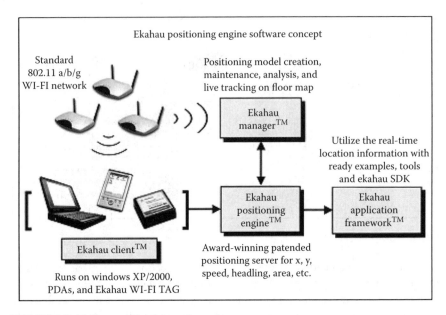

FIGURE 6.3 Ekahau positioning engine software components.

is minimal (tens of bytes per location request), thus making it very adaptable to any type of device while keeping network overhead at a minimum.

Ekahau T201 tags do not need any proprietary antenna or radio infrastructure on-site, but can work with any standard 802.11b and g AP brands, thus enabling pure Wi-Fi-based real-time tracking of people, equipment, and assets. In addition, unlike radio frequency identification (RFID)- or infrared (IR)-based solutions, the T201 tag is not required to be in the close proximity of a reader gate or scanner. This allows the T201 tags to be used for continuous location tracking as long as the tag is within the Wi-Fi coverage area, which can encompass an entire building or campus.

The Client's controller properties dialog can be used for:

- Viewing the Client's status
- Starting and stopping the Ekahau Client Service
- Configuring the allowed or restricted Positioning Engines
- Enabling or disabling the writing of an Ekahau Client log file

Ekahau Positioning Engine

Ekahau Positioning Engine 3.1 (EPE) is a Java-2 based platform and network-independent software architecture that runs on a desktop PC or (PC/Unix) server and calculates the client device (x, y), floor, heading, speed, area name, etc. The Ekahau Positioning Engine is the heart of the Ekahau system and where the complex location calculations are produced.

Ekahau Manager

Ekahau Manager is a stand-alone application tool for creating positioning models of the area, saving them in the Positioning Engine, drawing logical areas (or locales), testing live positioning, and analyzing positional accuracy. Logical areas are user-drawn areas (polygons) that can have a name (syntax-independent) and other location-based properties. Logical areas are used to determine whether a client device is within a given area.

Specifically, the Manager is for:

- Loading and configuring location maps
- Performing site calibration
- Restricting/allowing networks and APs
- Drawing and editing tracking rails (provide improved positioning accuracy)
- Displaying device locations and properties for administrative purposes
- Drawing and editing logical areas (used to determine whether a client device is within a given area — proximity)
- Saving and loading positioning models
- Sharing work by merging positioning models
- Analyzing positioning accuracy

Creating positioning models involves doing site calibration (collecting sample points every 3 to 5 m), which we outline in Chapter 8.

Ekahau Planner

Ekahau Planner is a planning tool for Wi-Fi network design and deployment. It can be used to intelligently plan initial AP network placement, even before visiting the installation site. A drag-and-drop graphical user interface (GUI) is included for access point placement on a facility floor map, with instant visual results on RF propagation and a variety of performance parameters. (Note that Ekahau Planner is part of the Ekahau Site Survey Tool.)

Ekahau Application Framework and SDK

This is a set of helpful tools and easy programming interface for authorized applications to quickly utilize EPE location information. The Ekahau Software Development Kit (SDK) contains the SDK Java package, Javadoc, and code examples for quickly connecting to the Positioning Engine and reading location information. Programmers using other than Java language can alternatively use the Ekahau YAX protocol (a socket-based prototocol) to read the location information via TCP socket. Both options utilize TCP sockets to connect to the Positioning Engine. Figure 6.4 shows how the various components interact.

The application layer consists of end-user applications for accessing the location information. The applications may be either custom built, provided by a third party, or one of three applications in the Ekahau Application Suite:

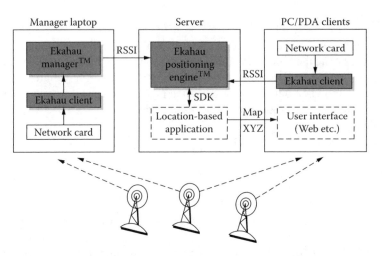

FIGURE 6.4 Ekahau system components.

- Ekahau Finder is a web application for querying the locations of people and assets from web browsers. The query results display the name of the area where the object is located and a view of the object on a map of the area.
- Ekahau Tracker is an application for tracking objects. It provides a real-time view of the tracked area with selected objects displayed graphically on a map of the area.
- Ekahau Logger is used for storing the historical tracking information to a database, for example, for analyzing the location data or recording procedures.

As an example, the Ekahau Finder application can be deployed on a server platform that runs the application, the application server, and the web server. When deploying Ekahau Finder, either the default application and web server that are delivered with the package or an existing application and web server can be used. To run Ekahau Finder, the Ekahau Positioning Engine needs to be installed, which provides the location information needed. Figure 6.5 shows the system requirements for this kind of application.

Another application/service example is the Hawk Tour project that we feature in Chapter 8, where the end-user application is a Java application that automatically presents a scalable map of the local area and links points of interest in the vicinity.

There are two ways for reading client location — (x, y), floor — speed, heading, and logical area information ("conference room"). These are using the Ekahau Java SDK or the language-independent Ekahau YAX TCP (if not using Java, the YAX protocol can be used to read the location info via TCP socket). For example, in the Hawk Tour application, the location coordinates are provided to the end-user application via location service built with Java and deployed to an Apache Tomcat server. This location service queries coordinates from Ekahau and provides them to the application using the Simple Object Access Protocol (SOAP).

System requirements

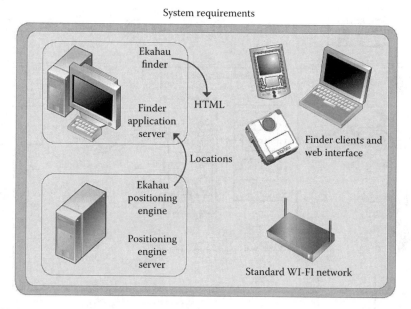

FIGURE 6.5 Ekahau system requirements.

Yet another example is the Accenture Positioning Enabled Data Management Solution, which combines the core knowledge of Accenture, Ekahau, and Solid. The Solid FlowEngine controls data distribution between PEDM DB and Devices (e.g., pushes data to terminal or synchronizes data) and posts positioning information events to PEDM Core Applications. The PEDM includes Core Applications as a base for Value Added PEDM Applications. Examples of core applications include route diagrams, rules to combine position, and data to assign (or de-assign) services to a mobile device, and compose data from different sources (e.g., from Enterprise DBs, document storages, or mobile devices). Accenture has interface components to many Enterprise Business Systems, e.g., SAP, Siebel, Oracle, and J2EE, and Microsoft technology-based systems.

A further example is the integration of Ekahau with IBM's WebSphere application server. WebSphere Everyplace Access (WEA) is a platform for mobile applications. Part of WEA is location-aware services (LAS) middleware, which provides API for application developers. Using LAS, developers can choose the most meaningful location provider for their application, freeing developers from learning yet another API for location data. WebSphere Everyplace Access has location-aware services middleware (Atlas) that provides standard API for various location providers and services (such as device position, geocoding, routing, mapping, etc.). Each location provider has implemented service adapters for their system. The Ekahau adapter has implemented a device position adapter interface. Location-aware portlet applications use adapters to get location data, and they can dynamically change location provider based on location, quality of service, etc. Portlets can be targeted for users who need location-aware services in their work or users who monitor WLAN clients or otherwise use the location of other WLAN clients, rather than

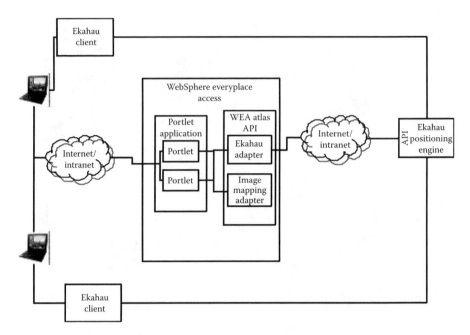

FIGURE 6.6 IBM WebSphere with location-aware service. (From Ekahau Positioning Engine: WLAN Location Services for WebSphere Everyplace Access, v. 0.2.1.)

their own. The WEA Atlas adapter utilizes Ekahau's own Yax protocol and allows portlet developers to use familiar API to implement portlet applications (Figure 6.6).

As an application server example, ESRI's ArcIMS geographic information systems (GIS) server can use the Positioning Engine to retrieve (x, y) positions to build indoor tracking applications, information services, and event-type emergency services applications. In addition, GIS enterprise users can also use Ekahau's Site Survey Software to plan and optimize Wi-Fi networks faster and deliver better connectivity for wireless data and voice. Ekahau's ability to extract location from any standard Wi-Fi network with a software-only solution allows existing GIS enterprise users to build a new breed of indoor GIS applications.

INTEL PLACE LAB AND SKYHOOK WPS

Place Lab and Skyhook's WPS use the map-based pinpointing approach. Once a user has Wi-Fi, he will not have to buy extra hardware to use Place Lab, and the software can be downloaded for free from the Place Lab web site. Increasingly, laptops, cellphones, and personal digital assistants (PDAs) are being sold with Wi-Fi capability already installed for around an extra $30. The basic idea is that every wireless access point broadcasts a unique ID, which can be used to tell it apart from other access points. This identifier is pretty useless to most of us; however, it can also be used as a landmark to identify a particular location. All your computer needs is a sort of address book, so when it sees access point A, it will be able to match it with the physical location.

These systems have no line-of-sight requirements, are accurate to within 20 m, and can be used indoors or outdoors to determine location in seconds. Skyhook does not need to pay for the network of millions of access points it uses. WPS is an all-software solution, which means it can be rapidly distributed and deployed. WPS takes advantage of the fact that almost all access points are configured to transmit their unique IDs whenever a Wi-Fi client sends out a scan request. Skyhook exploits this capability (which does not drain the resources or compromise the security of access points) by combining it with the data collected by war drivers, who use global positioning systems (GPS) and laptops to discover and map out access points in towns and cities. Once installed on a laptop or mobile device, the WPS software measures the time it takes for signals from nearby access points to return a scan request, and then uses triangulation to pinpoint the location. Because WPS is interoperable with any application that can use GPS, it can be put to use right away in applications such as Microsoft Streets and Trips, and Delorme Street Atlas. The software, which is now available to application developers and device manufacturers, uses 802.11 radio signals emitted from wireless routers to determine the precise location of any Wi-Fi-enabled device, whether it be a PC, laptop, PDA, Tablet PC, Smartphone, or RFID tag. The way it works is that the company has compiled a database of every wireless access point in a given a city. It did this by having people literally drive the streets "listening" for 802.11 signals. Using the unique identifier of the wireless router, it notes in the database where the access point is located. When a mobile user running the Skyhook client pops up in a neighborhood, the software scans for access points. It then calculates a user's location by selecting several signals and comparing them to the reference database. The more densely populated the area is with Wi-Fi signals, the more accurate the software is at locating the device. The precise location of any Wi-Fi-enabled device, including a PC, laptop, PDA, Tablet PC, Smartphone, or RFID tag, can be determined without adding any new hardware. WPS includes a reference database of over 1.5 million private and public access points along with their locations within these metropolitan areas. The WPS client software utilizes this reference database to calculate a device's location to within 20 to 40 m. Vendors of location-based services and applications can now appeal to a large number of users and devices by adding a reliable and easy-to-use location system without having to alter their systems or provide new hardware. The main difference between Place Lab and the other Wi-Fi-based positioning systems is that Place Lab did not base its concept on the limited-extent venues, where either the access point locations were known or extensive radio site surveying was critical (e.g., Ekahau, radar).

Like radar, Place Lab uses a device's embedded 802.11 interface, but does not rely on precalibrated fingerprints. Instead, it predicts location via the known positions of the access points detected by the device. The positions of these access points are provided by a database cached on the same device; this cache in turn can be filled from a variety of access point databases. This means that the Wi-Fi radio surveys rarely contain enough data from any one location to compute meaningful statistics, thus eliminating the possibility of the sophisticated probabilistic methods used in Haeberlen et al. (2004), Krumm and Horvitz (2004), and Ladd et al. (2002).

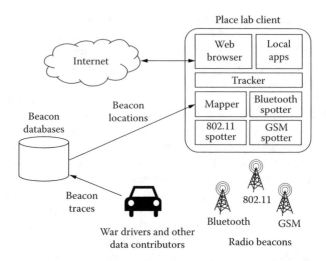

FIGURE 6.7 Key components in the Place Lab architecture. (From Place Lab: Device Positioning Using Radio Beacons in the Wild.)

A problem for 802.11-based location systems is the lack of 802.11 in less populated areas. To address this, Place Lab uses Global System for Mobile Communications (GSM) cell towers and fixed Bluetooth devices, as well as 802.11 access points. These systems can provide accuracy to within 20 to 30 m. But with improved algorithms that take into account, say, the height of the base station above the ground, or the building materials in the vicinity, GPS-like accuracy is possible in an area as densely served with Wi-Fi as downtown. And because Wi-Fi routers are often deployed closer together in cities than cell towers are, it can also be more accurate than cellular-based location systems. However, Wi-Fi routers do not necessarily stay in one position forever, like satellites or cell towers do. People often move and take their Wi-Fi routers with them. Skyhook software is able to detect when a Wi-Fi access point has been moved, and it makes note of the change in the database. Once a year the company plans to update the database by taking to the streets again to find new Wi-Fi signals.

A key observation here is that many developed areas have wireless hot spot coverage so dense that cells overlap. By consulting the Place Lab directories, which will map wireless hot spot multiple-access control (MAC) addresses to physical locations, mobile computers and PDAs equipped with Wi-Fi can determine their location. Furthermore, by keeping cached copies of these directories, computers can calculate their locations locally without transmitting information to any other computer. At this basic level, if one's personal device does not use the same wireless hot spot for communication, then this technique yields a totally private positioning technology. In these cases, one can still interact with cached offline content.

The Place Lab architecture consists of three key elements (Figure 6.7):

1. Radio beacons in the environment
2. Databases that hold information about beacons' locations
3. The Place Lab clients that use these data to estimate their current location

Place Lab was built using open-source and other redistributable components. The license agreements for all these components can be obtained from the Place Lab web site (http://placelab.org/toolkit/doc/complic.html). These components include the Hypertext Transfer Protocol (HTTP) proxy, written by Johannes Plachy; JDBM, developed by Alex Boisvert; and the Bluetooth Spotter, which uses BlueCove (http://sourceforge.net/projects/bluecove/), an open-source implementation of JSR 82. Communication with serial GPS units is handled through rxtx (http://rxtx.org/), Hsqldb (http://hsqldb.sourceforge.net/), and MySQL Connector/J (http://www.mysql.com/products/connector/j/). Apache Axis (http://ws.apache.org/ axis/) and the Xerces2 Java Parser (http://xml.apache.org/xerces2-j/) are used to interact with SOAP web services. The Place Lab distribution also includes SWT and the J9 run time, both from IBM.

How It Works

Radio Beacons

Place Lab relies on commodity hardware such as 802.11 access points and 802.11 radios built into users' devices (commodity laptops, PDAs, and cell phones) to provide client-side positioning. Positioning works in that a client device listens for the cell IDs of fixed radio beacons in their area — wireless networking sources like 802.11 access points, fixed Bluetooth devices, and GSM cell towers (collectively referred to as radio sources' *beacons*) — and uses a precomputed map (a DB with cached physical locations of radio sources in the area) to localize itself by means of referencing the beacons' positions in a cached database. Specifically, the beacons employ protocols that assign beacons a unique or semiunique ID. For example, 802.11 APs and GSM cell towers typically broadcast a beacon message that includes a unique identifier (the access point's MAC address or the cell tower's cell ID). Mobile devices associate the MAC address or the cell ID of the Wi-Fi AP or cell tower, respectively, that they connect to and look up the position in a MAC address or cell ID DB. Hearing this ID greatly simplifies the client's task of calculating her position. Figure 6.8 shows how Wi-Fi positioning uses client-received AP MAC identifiers and signal strengths to privately compute a position, which may also have user-meaningful places associated with it.

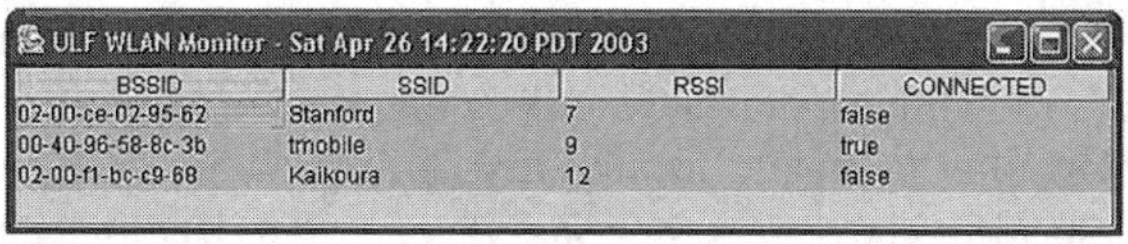

BSSID	SSID	RSSI	CONNECTED
02-00-ce-02-95-62	Stanford	7	false
00-40-96-58-8c-3b	tmobile	9	true
02-00-f1-bc-c9-68	Kaikoura	12	false

ULF WLAN Monitor - Sat Apr 26 14:22:20 PDT 2003

FIGURE 6.8 Wi-Fi positioning using Place Lab. (From Challenge: Ubiquitous Location-Aware Computing and the "Place Lab" Initiative.)

Mapping Service

Place Lab clients compute their location entirely on their own devices, but rely on a mapping service to provide them with (segments of) a database of known beacon locations. (While the Place Lab mapping service was initially deployed as a centralized service, concerns about privacy and organizational control have required transitioning to a decentralized architecture.) When a new Place Lab client starts up (or when an existing client moves to a new region), it downloads to its local disk a segment of the "mapping database" for its region (typically its surrounding metropolitan area or even the entire state). Whenever a user turns on her notebook in the presence of a beaconing access point, the Place Bar looks up the MAC addresses of nearby hot spots in a cached directory and determines the user's location. The Place Bar can then connect users to web content relevant to that location, such as a restaurant, hotel, or microtravel blog sites; services, like proximate Yellow Pages; and a wave of place-enhanced pages, such as a matchmaker, that compares your likes and dislikes with those of other people in the same hot spot zone.

Coverage and accuracy are dependent on the number and type of beacons in range of the client device. Place Lab devices need only interact with radio beacons to the extent required to learn their IDs. Clients do not need to transmit data to determine location, nor do they listen to other user's data transmissions. In the case of 802.11, receiving beacons can be done entirely passively by listening for the beacon frames periodically sent by access points. These beacon frames are sent in the clear and are not affected by either WEP or MAC address authentication. Other technologies, like Bluetooth, require clients to initiate a scan in order to find nearby beacons. Due to restricted programming interfaces, detecting GSM cell IDs requires handsets to associate with nearby cell towers, as they normally do when carried around and not on a phone call. Some urban areas already have wireless hot spot coverage so dense that cells overlap.

Due to the minor amount of calibration required, Place Lab's location estimates are less accurate than those produced by systems like radar. In residential and urban settings with GSM coverage and moderate 802.11 density, Place Lab produces location estimates with 20 to 25 m of accuracy, almost a factor of 10 worse than systems that maintain a more detailed radio map.

Location Databases

A precondition with Place Lab is the fact that a database with beacons' positions must be available, which serves the beacons' geographical location coordinates (e.g., World Geodetic System 1984, or WGS84) to client devices. If Place Lab knows nothing about a beacon, being in range does not improve its location estimates. Place Lab requires the following capabilities: routing data to and from the individual Place Lab servers, replicating to ensure availability of the data, and efficiently gathering data relevant to specific geographic regions. Place Lab's beacon database is formed by merging together publicly available information about beacon locations *and* user-contributed data collected by *war driving*. War driving is the process of using software on Wi-Fi- and GPS-equipped mobile computers and driving or walking through a neighborhood collecting traces of Wi-Fi beacons in order to map out the locations of Wi-Fi access points. A typical war drive around a single city neighborhood takes less than an hour. Contrast this with the typical calibration time for a single in-building positioning system, which can require many hours of careful mapping. War driving is already a well-established phenomenon, with web sites such as wigle.net gathering war drives of over 1.4 million access points across the entire U.S. Each day the RF beacon DB is improved by submissions from thousands of clients all over the world. The service compares the records in each submission and discards and statistically combines entries. The service is continuously updated, combined from a huge number of sources, and made instantly available offline or by high-speed servers. Multiple beacon databases are allowed. Place Lab does not specify whether beacon databases are private or public, how clients authenticate with a database, or how many databases a client should load data from. At the time of writing, Place Lab clients have had access to location information for approximately 2.2 million radio beacons (most of which are 802.11 access points). These mostly come from wigle.net. Other sources include databases for the University of California–San Diego and the University of Washington, as well as some GSM tower locations imported from the Federal Communications Commission's (FCC) database. Intel Research also maintains its own database that, at the time of writing, has had location estimates for around 40,000 GSM, 802.11, and Bluetooth beacons. The limitation of this is that even if large maps were gathered, users may move outside the mapped area (this goes against the principle of always-on service). If, in the future, such mapping databases are well populated, a client-based location service such as Place Lab would be suitable. Or, if a client device knows the physical location of many radio beacons, it can triangulate its own position by listening for the set of beacons that are transmitting in its immediate vicinity.

Place Lab Clients

Clients use live radio observations and cached beacon locations to form an estimate of their location. The client component characterizes radio and uses DB to instantly provide longitude and latitude. For privacy and reliability, large portions of the radio beacon database will be cached on clients. To make the client both extensible and portable, client functionality is broken into three logical pieces: spotters, mappers, and trackers.

Spotters are the eyes and ears of the client and are responsible for the observing phenomenon in the physical world. Place Lab clients typically instantiate one spotter per radio protocol supported by the device. For example, a laptop running Place Lab might have a Bluetooth and an 802.11 spotter, while a cell phone might run a Bluetooth and a GSM spotter. The spotter's task is to monitor the radio interface and share the IDs of the observed radio beacons with other system components. An observation returned by a spotter is of little use if nothing is known about the radio beacons.

The job of the *mapper* is to provide the location of known beacons. This information always includes a latitude and longitude, but may also contain other useful information, like the antenna altitude, the age of the data, a learned propagation model, or the power of the transmitter. Mappers may obtain these data directly from a mapping database, or from a previously cached portion of a database. This cache can contain beacons for a large area or may, due to capacity concerns, just contain information for a single city.

The *tracker* is the Place Lab client component that uses the streams of spotter observations and associated mapper data to produce estimates of the user's position. The trackers encapsulate the system's understanding of how various types of radio signals propagate and how propagation relates to distance, the physical environment, and location. Trackers may use only the data provided to them by the spotter and mapper, or may use extra data like road paths and building locations to produce more accurate estimates. For example, Place Lab includes a simple tracker that computes a Venn diagram-like intersection of the observed beacons. This tracker uses very few resources, making it appropriate for devices like cell phones. Place Lab also includes a Bayesian particle filter (Doucet and de Freitas, 2001) tracker that can utilize beacon-specific range and propagation information. While computationally more expensive, the Bayesian tracker provides about a 25% improvement in accuracy and allows Place Lab to infer richer information, like direction, velocity, and even higher-level concepts, like mode of transportation (walking, driving, etc.). For more information on the intricacies and advantages of using probabilistic Bayesian filters in location systems, see Hightower and Borriello (2004) and Patterson et al. (2003).

Communication Protocols

Place Lab supports five ways of communicating location information to applications: direct linking, daemon, web proxy, JSR 179, and NMEA 0183. With direct linking, applications may link against the Place Lab Java library and invoke a single method to start the location-tracking service. Daemon is for lighter-weight interactions. Place Lab can be run in daemon mode and applications can query Place Lab via HTTP. This HTTP interface allows programs written in most languages and styles to use Place Lab. Regarding web proxy, Place Lab supports location-enhanced web services by augmenting outgoing HTTP requests with extension headers that denote the user's location. By setting their web browser to use the Place Lab daemon's web proxy (in the same way one uses a corporate firewall's proxy), web services that understand our HTTP headers can provide location-based service to the user. See more on web proxies in Chapter 8. JSR 179 is used to support existing Java location-based applications. This specification defines a J2ME Optional Package that enables mobile location-based applications for resource-limited devices. The API is designed to be a compact and generic API that produces information about the present geographic location of the terminal to Java applications. This API covers obtaining information about the present geographic location and orientation of the terminal and accessing a database of known landmarks stored in the terminal. The location API for J2ME is designed as an optional package that can be used with many J2ME profiles. The minimum platform required by this API is the J2ME Connected Limited Device Configuration (CLDC) version 1.1 (v1.1). The API can also be used with the J2ME Connected Device Configuration (CDC). (Note that due to using the floating point data types in this API, it cannot be used with CLDC v1.0.)

Classes pertain to address information, coordinates, landmarks, location, orientation, proximity, etc. The `AddressInfo` class includes both outdoor- and indoor-related address fields. Outdoor ones include city, county, zip code, etc. Indoor fields include building floor, building name, building room, and building zone.

The Landmark class represents a landmark, i.e., a known location with a name. A landmark has a name by which it is known to the end user, a textual description, coordinates, and optionally address info. The Location class represents the standard set of basic location information. This includes the time-stamped coordinates, accuracy, speed, course, and information about the positioning method used for the location, plus an optional textual address. The Orientation class represents the physical orientation of the terminal. Orientation is described by azimuth to north (the horizontal pointing direction), pitch (the vertical elevation angle), and roll (the rotation of the terminal around its own longitudinal axis). It is not expected that all terminals will support all of these parameters. If a terminal supports getting the Orientation, it must provide the compass azimuth information. Providing the pitch and roll is optional. Most commonly, this class will be used to obtain the current compass direction.

The `ProximityListener` interface represents a listener to events associated with detecting proximity to some registered coordinates. Applications implement this interface and register it with a static method in LocationProvider to obtain notifications when proximity to registered coordinates is detected. This listener is called when the terminal enters the proximity of the registered coordinates. The proximity is defined as the proximity radius around the coordinates combined with the horizontal accuracy of the current sampled location. With NMEA 0183, Place Lab provides a virtual serial port interface that can mimic an external GPS unit by emitting NMEA 0183 navigation sentences in the same format generated by real GPS hardware.

MICROSOFT RESEARCH RADAR

Radar is an in-building location-aware tracking system based on the 802.11 Wave-LAN wireless networking technology. It allows wireless LAN-enabled mobile devices to compute their location based on the signal strength of known infrastructure access points (APs). RF-based systems mainly use IEEE 802.11 wireless LANs and attempt to deal with the noisy characteristics of wireless radio; multipath fading is the major reason for this noise. Radar is probably the first example of a positioning system using an IEEE 802.11 network. The system leverages the existing wireless LAN infrastructure of the building, without setting up additional location-tracking components. The system calculates the position coordinates of a device either by empirical methods based on comparison (fingerprinting), with previously measured locations mapped on a radio map (in essence, comparing the current measurement of the device with a database of previous measurements to see if it matches), or by using a mathematical model of indoor radio signal propagation (the map is measured in advance for its radio propagation properties). Radar overcomes the noisy environment by creating a radio map in the offline phase by collecting signal strength samples for each user location and orientation. In the real-time phase, each access point measures signal strength of the mobile terminal and searches through the radio map database to determine the location of the mobile terminal. The radio map-based method is an empirical method. A mathematical method using a mathematical model of indoor RF propagation and floor layout information instead of a radio map has also been proposed for radar. Hence, the system implements a location service utilizing the information obtained from an already existing RF data network. The same network is used for the control signaling (e.g., transmitting the database information). The system is able to estimate a user's location to within 2 to 3 m (about the size of a typical office room) of his or her actual location (with 50% probability). This is achieved using radar's scene analysis. Another implementation of radar uses lateration, which has a 4.3-m accuracy at the same probability level. Although the scene analysis version provides greater accuracy, significant changes in the environment, such as moving metal file cabinets or large groups of people congregating in rooms or hallways, may change the radio fingerprint and force the system to reconstruct the predefined signal strength database, or even create an entirely new database. In addition to the user's location, radar can also record the direction and orientation (one of north, south, east, or west) that the user is facing at the time the measurement is made. Also, radar allows for the user to indicate his or her current location by clicking on a map of the floor. The user's coordinates (x, y) and time stamp are recorded (together with a measurement of the current radio fingerprint). In case privacy is a concern, the architecture of radar enables a mobile device to track its own location without other nodes in the system being aware of it, since it is only reusing the signal from the wireless LAN transmitters. In the extreme, a mobile device can turn off data connectivity and use its wireless interface (in conjunction with radar) solely for the purpose of tracking its own location (provided the database of fingerprints has been downloaded first). Other than the signal strength values derived from beacons, the mobile device only needs the radio map and the layout map of the building, which it can download once, e.g., the first time it enters the building.

Some of radar's limitations are the following:

1. Object tracking. It must support a wireless LAN (the network interface card (NIC)), which may be impractical on small or power-constrained devices.
2. Effect of multifloored buildings (or three dimensions). Signal aliasing between points on adjacent floors could cause the system to place the user on the wrong floor.
3. Signal generated by the mobile host might be obstructed by the user's body. While this may not be realistic given the antenna design and positioning for existing wireless LANs, it may be possible to approximate an ideal case with new antenna designs (e.g., omnidirectional wearable antenna).
4. Interference from other radio sources. The radio map database assumes that the signal from the Wi-Fi base stations does not change. However, either interference from other transmitters, or moving objects in rooms can change the signal. While it takes a great deal to invalidate the database, it is not impossible to do so.

ROSUM TV

Rosum's technology uses high-power, high-bandwidth, unmodified analog and digital broadcast TV signals to determine the position of a receiver. Receivers may be stationary or mobile and may be integrated into pagers, mobile telephones, PDAs, computers, vehicles, and people-tracking devices. Due to the high broadcast power level and relatively low frequency, TV signals penetrate deep into buildings and structures. Rosum technology could therefore be used to monitor the location of high-value assets or people, including parolees, soldiers, police, or even firefighters, as they go inside buildings.

All standard TV signals contain synchronization information that Rosum uses to measure the timing of the TV signal, a key step in determining the receiver's location. The Rosum system employs a distance measuring technique known as "multilateration," where signals from three or more transmitter sites, with known fixed coordinates, are measured by a remote device (the device being tracked). Distances to each transmitter site are then calculated, simultaneous equations are solved, and the location of the remote device is then determined. (The Rosum system, technically speaking, does not use triangulation where angles, rather than distances, to transmission sources are determined.) For the ATSC digital TV standard, Rosum uses the field synch and segment synch information; for NTSC, Rosum uses information included in the Vertical Blanking Interval, including the ghost cancelling reference (GCR) signal. In order to position, the receiver measures pseudo-ranges from TV signals broadcasted from three or more different TV towers. Using precision timing, it can be figured out how far a TV signal travels before it is picked up by a device equipped with the chips. Next, at the Rosum location server, the measurements are compared against other data that are collected by proprietary listening stations, and the device's position is calculated.

However, the technology does not provide reliable vertical information. This is due to the geometry of the TV towers; that is, signals arrive at the receiver from the horizon, and the difference in TV tower height is not sufficient to provide a meaningful vertical component to location. It can locate a person in an office tower but cannot determine what floor she is on unless the building is equipped with a Rosum pseudo-TV transmitter, which augments the local TV signal environment. Rosum has developed a high-accuracy three-dimensional positioning system. This system is deployed locally at the site and provides high-accuracy (room-level) location and tracking in the local area.

Rosum TV-GPS is a hybrid positioning system, based on augmenting GPS with its TV positioning system. Rosum combines pseudo-range information from GPS satellites and TV towers alike to provide a seamless indoor/outdoor location system. Outdoors, GPS signals are used primarily, with GPS performance. Indoors and in urban canyons the system becomes TV only, and accuracy of 50-m RMS is provided.

Rosum has developed algorithms to support other global TV standards, including PAC, SECAM, DVB, and ISDB.

ROSUM TV COMPONENTS

The Rosum TV-GPS and TV-GPS Plus location systems are made up of the following components (Figure 6.9):

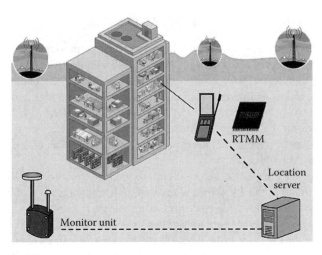

FIGURE 6.9 Rosum TV components.

- Rosum TV Measurement Module (RTMM) (positioning module contained in the positioning device)
- Location server (deployed at service provider site)
- Monitor unit (typically three or four per large metropolitan area)
- Communication channels (cellular SMS and packet data, Internet; proprietary radio for TV-GPS Plus)
- Pseudo-TV transmitter (TV-GPS Plus applications only; typically five or six per deployment)

ROSUM TV MEASUREMENT MODULE

The RTMM is a chip set that can be integrated into the user device to be located or tracked. The RTMM contains a TV tuner module, a digital signal processing module, and other supporting glue logic and memory. The radio chips and the modules (size is of a matchbook) that house them are placed into mobile devices. One of the main computational tasks of these devices is to filter out reflected signals from the direct path signal. These ghost images must be filtered out in order to provide reliable ranging to TV transmitters. This $40 component (the price is expected to go down when produced in large volume) receives the synchronization signal used by analog TVs to stabilize their vertical hold. Location servers installed in a building use this synchronization signal to determine the distance that the signal travels. Rosum's technology also depends on the installation of several nearby "monitoring units" which have sensitive antennas that monitor the timing and stability of TV signals and send calibration information to the location servers.

There are two versions of this chip set: one is DSP based and another is application-specific integrated circuit (ASIC) based. The RTMM receives aiding information from the location server and sends back to the location server pseudo-range measurements for TV channels in the area. The RTMM has been designed to process all standard analog and digital terrestrial broadcast TV signals based on simply changing the internal description of the signal(s) to be processed. The RTMM is common to both the TV-GPS and TV-GPS Plus systems.

LOCATION SERVER

The location server receives monitor data from monitor units in the region and forwards that information, in the form of aiding information, to RTMMs. The location server's primary role is to compute the position of each RTMM based on its pseudo-range measurements. Once a position has been computed, that information is either sent back to the RTMM to be displayed on the mobile device or is forwarded to a tracking application or service provider. The location server may be hosted by Rosum, or it may be hosted by the service provider or the end customer, depending on the application requirements and needs.

MONITOR UNITS (REFERENCE STATIONS)

The monitor unit is a small device that is deployed in a fixed location in the region in which RTMMs (and their host location devices) are deployed. Each monitor unit can be implemented as a small unit including a transceiver and processor, and can be mounted in a convenient location such as a utility pole, TV transmitters, or base stations. The role of the monitor unit is to use its highly sensitive antenna to monitor TV signals in the area, to analyze the stability and timing of these signals, and to report this information back to the location server.

PSEUDO-TV TRANSMITTER

Pseudo-TV transmitters (PTTs) are an option system component that are used to enhance the TV signal environment beyond what is provided by existing TV towers. PTTs can be used to increase accuracy and availability in the target deployment area. PTTs are deployed locally, around a building, campus, theater of war, or metropolitan area to infiltrate buildings or areas with additional TV signals that the RTMM can detect. PTTs broadcast standard or proprietary TV signals on unused TV channels as permitted by a FCC licensing agreement.

Rosum calls this system TV-GPS Plus. The system components for a typical TV-GPS Plus deployment are shown in Figure 6.10. Note that the receiver used in the local area is the same wide-area receiver. Hence, the user transitions from wide area coverage into local area enhanced coverage seamlessly.

For even greater accuracy, it is possible to install PTTs around the outside perimeter of a building to provide position fixing within areas as small as a single room. This could be useful for tracking people inside burning buildings. Firefighters could temporarily set down a number of Pseudo-TV transmitters around a burning structure and see the location of the people trapped inside (provided they are carrying a compatible device).

COMMUNICATION CHANNELS

Rosum uses the following channels for intrasystem data communication. The Internet is used for monitor-to-location server communication; SMS and GPRS/1xRTT are

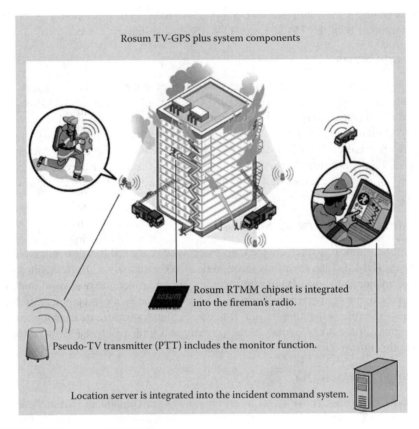

Rosum TV-GPS plus system components

Rosum RTMM chipset is integrated into the fireman's radio.

Pseudo-TV transmitter (PTT) includes the monitor function.

Location server is integrated into the incident command system.

FIGURE 6.10 Rosum TV-GPS components.

used for location server-to-RTMM communication. A proprietary-based mechanism for communication can also be used.

How It Works

The components of Rosum's TV signal-based positioning system are: a positioning chip contained in the MS, a location server, and a monitor unit (typically two to three per large metropolitan area). The communication channels include cellular SMS and packet data.

The positioning chip set can be integrated into the user device to be located or tracked. It contains a TV tuner module, a digital signal processing module, and other supporting glue logic and memory. It receives aiding information from the location server and sends back to the location server pseudo-range measurements for TV channels in the area.

The location server receives monitor data from monitor units in the region and forwards that information, in the form of aiding information, to the positioning chips. The location server's primary role is to compute the location of each positioning chip based on its pseudo-range measurements. (Note that the position calculation can be handset or network based.) Once a position has been computed, that information is either sent back to the positioning chip to be displayed on the MS or forwarded to a tracking application or service provider, as required by the specific application.

The monitor unit is a small device that is deployed in a fixed location in the region in which positioning chips (and their host MSs) will be deployed. The role of the monitor unit is to use its highly sensitive antenna to monitor TV signals in the region, to analyze the stability and timing of these signals, and to report this information back to the location server.

Pseudo-TV transmitters may be deployed locally to "light up" buildings or areas with TV signals that the positioning chip can detect. The result of using higher-powered local TV signals is greatly increased accuracy (room level).

Rosum TV-based positioning works by receiving, at the user terminal, a broadcast analog TV signal having a periodic component. It correlates the broadcast TV signal with a predetermined reference signal based on the periodic component, thereby producing a pseudo-range. The position of the MS is determined based on the pseudo-range and a location of the transmitter of the broadcast TV signal.

The TV transmitter location data may be stored at the MS or at the server. By one of a variety of tracking techniques, the MS measures the pseodo-ranges to each of a subset of visible transmitters. Pseudo-ranges to three spatially separated transmitters are sufficient to calculate the position (using the triangulation method). In order to compute the precise position of an MS, the precise timing of the TV synch code transmissions must be known. Monitor units at known locations are used to independently monitor the TV station clock offset. This clock offset information may be applied to the handset's or server's position calculation.

The following describes how the Rosum TV-GPS works. First, the monitor units analyze the broadcast TV signals and send channel stability and timing information to the location server. A mobile station either tracks the TV signals emitted by each TV transmitter and measures a pseudo-range for each of the TV transmitters, or transmits the measured pseudo-ranges to the location server. The location server then combines the information related to the transmission times and pseudo-ranges to compute the position of the mobile station. The position of the mobile station is then relayed back to the mobile station and used as positioning information for location-based service (LBS) applications.

In addition, each monitor unit measures, for each of the TV transmitters from which it receives TV signals, a time offset between the local clock of that TV transmitter and a reference clock. The reference clock could be derived from GPS signals. The use of a reference clock permits the determination of the time offset for each TV transmitter when multiple monitor units are used, since each monitor unit can determine the time offset with respect to the reference clock. Thus, offsets in the local clocks of the monitor units do not affect these determinations. The monitor unit function is incorporated into PTTs for TV-GPS Plus applications.

The A-GPS model is used for sending aiding data from the monitor unit to the server. The results are sent back to the handset or a third party. Potentially, the D-GPS model could also be used, which uses the monitor units that send aiding data back for transmission over the FM spectrum. A similar model, i.e., sending the aiding data over the TV spectrum, could be explored. This would involve a local-area monitor unit that would serve the role of the D-GPS reference station.

Second, the location server sends aiding information to the RTMM positioning module. This includes information on the set of channels the receiver should listen to, and information on where in frequency and time to find them. The information gathered at the receiver is not actually processed in the receiver. Rather, it is sent via a wireless connection to the Internet to a server that performs the necessary calculations to determine time of travel of each signal and then the corresponding distance. Stored in the server are the coordinates for each transmission site. The Rosum system clocks the time the signal leaves the transmitter and the time the signal arrives at the receiver in the mobile device. Because TV signals are not locked to a known time reference, the existing implementation of the Rosum system includes additional receivers at fixed locations which are timed to GPS. The receivers are small PCs. Each has two antennas, one for TV and one for GPS. The receivers scan the same frequencies as the mobile device and forward that data, as well as the GPS time stamps of when the signals were actually received, to the server. The server uses this information to calculate the timing offset used to make corrections in calculated transmission travel time of signals received by the mobile device. The monitor could be eliminated if a stable clock reference were added to each TV transmission (an estimated cost per transmitter would be around $5,000.) Another emerging technology could also be used to provide a stable reference, the RF watermark or transmitter ID signal making its way through the ATSC standardization process. (Candidate Standard CS/110A for the identification and synchronization of distributed TV transmitters.)

ROSUM HYBRID TV-GPS SYSTEM

To this point we have discussed only the Rosum TV system. Rosum TV-based positioning technology has been implemented as a hybrid system in that the mobile device includes both TV and GPS receivers. The Rosum system utilizes both TV and GPS signals and abstracts away the signal type so that if, say, two TV towers and two GPS satellite signals were seen, the system would generate a position fix. This hybrid technology will soon find a practical application, incorporated into Trimble's newest version of its TrimTrac locator system. Trimble will add radio positioning via TV signals to its GPS tracking system for enterprise-wide automobile fleets.

RTMM receives local TV and GPS signals, measures their timing, computes pseudo-ranges, and sends information to the location server. The following describes how the position of a mobile station is determined based on the pseudo-range and a location of the transmitter of the broadcast analog TV signal. The mobile station receives the analog TV signal and correlates signal with a predetermined reference signal, thereby producing a pseudo-range. Using TV signals, a pseudo-range can be obtained to the TV signal transmitter. When multiple such pseudo-ranges are known, and the locations of the transmitters are known, the position of the mobile station can be determined. Within the ATSC digital TV signal synchronization codes are used, such as the field synchronization segment and the synchronization segment within a data segment. Within the ETSI DVB-T and ISDB-T digital TV signals are scattered pilot carriers. Within the NTSC analog TV signal are the horizontal synchronization pulse, the horizontal blanking pulse, the horizontal blanking pulse and horizontal synchronization pulse taken together, the ghost cancelling reference signal, and the vertical interval test signal. However, in some regions, hills, buildings, other obstructions, or even the body of a user may block one of the TV signals. Alternatively, the mobile station may simply be located in a rural region too distant from the required number of TV transmitters. In such cases the remaining pseudo-ranges can be supplied using signals transmitted by mobile telephone base stations.

Rosum's hybrid strategy has led them to continue their work on incorporating additional signals into their system. Positioning is also possible using TV signals and mobile telephone signals. The TV signal from a TV transmitter is received at the mobile station, and a first pseudo-range between the mobile station and the TV signal transmitter is determined. Similarly, a second pseudo-range between the mobile station and the mobile telephone base station is determined based on the GSM signal received at the mobile station from a mobile telephone base station. The actual position of the mobile station is determined using both the first and second pseudo-ranges.

Figure 6.11 portrays the different positioning sources for either network- (server-) based or handset-based positioning.

Next, the location server uses the measurements from the mobile receiver and monitor units, whose locations are known, to calculate the position of the device. The location server then sends this location information back to the device (in the case of a navigation application) or to a tracking application. Figure 6.12 portrays this description.

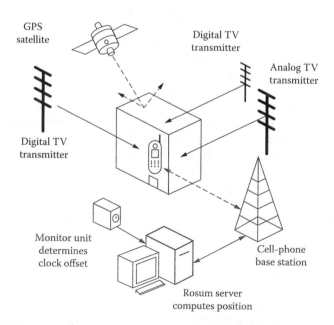

FIGURE 6.11 Network- and server-based positioning using Rosum TV.

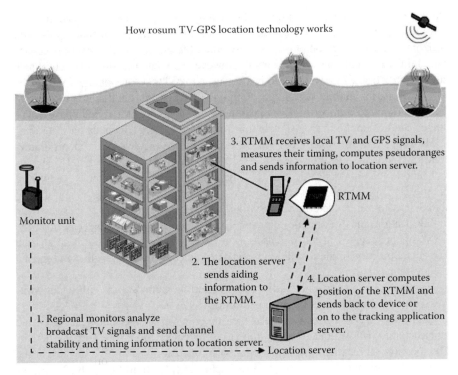

FIGURE 6.12 How Rosum TV-GPS location technology works.

AEROSCOUT

AeroScout's Wi-Fi positioning technologies use the wireless infrastructure to locate any standard 802.1b and g mobile station, including laptops, PDAs, bar code scanners, and RFID readers, in addition to battery-powered AeroScout tags attached to people or any other assets and equipment.

The AeroScout system uses both TDOA and RSSI algorithms to determine location, depending on the environment and application. TDOA is used for large, open, indoor environments, such as manufacturing, and outdoors; RSSI is used for tight indoor environments. For TDOA installations, AeroScout location receivers are used as long-range Wi-Fi readers; for RSSI installations, the same AeroScout location receivers or existing standard Cisco access points are used as readers.

AEROSCOUT COMPONENTS

Figure 6.13 shows AeroScout's system architecture and components.

AeroScout Wi-Fi Tags

The AeroScout Wi-Fi tags are long-life active RFID tags using the wireless LAN communications standard. AeroScout's T2 Tag model enables the wireless network infrastructure to locate people and assets otherwise not connected to a wireless network.

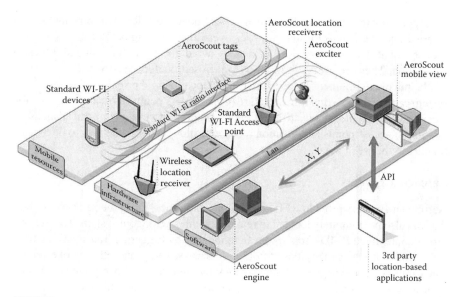

FIGURE 6.13 AeroScout system architecture.

Unlike traditional passive RFID devices, AeroScout's solution does not require dedicated readers, but rather allows the WLAN to act as a long-range reader, locating wireless devices and active RFID tags anywhere within the network.

AeroScout T2 Tags can be accurately located or detected by access points or readers at long range (several hundred feet). They can also be triggered at short range by an AeroScout Exciter, which triggers them to transmit as they pass through a defined area. Exciters have up to a 6.5-m (20-ft) range, enough to cover wide gate areas. In wider gates, up to eight exciters can be chained and operated synchronously as one exciter, thus covering gates of up to 275 ft (84 m).

These battery-powered tags have a long battery life (up to 5 years). Tag battery life can be extended further by switching them off automatically when they leave a defined tracking area through a gate or doorway.

The tags come with a wireless or serial programming interface for remote programming and telemetry applications. Through the use of a clear-channel assessment mechanism, they are able to avoid interference with standard Wi-Fi data communications.

AeroScout Location Receivers

The location receivers provide robust location measurement capabilities based on time difference of arrival (TDOA). The location receiver measures the time of arrival of standard 802.11b messages to the nanosecond, and sends that information to the AeroScout Engine to process and determine precise location. Each location receiver can process over 300 location measurements per second, enough to satisfy even the most demanding applications.

Approximately the size of a small access point, location receivers include an external antenna connector and a power-over-Ethernet adaptor. Location receivers can be connected to a wireless bridge to avoid the need for a wired Ethernet connection. They are available in NEMA-enclosed outdoor versions.

In indoor environments where RSSI-based location is appropriate, the AeroScout system can use the existing infrastructure (Cisco access points) as readers, so location receivers are not necessary. Mixed indoor–outdoor environments can also use a mixture of access points and location receivers, allowing the same tags to be located seamlessly between areas.

AeroScout Exciters

AeroScout Exciters provide RFID detection capabilities, using the same Wi-Fi tags that can also be accurately located in real time by the AeroScout system. The exciters provide integrated RFID choke-point functionality by triggering transmissions from the AeroScout tags as they pass through doorways, gates, and other choke points. The exciter can also trigger T2 Tags as they pass through a choke point to transmit a message that is received by an AeroScout location receiver. This provides instant knowledge that a tagged asset or person passed through a gate, doorway, or some other tightly defined area. AeroScout Exciters also enable tag telemetry functions, and can be used to modify tag behavior wirelessly. Telemetry functions provide the ability to use the exciter to store messages on the tag for later transmission. Message transmission can later be triggered by other exciters, enabling sophisticated process control functions. Exciters can trigger a tag to send (or store) up to 10 bytes of data and transmit 1 of 10 prestored messages.

AeroScout Engine

This is a core software component that manages the collection and processing of location data, including the use of sophisticated algorithms to combat indoor multipath. The AeroScout Engine (Positioning Server) uses TDOA or RSSI algorithms to determine location (1 to 5 m location detection accuracy with TDOA) of devices. The TDOA mode avoids the use of client-side software drivers, enables accurate outdoor and open-area location, and eliminates the need for radio map calibration processes, which we discuss in Chapters 4 and 8.

The AeroScout Engine includes the following main components:

- AeroScout Positioning Server: Core component in charge of the TDOA location and presence calculations and measurements designed to maintain various positioning schemes (see below).
- AeroScout System Manager: A built-in graphical installation tool that makes the system setup simple. Administration and configuration are centralized in the AeroScout System Manager and allow the importing of a variety of map file types (JPEG, GIF, and BMP), the setup of the system topology, and remote configuration of location receivers.

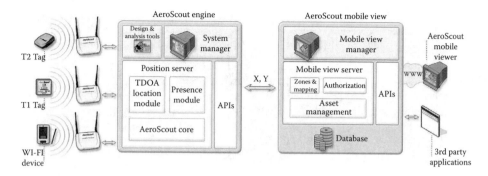

FIGURE 6.14 AeroScout software components.

- AeroScout positioning API: An open eXtensible Markup Language (XML) interface that application developers can use to create a wide variety of wireless location-based applications.
- AeroScout analysis and design tools: A set of sophisticated tools for system design, planning, and analysis. These include utilities such as system accuracy analysis, synchronization analysis, and system coverage reports for (pre-)installation purposes.

The above-mentioned components are portrayed in Figure 6.14.

AeroScout supports various positioning schemes (location modes) in a single system. These include absolute location based on TDOA, symbolic location based on RSSI, presence and choke point, and telemetry.

Figure 6.15 shows TDOA algorithms triangulating the precise location of Aero-Scout tags and Wi-Fi devices.

Figure 6.16 shows localization based on presence using active RFID tags. This is adequate for environments where detecting the presence of an asset within a set area (or lack thereof) is critical; a single location receiver or access point can detect

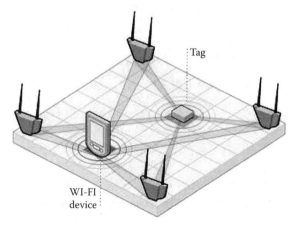

FIGURE 6.15 Triangulation using AeroScout tags.

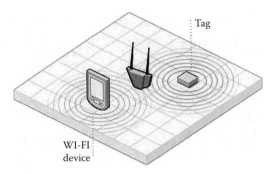

FIGURE 6.16 Localization based on presence using AeroScout tags.

tags and Wi-Fi devices. Examples include automated inventory counts and sorting cars on a lot by the zone in which they are located.

Figure 6.17 shows choke-point detection and telemetry: by using AeroScout Exciters, customers can detect a tag's passage through a defined area, such as a gate or doorway. This provides instant knowledge that an asset has entered or exited, and can be used to trigger events such as alarms, tag behavior modification, or telemetry data retrieval.

The following describes how the software works. The AeroScout Engine (or Cisco 2700) processes Wi-Fi messages coming in from the hardware and determines the raw location of a tag. These raw data will be of the form "Tag with Mac address 000 is at location 'X, Y' on map Z." These raw data are then sent to MobileView, where the intelligence is added. The MobileView end user has created zones within the map (using a web-based tool), has a database that associates each tag with a certain asset (and category of asset), and has created alerting rules based on asset location. In other words, in MobileView, the raw "Tag with Mac address 000 is at location 'X, Y' on map Z" is compared to the database and list of rules and becomes something usable, like "Infusion Pump 341 has just exited the patient rooms on the west wing of floor 4. Send an automatic pager alert to the floor 4 administrator on duty." MobileView also has a web-based user interface where users can search for

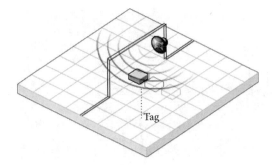

FIGURE 6.17 Localization based on choke-point detection and telemetry using AeroScout tags.

any asset (or asset category, or MAC address, etc.) and view where the asset is on a map. The maps have zoom and pan functions, and the assets show up as icons assigned by the user. MobileView also has advanced functions for reporting on asset locations, history, and events. Users can create a hierarchy of zones and subzones: the engine will determine an item's ("Tag A") location through either TDOA or RSSI with a certain level of accuracy. Let us say it determines that the item is at (x, y) coordinates (50, 50) on the map. Now, in MobileView, let us say that the user has created a zone ("Zone B") covering the square with corners (0, 0), (0, 100), (100, 0), and (100, 100). When the location report comes from the engine ("Tag A at (50, 50)"), MobileView compares Tag A with the database to see with which item it is associated ("Asset C"). MobileView puts Asset C on the map at (50, 50). Since that lies within the boundaries of Zone B as defined above, any alerts created by the user (i.e., "Send a text message whenever an asset enters Zone B") will be triggered. You can also create subzones; for example, Zone D could be the square between (25, 25), (25, 75), (75, 25), and (75, 75). Since this is all within the boundaries of Zone B, anything that applies to D will also apply to B (but not the other way around — an item at (10, 10) will be inside B but outside D). These zones are all created graphically with a drawing tool on top of the picture of the map in MobileView.

You configure the map in the engine by inputting the real-world locations of fixed receivers/access points. For example, if there is an access point at the far southwest corner of a building, you might input that access point at location (0, 0) on the map. Another AP 100 m due north could be entered at (0, 100) (depending on the scale you have chosen). Given the fixed locations of the readers/APs, the system algorithms determine the locations of tags relative to those fixed readers.

BLIP SYSTEMS

BLIP System's core product is BlipNet, which is a managed Bluetooth network offering access to LAN/WAN via Bluetooth. A connection can be established in three ways using BlipNet. One way is to register an SPP (Serial Port Profile) service on a BlipNode, the same way as registering an RMI service. When the service is registered, it is possible for clients in range of the BlipNode to search for the service and connect to the server using this service. This connection is obtained as the BlipNode informs the server application about the client, with an SPP event, which contains a BluetoothSerialPort for further communication.

Another way to establish a connection is to continuously send inquiries from the BlipNodes, to discover devices in range and then filter the Bluetooth device found, for example by searching for those capable of receiving objects. These inquiry events include the specific BluetoothAddress of the Bluetooth device and, when this is known, it is possible to use one of multiple protocols for establishing a connection.

The third and final way to establish a connection is to let the Bluetooth device establish a LAN connection and send an HTTP request for a specific URL. The BlipServer can then verify the device's BluetoothAddress based on such a request and pass connection data on to the server application. The Bluetooth device can establish a LAN connection using BlipNet, without any server side development, as this is a part of BlipNet.

BLIP Systems Components

A BlipNet network consists of four different kinds of elements: Bluetooth devices (e.g., mobile phones), the BlipNodes, a BlipServer (with BlipManager), and a server application. The BlipNet API provides an interface to the BlipServer for setting up the BlipNodes and establishing a connection to Bluetooth devices. When a Bluetooth device is in range of a BlipNode that is connected to a BlipServer, it is possible for both Bluetooth device and server to initiate and establish a communication link.

BlipNode

The BlipNode is a Bluetooth AP to which terminals such as mobile phones and PDAs may be connected for receiving information. It enables Bluetooth devices to access all services on a LAN. BlipNode requires a BlipServer to operate. The connection to the BlipServer can be established via local LAN, the Internet, or a Bluetooth link to another BlipNode. Access to devices and the BlipServer is enabled via a Bluetooth-enabled serial port.

A note about the performance of actions on a BlipNode: In case the application wants to change some of the settings in a node, or push objects via the node, the application must perform some extra steps to gain access to the node. These steps ensure that only one application at a time has the right to push to terminals, change certain configuration items, etc. When an application has acquired a BlipNodeHandle and invoked the lock method, that application has exclusive access to the physical BlipNode until the handle is released.

The BlipNode requires a Dynamic Host Control Protocol (DHCP) server on the LAN for allocation of IP addresses. After the BlipNode has received an IP address, it is possible to set the address to a static one via a Telnet command. If the BlipNode is applied for the LAP profile, a DHCP server is always required for allocation of IP addresses to the connecting terminals. The BlipNode acts as a DHCP relay agent. If the BlipNode is applied for the PAN profile, it is the connecting client's own responsibility to allocate an IP address. This implies that it is feasible to run without a DHCP server on the LAN, in this specific case, if both the BlipNode and the connecting devices are configured with a static IP address. A DHCP server is provided within a standard Red Hat Linux distribution or Microsoft Windows 2000 Server Edition.

BlipServer

The BlipServer is the core component in the BlipNet architecture. Figure 6.18 shows the BlipServer architecture.

The BlipServer configures, monitors, and controls the BlipNodes in the BlipNet. It provides a feature-rich open Java API (BlipNet API), which makes it possible for third-party developers to interface with the BlipNet and create custom applications. The API enables positioning of Bluetooth devices. It also provides the following:

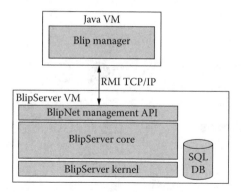

FIGURE 6.18 BlipServer.

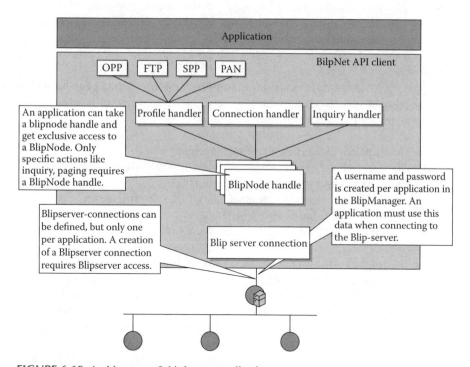

FIGURE 6.19 Architecture of third-party applications.

control of link establishment, pushing and receiving of objects or serial data, and the establishment of IP-based connections. The API is built on top of Java RMI to enable easy development of portable networking applications. Figure 6.19 shows the architecture of a third-party application.

Connection of an application to the BlipServer works as follows: The BlipNet API is IP based and can be accessed from remote PCs (depending on firewall settings). To ensure that the system administrator has full control of the BlipNet,

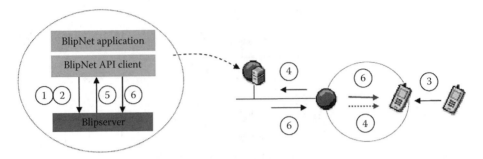

FIGURE 6.20 BlipNet API used for push of objects.

applications can only gain access to nodes if they have a password and a username. This ensures that unauthorized applications cannot connect remotely.

The BlipNet API is network based. The networking concept enables a range of applications that could not be constructed using a stand-alone system. Although the system is network based, it is of course still possible to write applications using a single node.

Figure 6.20 shows an example of usage of how the BlipNet API can be applied for the push of objects. The interaction between an application, BlipNet, and a terminal-entering range is portrayed:

1. The BlipNode is configured to track devices.
2. The application subscribes to information from the BlipServer.
3. The terminal enters the range of the BlipNode.
4. The BlipNode detects the terminal and informs the BlipServer.
5. The BlipServer informs the application.
6. The application invokes establishment of a link and pushes an object.

Via the BlipNet API, it is possible to start or stop a discovery procedure on a specific BlipNode. This feature can also be used for device-tracking applications. Via the BlipNet API, an application can subscribe to the tracking information. This information could be applied to open doors when users approach, showing user-specific content on a nearby screen when in range, etc. Via the BlipNet API, it is possible to start and stop inquiry on a specific BlipNode on the fly.

The inquiry responses will also contain the RSSI value from the found devices. As a method for establishing the best possible link between Bluetooth devices and BlipNodes, every BlipNode has the possibility to measure RSSI of devices in range. Whenever a Bluetooth device enables its Bluetooth service, it starts sending out signals to find other Bluetooth devices, and this triggers an Inquiry event on the BlipNodes, which contains the RSSI measurement of the signal received. These RSSI measurements are measured in integers in the range -90 to -20, where the higher values mean better signal strength. Alternatively, it is possible to measure RSSI over an ACL13 link, but this only gives values in the range of -10 to 20, and is affected by something BlipNet calls "a golden receiver range," which returns 0 in a certain range. This makes positioning using an ACL link very imprecise, as it is only possible to divide the measurements into three zones of length. Despite this, Blip Systems has already implemented positioning using this ACL connection, and it works to a precision of ±5 m.

There is also the BlipNet web API that gives access to the standard BlipNet API from a web-based application running in the BlipNet web container.

The connection procedure for a new application that wants to subscribe on events from the BlipNet is like this:

1. A login is defined via the BlipManager.
2. The login is inserted into the application.
3. The application constructs a BlipServerConnection using the login data entered in the BlipManager.
4. At this point the application can obtain all information and all events from the system.

IP traffic from PCs and PDAs accessing the LAN via Bluetooth bypasses the BlipServer and goes directly on the LAN. Serial data from devices like RS232 devices and MIDP Java applications on mobile phones are routed via the BlipNode to the BlipServer, where they can be accessed from third-party applications. OBEX data from devices like Mobile phones are routed via the BlipServer, where they can be accessed from third-party applications.

Figure 6.21 shows routing of data streams. IP traffic is routed directly to the LAN. Serial and OBEX data are routed to the BlipServer.

BlipServer modules include the mobility module, positioning module, push module, and beacon module.

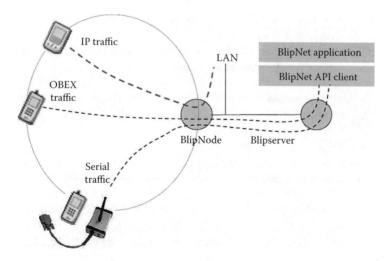

FIGURE 6.21 Routing of data stream.

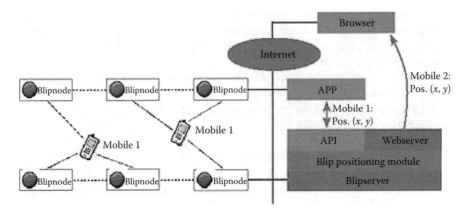

FIGURE 6.22 BlipNet positioning module.

The mobility module provides seamless connectivity. BlipNet connects automatically, handover between BlipNodes when required. The wireless BlipNode enables operation of BlipNodes without an Ethernet cable. Via this module, it is possible to configure the topology of the meshed network.

The positioning module is the Positioning Engine. Distance between BlipNode and the item to be tracked can be up to 100 m. The BlipNet positioning module is an extra module to the BlipServer. The module is configured via an easy-to-use graphical user interface. Installation and deployment are simple due to advanced calibration techniques. The positioning module requires the web module. Figure 6.22 portrays the positioning module.

The positioning module has these interfaces:

- A JAVA API, which enables easy integration between the Blip positioning module and other Java applications.
- HTTP- and XML-based APIs (via the web module) enable easy integration with other web-based applications.

The push module pushes content to devices, including web-based GUIs. Administrators can control the push of content in terms of:

- Content based on location
- Content as a function of time
- Content as a function of previously received content (flow)

The beacon module handles communication with BlipBeacons and, on mobile phones, the Beacon Tracker application. This module is required to implement a BlipNet Beacon solution. Figure 6.23 shows the architecture of the BlipNet Beacon. The Beacon tracker connects to the beacon module on the BlipServer.

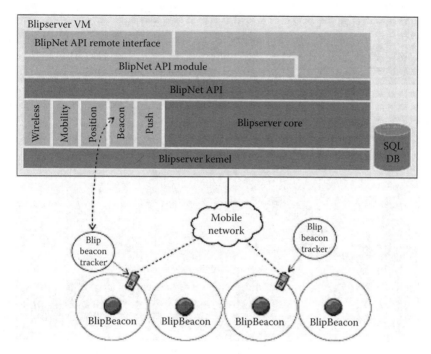

FIGURE 6.23 Architecture of BlipNet Beacon.

The web module consists of a web server integrated with the BlipServer. The module enables access to the BlipServer via HTTP and allows the writing of web pages or web applications that take full advantage of BlipNet. Functionality of the BlipNet API can be accessed from Java Server Pages (JSP), Servlets, Java Server Faces (JSF), or WebServices (WS), allowing BlipNet server-side web application development.

The BlipNet WebServices is a premade set of WebServices to configure and control BlipNet. This allows other WebService-enabled services, like existing web solutions, Microsoft.NET applications, and WebService-enabled mobile phones and PDAs, to use the services offered by BlipNet. Figure 6.24 shows the BlipServer architecture, including modules. The orange part of the figure is the optional modules. The green parts are third-party applications.

BlipManager

This is a GUI of the BlipServer for configuration and management of BlipNet. It connects to the BlipServer. The BlipManager allows configuration and monitoring of the entire BlipNet. The BlipManager connects to the BlipServer via TCP/IP. The BlipManager is implemented as a thin client. It does not have any local storage. All configuration changes are made directly to the BlipServer.

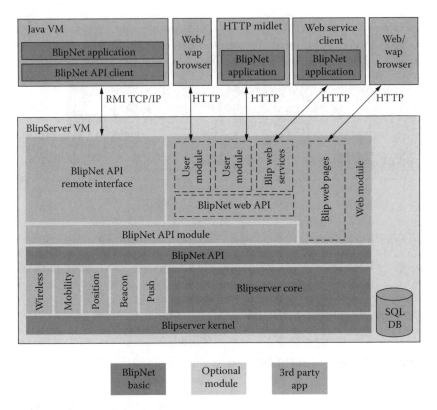

FIGURE 6.24 BlipServer architecture.

REFERENCES

Doucet, A. and de Freitas, N., Ed., *Sequential Monte Carlo in Practice*, Springer-Verlag, New York, 2001.

Haeberlen, A., Flannery, E., Ladd, A.M., Rudys, A., Wallach, D.S., and Kavraki, L.E., Practical robust localization over large-scale 802.11 wireless networks, in *Proceedings of ACM MobiCom'04*, Philadelphia, September 2004.

Hightower, J. and Borriello, G., Particle filters for location estimation in ubiquitous computing: a case study, in *Proceedings of the Sixth International Conference on Ubiquitous Computing (Ubicomp)*, Springer-Verlag, New York, 2004.

Krumm, J. and Horvitz, E., Locadion: inferring motion and location from Wi-Fi signal strengths, in *Proceedings of International Conference on Mobile and Ubiquitous Systems: Networking and Services (MobiQuitous'04)*, 2004.

Ladd, A.M., Bekris, K.E., Rudys, A., Marceau, G., and Kavraki, L.E., Robotics-based location sensing using wireless Ethernet, in *Proceedings of ACM MobiCom'02*, September 2002.

Patterson, D.J., Liao, L., Fox, D., and Kautz, H.A., Inferring high-level behavior from low-level sensors, in *Proceedings of the Fifth International Conference on Ubiquitous Computing (Ubicomp)*, Springer-Verlag, New York, 2003, pp. 73–89.

7 Modeling Location

Most positioning systems allow independent determination of position, yet this information must be communicated in order to be shared. It is difficult to exchange information between these systems today (being self-contained and vertically integrated). This would require data conversion, among other things. Nor is it possible to notify the applications on one system about information changes based on the information sensed by another system.

The world model described in Chapters 4 and 5 is both represented and queried by a data location modeling language. In principle, a location modeling language could equally well represent a symbolic (relative) position as a geometric (absolute) position. Existing standard languages, however, are mostly applicable to the geometric location model. Since they are based on the eXtensible Markup Language (XML), they are extensible and can be adapted for symbolic location models as well. The location modeling languages can be regarded as application programming interfaces (APIs) for the location data, which are also used as the communication exchange protocol for the information. They overlap to a very large degree and can be combined to create more precise or adapted models. Manipulating the models is often a matter of manipulating XML, which is out of the scope of this book, but we will touch on it at the end of the chapter and in later chapters.

LOCATION MODELING LANGUAGES

A location can be described in terms of a set of points (with probability intervals to compensate for the measurement error). In the most extreme case, it is a two-dimensional point on a surface (the surface of the Earth). However, the Earth is not a smooth globe, and in global positioning representation, there has to be a model to relate the position of the surface to what the surface is. There are other books specifically about this (e.g., Johan Hjelm's *Developing Location-Based Services*).

The points of a position can be connected to form polygons. These, in turn, can be used to describe other shapes. When using a geometric location data model, there are a number of shapes that can be used to represent a geographic area that describes where a mobile user is located. Additional shapes are required for more advanced location-based services.

But it is not sufficient only to manage shapes. The language also has to manage their relations; i.e., it is not sufficient to describe how to manage the data — you also have to manage the data model. Using a query language to decouple the application from the internal model representation allows the hiding of the different levels of detail of models, as well as the necessary model representations.

The shapes can easily be identified using high school mathematics. How to describe them in XML, however, is not automatically determined in the same simple way. There are a number of standards bodies for geographic data (and hence representation languages for position information), which can be used for advanced location-based services such as routing, geocoding, coordinate conversion, and map display. These bodies are the Open Mobile Alliance (OMA) Location WG (formerly the Location Interoperability Forum, or LIF), the Open Geospatial Consortium (OGC), and the International Standards Organization (ISO) TC211 working group. The current public XML specification defining geography from these groups is Geographic Markup Language (GML). In addition, the Scalable Vector Graphics (SVG) standard from the World Wide Web Consortium (W3C), and the GeoVRML language for virtual reality models from the XVRML consortium are used for the representation of maps.

An additional complication is that position measurements are not absolute. Measuring a local position can be done with higher precision than global positions, since the sensors are closer together and the environment is more controlled. Some location representation languages have functions to represent the inaccuracy of measurements, mostly in terms of probabilities. You define a shape, within which the position is located with a certain probability. The size of the shape depends on the measurement system. In global positioning systems (GPS), for instance, the probability that a terminal is to be found within 50 m of the positition given is almost 100%.

Languages for modeling location information must conform to a number of requirements, which come from their aspects as both representation languages and APIs to data models. Overall, the requirements for a modeling location language are:

- Model the world independent from a particular application.
- Describe the geometry of objects relative to different coordinate systems.
- Express symbolic information, i.e., names and descriptions of objects.
- Express certain relations (i.e., *inside, overlaps, includes, excludes,* and *closest*) between objects.
- Support the description of objects in different levels of detail/scale.

In our conceptual model, the world model is stored in a database. A modeling language is used to describe the model, and could also be used to query the model — if it has the appropriate functions. It is possible to model the location in one language and use the query language of the database to retrieve the individual values of the attributes. The Structure Query Language (SQL), for instance, has extensions that are intended to handle location models. These, however, cannot handle symbolic location models, only geometric models.

The Open GIS Geography Modeling Language can be used to describe properties of geographic features, which includes geometry. It can also be used to model symbolic information (attributes of features) and to encode spatial relations, which can be either symbolic or geometric.

MOBILE LOCATION PROTOCOL

The most frequently used API for mobile location systems is the Mobile Location Protocol (MLP). It is standardized by the Open Mobile Alliance (formerly by the Location Interoperability Forum, which became a part of OMA in 2003). Not strictly speaking a navigation language (and not a protocol either), MLP has elements that enable the positioning system to return values for direction and velocity. The DIRECTION element expresses position in degrees (with north as 0). The VELOCITY element expresses the speed of the mobile device in meters per second. Both are only present if the positioning method can be used to calculate it (positioning using cell ID does not make it possible to calculate the velocity, for example).

Location positioning can be of two types: relative (i.e., in the vicinity) or absolute (i.e., (x, y) coordinates). This use case is about absolute positioning since the LPS MIT Cricket gives (x, y) coordinate positions. However, other LPS are designed to represent position as relative with respect to other objects or entities.

In terms of the positioning infrastructure, the requirement is that the user is located (or tracked) in real time at all times when needed. Communication infrastructure providers can benefit from standardized location positioning protocols like the LIF MLP API. If the indoor and outdoor location worlds (or subsets of either world) use the same standard for modeling and encrypting user positions, this would enable seamless positioning handover from one region or zone to another, without the need for data format conversations and reference system transformations.

Moreover, the LIF MLP API enables integrating positioning based on microlocation with network positioning that allows for a centralized architecture, allowing a one-stop shop environment for developers.

LIF developed, and OMA maintains, the Mobile Location Protocol (MLP) for communication infrastructure providers and the software (services) and infrastructure providers (i.e., mobile application developers) to use positioning infrastructures.

The LIF MLP API is used to seamlessly integrate location from the communication network that is position enabled (i.e., Global System for Mobile Communications, or GSM, and Wi-Fi cell ID-based positioning). Specifically, this API serves as the interface between a location server and a location-enabling server, which in turn is interfacing with the application server.

The API describes the request–response that gathers position information (i.e., (x, y) coordinates) from the mobile positioning center (MPC)/GMLC servers. Moreover, it defines the core set of operations that a location server should be able to perform. The location-enabling server (middleware) effectively acts as an LIF-MLP pass-through, and subsequently passes on requests from the location-based service (LBS) application to the MPC/GMLC for location gathering. There is also a function to set a privacy flag in the Home Location Register (HLR) database. If the flag is set, the user has requested not to be positioned, and it can be overridden only by an emergency request.

There are two main functions for retrieving location information: immediate and deferred (for example, triggered by a timer). The LIF API assumes that the location information is delivered to the client (or the application server) as a result of a query.

The GEO_INFO element stores the position information. Position can be in (x, y) coordinates, for example.

The API is formally defined in a number of XML document type definitions (DTDs). They give the type definitions for XML elements that are to be sent in the different documents that comprise the messages. Because there are a number of common structured elements among different services, the DTD that defines a single location query service is composed only by the definition of the root element and the inclusion of the necessary common DTD. In effect, the DTDs define data structures for the Hypertext Transfer Protocol (HTTP) methods.

The LIF API defines a number of geometric shapes used in defining a geographical position. When requesting and reporting a position, the result can be a point with (x, y) coordinates. To assess the applicability of the LIF API for the indoor world, it is necessary to apply it to both absolute and relative positioning. Indoor LPS operating on absolute positioning like MIT Cricket and radar can supply (x, y) coordinate location for which the LIF MLP API can be used to model position location. However, LPS such as Active Badge give a symbolic location ("in Room 101"), and this is where the LIF MLP API might be lacking, in terms of modeling relative positioning.

To explore the matter of modeling relative positioning using the LIF MLP API, consider the following: Since it is rare to have a precise position like an (x, y) point coordinate, the LIF MLP API can describe position in terms of inaccuracy as a circle or a polygon (or some other shape), with a radius that describes the inaccuracy. Specific elements of the LIF MLP API that could be considered to model this include CircularArea, CircularArcArea, and Polygon. The CircularArea element can be used to define the Wi-Fi access points' range within which the user is positioned. These elements represent geometric shapes that have absolute positions (e.g., an arc is made up of points that have (x, y) coordinates), and hence are not suited to model symbolic locations like "User1 is in Room 101" as she is accessing the Wi-Fi access point (AP) that is located in Room 101. It would seem that symbolic locations like "in the room" or "in the hallway" would need to be defined for the LIF MLP standard to define relative positions in a common way.

The LIF MLP API can also be used to set quality requirements on the position information by using the LEV_CONF attribute, which indicates the probability in percent that the mobile station is located in the position area that is returned. It is a percentage value associated with the accuracy. If a location positioning system can determine that a user is in a circle sector that is long and narrow, by measuring the circle arc in which the user is located and finding that it is narrow and broad, or determining that the user is in a cell, there is a trade-off between the size of the inaccuracy area and the accuracy required. It is not very useful to know whether a user is within a certain cell if it is large, but it will help to know whether the user is somewhere in a circle sector.

The TIME element can be used in calculating position based on the time of arrival (TOA) geolocation method. The location can be one of four different types: current, last known, current or last known, or initial emergency call location.

In addition to the X and Y elements, LIF MLP API also includes the Z element for two dimensions.

The assumption is that the location-enabling server initiates the dialog by sending a query to the MPC, and the MPC responds to the query. There is also a second-use case for triggered requests, where the request is responded to during a period or while the user is in a certain area.

The protocol used is HTTP, and in the context of MLP the client is the LCS client and the server is the location server (GMLC/MPC). There are two ports: port 700 for secure transactions (using Secure Sockets Layer (SSL)/Transport Layer Security (TLS)) and port 701 for transactions without encryption (for cases where the system is operating within a trusted domain). The standard port for HTTP is port 80, but as these ports are unreserved, any function can be mapped to any port number.

There are two basic ways of requesting a position: an immediate request that is serviced immediately and a triggered/deferred request that can be triggered by a timer (perhaps several times or periodically).

The application server requests a position (called a location query service in the standard) by issuing an HTTP POST request toward the location server. The location query services are invoked by sending a request using HTTP POST to a URI. It is possible to send both immediate requests and deferred requests, where the result can be delivered later, and periodic requests, where the delivery of the location information is triggered at a predetermined time period (or position). The difference between a deferred request and a triggered request is that the trigger can be set off at a later occasion, whereas the deferred request should be fulfilled as soon as possible. The deferred request is also fulfilled only once, and the triggered request is fulfilled only when the trigger criteria are fulfilled.

If the request is a deferred request (triggered or periodic), the result is delivered to the client through an HTTP POST operation issued by the MPC. This situation implies that the client must be able to receive HTTP POST requests and be able to give a valid response. The response to the invocation of a location query service is returned by using an HTTP response.

The POST message contains a document, which in turn contains the requirements for the position information and the information about who triggered it. Triggers can be both the user requesting a position and an SMS being sent or received, as well as moving into another cell or making an emergency call. There are also possibilities for users outside the mobile network (for instance, a fleet management center) to request positioning not just for one mobile terminal, but also for a number of them (either by specifying the numbers or specifying a range of numbers).

The LIF API also has functions to manage the privacy flag in the Home Location Register. If the flag is set, the user has requested not to be positioned, and it can be overridden only by an emergency request. The status of the privacy flag is determined by using the PFLAG_STATUS element, which can be set in the status attribute to three different values: ON, which means no location requests can be made (although this setting can be overridden by emergency location requests); OFF, which means that the privacy flag is off and all location requests can be performed; and UNDE-FINED, which means that the operator can determine the status depending on the local legal requirements (for example, in some countries, the flag can be set to OFF by default except for emergency positioning). It is a unclear how this will be implemented in the 3G systems, however.

The LIF API is formally defined in a number of XML document type definitions (DTDs). They give the type definition for the XML elements that are to be sent in the different documents that comprise the messages. Because there are a number of common structured elements among the different services, the DTD that defines a single location query service is composed only by the definition of the root element and the inclusion of the necessary common DTD. The Mobile Location Position (MLP) is distributed over a set of common DTDs that define the core elements. In effect, the documents and DTDs together define data structures for the HTTP methods.

It is possible to trigger a location report when a mobile station enters or leaves an area. The area has to be defined by using the shape elements (more about that later) or a cell ID. Because the user is registered when he or she enters or leaves a cell anyway, it is more likely that that is how it will be used. (If the area is a polygon inside a cell, it will still require the position to be measured regularly to determine where in the polygon the mobile station is or if it has left it.)

Triggered reports can also be set based on timing. There is a special message element for both start and stop times for the request, which describes when the tracking of the user should start but can be sent at any time before the start (or stop) of the request.

The LIF API defines a number of shapes used in defining a geographical position. These go back to Chapter 4, where the different shapes and coordinate systems used to describe the Earth are discussed. When requesting and reporting a position, the result can be a point, but it is very rare that there are such precise positions. Instead, the position inaccuracy can be described as a circle (or some other type of shape, as discussed in Chapter 4) around the point with a radius that describes the inaccuracy. Requests can be made for different accuracies; the response will contain the accuracy that can be delivered.

It is possible to set quality requirements on the position information by using the LEV_CONF attribute, which indicates the probability in percent that the mobile station is located in the position area that is returned (this information can also be used by responding entities, it seems). It is a percentage value. This element might sound spurious, but it has to do with the accuracy. If a system can determine that a user is in a circle sector that is long and narrow, by measuring the circle arc in which he or she is located and finding that it is narrow and broad, or determining that he or she is in a cell, there is a trade-off between the size of the inaccuracy area and the accuracy required. It is not very useful to know that a user is within a certain cell if it is large, but it will help the user know whether he or she is somewhere in a circle sector. At least he or she will know where he or she is going. Needless to say, the accuracy will depend on the application. If the requesting system requires an accuracy that is higher than the best value it can give by using any other positioning method, it will have to return the cell.

The quality of position is not just dependent on the position, however, but also on the time when the position was measured. The RESP_REQ element can have three different attributes (in the resp_req_type attribute) that will set the quality level (nothing is said what quality level will apply if this attribute is missing). The three attribute values are NO_DELAY (where the initial or last known location of the user will be sent back), LOW_DELAY (where the current location with minimum delay is returned with a best effort to fulfill the requirements), and DELAY_TOL (where the response is delay tolerant). The default value is DELAY_TOL.

The location can be one of four different types — current, last known, current or last known, or initial emergency call location — which is defined in the LOC_TYPE element and which type is set in the loc_type_type attribute. It can have four values: CURRENT (which means the current location), LAST (the last known location), CURRENT_OR_LAST (current or last known location), and INI-TIAL (the initial location when the user makes an emergency call). All these have to be qualified by the time and accuracy of the position, of course.

To know what is current, last known, and so on, you need to have a rather precise idea about the time of your measurement. You can perform this task by using the TIME element, but there are several other elements that are used to describe timing in the LIF API. The RESP_TIMER element allows you to set the time in minutes and seconds for an interval during which the location should be obtained and returned to the requestor. The time of the requisition of a position measurement is declared by using the REQ_TIME element. This time does not have to be the same as the time the position was measured (which can be true, for instance, in the case of triggered positions).

Calculating velocity and direction is, as we noted in Chapter 3, rather tricky. There are elements in the LIF API, however, that enable the positioning system to return exactly those values. The DIRECTION element expresses position in degrees (with north as 0). It is only present if the positioning method can be used to calculate it. VELOCITY is the speed of the mobile station in meters per second (so this value might have to be recalculated to kilometers per hour, miles per hour, or knots as required). It is also present only if it can be calculated by using the positioning method (positioning using a cell ID does not make it possible to calculate the velocity, for instance).

Altitude can also be measured by some positioning methods (for example, GPS, as we noted in Chapter 3, but not the pure network-based positioning methods). The altitude is expressed in the ALT element, which gives the altitude of the mobile station in meters (above mean sea level). The ALT_ACC element expresses the accuracy of the requested altitude in meters.

The LIF API has a special element for position information: the GEO_INFO element. Position can be in UTM or longitude–latitude (LL) and (x, y) coordinate formats, and it is possible to define different data. In addition, there are a number of elements and attributes to describe shapes, as we discussed in Chapter 3.

The LIF API relies heavily on the mobile system to identify the user. There are no alternative functions to provide an identity for a mobile station (you could, for instance, imagine IP numbers being used for this purpose — as they indeed are in the WAP Location Framework). There are six different types of identities for mobiles, as shown in Table 7.1.

TABLE 7.1
Different Types of Mobile Identity Numbers in the LIF API

AbbreviationType of Identity Number	
IMEI	International Mobile Station Equipment Identity
IMSI	International Mobile Subscriber Identity
MSISDN	Mobile Station International Integrated Services Digital Network (ISDN) Number
MIN	Mobile Identification Number
MDN	Mobile Directory Number
EME_MSID	Emergency MSID

Which of these number types applies depends on which network technology is used and whether the mobile is making an emergency call.

Among the identity-related functions in the LIF API is also CELLNO, which is a unique cell number that will identify the cell within a network (remember that there is a global numbering plan for mobile telephony that allocates numbers to networks so that all networks can really have unique numbers). The number is a string of characters; for instance, the Location Area Code (LAC) plus the cell ID (CI) in GSM.

The VLRNO is another element that identifies a network entity; in this case, the visited location register number (which in GSM and UMTS is known as MSIN in the Global Title). The Global Title is the same format as an MSISDN; in other words, CC + NDC + MSIN.

Because the LIF API is supposed to work with the mobile positioning center, the identity is somewhat restricted. From the point of view of the mobile system, the only identities that exist are the identities of mobile phones, although there is an ID element in the LIF specification, which is a string defining the name of a registered user performing a location request. In an answer, the string represents the name of a location server. There is, however, a password element in the standard as well. The identity of a mobile phone resides on the subscriber identity module (SIM) card of the mobile station. The system also has a function to position a range of identities, however (which is possible because the MSISDN is a number) — something that can be useful when working with applications like fleet tracking, where the phones in the cars of the fleet can have numbers in the sequence. The MSID can be of several types that are declared in the msid_type attribute of the MSID element. The msid_enc attribute contains the encoding of the identity, ASCII (which is the default value), base 64, or encrypted.

In some countries, the network operator (where the location server is assumed to be placed) is not allowed to send the private information of a mobile station (like the MSISDN) to an MPC over the network. It might be allowed to send encrypted information, though, because only the network operator will be able to decode this information. The msid_enc attribute is intended to designate that this task is done, and the CRP value shows that the address has been encrypted (there are two other values, B64 for base 64 and ASC for ASCII).

Mobile Country Code (MCC) and Mobile Network Code (MNC) are two numbers that are part of the global numbering plan and are used to identify the country and the network where the call originates. The Network Destination Code (NDC) specifies where the call is directed.

The Emergency Services Routing Digits (ESRD) number is a special number that is used for emergency services. The ESRD is a telephone number in the national numbering plan that can be used to identify an emergency services provider and its associated location services client. The ESRD also identifies the base station, cell site, or sector from which an emergency call originates. The Emergency Services Routing Key (ESRK) is another telephone number in the national numbering plan that is assigned to an emergency services call for the duration of the call. The ESRK is used to identify (for example, route to) both the emergency services provider and the switch that is currently serving the emergency caller. During the lifetime of an emergency services call, the ESRK also identifies the calling subscriber. ESRD and ESRK can be of two different types: NA for North American and EU for European, plus Other for other expected formats (which regional or national standards bodies might define).

Emergency positioning is different from other positioning requests in that it can be triggered by an emergency call being made. When it is made, the EME_EVENT element is triggered and the eme_trigger attribute determines whether the emergency call started (the eme_org value) or if the call is disconnected (the eme_rel value).

When an emergency call is made, you know it is urgent. But positioning requests other than those that are emergency call related might also need to be served urgently. To that end, there is a PRIO element in the API that can be used to determine whether the current request should be served before others. This element can be of two different types, high and normal, which are expressed with the prio_type attribute value. The default value is normal.

Each of the different methods is expressed in a separate XML document, which is sent as a POST to the MPC from the client. Below is an example of a request for immediate location information (for a range of mobile phones) in the IDMS3 format. It contains a request for positioning of a number of mobile stations (the MSID_RANGE element), and the quality of position should be accurate within 1000 m altitude and low delay:

```xml
<?xml version = "1.0"?>
<!DOCTYPE SLIR SYSTEM "MLP_SLIR.DTD">
<SLIR ver="1.0">
  <CLIENT>
      <ID>TheUser</ID>
      <PWD>The5PW</PWD>
  </CLIENT>
  <ORIGINATOR>
      <ID>TheASP</ID>
  </ORIGINATOR>
  <MSIDS>
```

```
        <MSID>461011334411</MSID>
    <MSID_RANGE>
        <START_MSID>461018765710</START_MSID>
        <STOP_MSID>461018765712</STOP_MSID>
    </MSID_RANGE>
  </MSIDS>
  <EQoP>
    <RESP_REQ resp_req_type="LOW_DELAY"/>
    <HOR_ACC>1000</HOR_ACC>
  </EQoP>
  <GEO_INFO>
   <FORMAT>IDMS3</FORMAT>
  </GEO_INFO>
  <LOC_TYPE loc_type_type="CURRENT_OR_LAST"/>
  <PRIO prio_type="HIGH"/>
  </SLIR>
```

The response to the request is sent in a Standard Location Immediate Answer document. Each document can contain a number of responses with positions for different mobile stations:

```
<?xml version="1.0"?>
<!DOCTYPE ELIA SYSTEM "MLP_SLIA.DTD">
<SLIA ver="1.0">
 <POS>
  <MSID>461011334411</MSID>
  <PD>
        <TIME>20000623134453</TIME>
        <SHAPE>
            <CIRCLE>
                <POINT>
                <LL_POINT>
                        <LAT>301628.312</LAT>
                        <LONG>451533.431</LONG>
                </LL_POINT>
                </POINT>
                <RAD>240</RAD>
            </CIRCLE>
        </SHAPE>
  </PD>
 </POS>
 <POS>
```

```
            <MSID>461018765710</MSID>
            <PD>
                    <TIME>20000623134454</TIME>
                    <SHAPE>
                            <CIRCLE>
                                    <POINT>
                                            <LL_POINT>
                                                    <LAT>301228.302</LAT>
                                                    <LONG>865633.863</LONG>
                                            </LL_POINT>
                                    </POINT>
                                    <RAD>570</RAD>
                            </CIRCLE>
                    </SHAPE>
            </PD>
    </POS>
    <POS>
     <MSID>461018765711</MSID>
     <PD>
            <TIME>20000623110205</TIME>
            <SHAPE>
                    <CIRCLE>
                            <POINT>
                                    <LL_POINT>
                                            <LAT>781234.322</LAT>
                                            <LONG>762162.823</LONG>
                                    </LL_POINT>
                            </POINT>
                            <RAD>15</RAD>
                    </CIRCLE>
            </SHAPE>
     </PD>
    </POS>
    <POS>
     <MSID>461018765712</MSID>
     <POSERR>
            <RESULT resid="10">QOP NOT ATTAINABLE</RESULT>
            <TIME>20000623134454</TIME>
     </POSERR>
    </POS>
```

```
<GMT_OFF>+0200</GMT_OFF>

<RESULT resid="0">OK</RESULT></SLIA>
```

There is much that can go wrong when retrieving the location of a mobile terminal, however. When an error message is returned, the position gateway will return a result code with it to describe what went wrong. The RESULT element will contain the result code in the resid attribute. In Table 7.2, the result codes and the descriptions of what they mean are listed. The numbering is not consistent because several codes have been removed in the draft version.

OGC Languages: SensorML, GML, Etc.

The Open Geospatial Consortium (OGC) has developed a suite of specifications for modeling and markup languages. It has also developed a set of specifications to simplify the development of services. Together, these form the OGC services framework. To a developer, these form the interfaces between the information, the services, and the network.

Within the OGC services framework, there is a range of services and content protocols, which include:

- **Location content access services** (i.e., web map service, web feature service): These provide access to repositories of geospatial data.
- **Geocode, geoparse, and gazetteer services**: Determine the geographic location for addresses (a store inside a shopping mall), landmarks (i.e., ATMs), places, and other textual/coded location descriptors.
- **Coordinate transformation services**: These provide the coordinate transformations between various projection coordinate systems.
- **Discovery of location services and content holdings** (i.e., basic service model and catalog services): Used to discover location services and location content.
- **Portrayal services** (i.e., Style Layer Descriptor (SLD) and legend): Provide for the customization, tailoring, and understanding of the display of geospatial information.
- **Location content encoding and transport protocols** (i.e., Geographic Markup Language (GML), Location Organizer Folder (LOF), and Geolink): These content specifications apply to the encoding and transport of collections of related location content.

In the following sections, we will look at some of these components.

OGC Sensor Model Language

As already mentioned, local positioning systems include many types of sensors, including infrared (IR) proximity badges, passive RFID tags, ultrasound ranging tags, etc. These sensors typically have proprietary means for management, control, and access to data. Each sensor driver collects and classifies the data produced

TABLE 7.2
The Result Codes in the LIF API

Result Code	Slogan	Description
0	OK	No error occurred while processing the request
1	SYSTEM FAILURE	The request cannot be handled because of a general problem in the location server or the underlying network
4	UNKNOWN SUBSCRIBER	The user is unknown, i.e., no such subscription exists
5	ABSENT SUBSCRIBER	The user is currently not reachable
6	POSITION METHOD FAILURE	The location service failed to obtain the user's position
101	CONGESTION IN LOCATION SERVER	The request cannot be handled due to congestion in the location server
102	CONGESTION IN MOBILE NETWORK	The request cannot be handled due to congestion in the mobile network
104	TOO MANY POSITION ITEMS	Too many position items have been specified in the request
105	FORMAT ERROR	Parameter in the request has invalid format; the invalid parameter is indicated in ADD_INFO
106	SYNTAX ERROR	The position request has invalid syntax; details may be indicated in ADD_INFO
107	PROTOCOL ELEMENT NOT SUPPORTED	The element specified in the position request is not supported by this implementation; the element is indicated in ADD_INFO
108	SERVICE NOT SUPPORTED	The indicated service is not supported in the location server
201	UNKNOWN SUBSCRIBER	The user is unknown, i.e., no such subscription exists
202	HLR PRIVACY FLAG SET	The privacy flag of the MSID in the HLR is set to ON, meaning that location requests cannot be performed on the MSID
207	MISCONFIGURATION OF LOCATION SERVER	The location server is not completely configured to be able to calculate a position

into measurements of type distance, angle, proximity, or position. In addition, each measurement has an uncertainty model derived from the physical characteristics of the sensor and the environment. These data are then stored in the location models. Currently, these systems are not accessible through online services, which would allow for dynamic updating of information within a distributed computing environment.

To create a standard solution for online access to sensor systems, the OGC has developed a modeling language, SensorML. In particular, it is intended for any application that wants to access sensor assets through a set of web services.

With improvements in sensor, communications, and Internet technologies, it is feasible to talk about a Sensor Web consisting of loosely coupled processing components. The OGC is promoting the concept of the Sensor Web, which is a set of online web services that provide access to sensor assets. These services provide the means to determine collection feasibility (i.e., allows determination of whether sensor assets are available to meet specific needs) and to collect observations from available sensors in a standard way. One of the capabilities of Sensor Web is the means to model sensors and encode information about sensors (SensorML).

Open GIS Web Services

Enabled by the advancements in web services in general, the Open GIS Web Services (OWS) architecture is rapidly manifesting itself. Various groups within OGC are working on service categorization (data, processing, and registry/catalog services), data encodings (SLD, GML), and service chaining (WMS/SLD/CPS), which, taken together, show how to do service chaining in a web services environment. Within this work, general web services technologies have been critical: examples include Web Services Description Language (WSDL) for service description, Universal Description, Discovery, and Integration (UDDI) for service discovery, Simple Object Access Protocol (SOAP) for passing XML-encoded data, and IBM WSFL and MS XLANG for web service composition and process languages for orchestrating web services.

Several of these interfaces will prove critical in developing indoor location-based services. These include the WMS interface, the WFS interface, and, although strictly speaking not a web services specification, the Open GIS Catalog interface.

The WMS Interface

For map retrieval and display, OGC's Web Mapping Service specification can be used.

Using OGC's WMS request syntax, a map request can contain a GetFeatureInfo call, meaning that even though the map is in a raster image format, as is often the case with floor plans, information about a feature can be obtained even though it is part of the map image.

The Open GIS Web Map Server (WMS) specification explains how a map image server should respond to basic queries such as GetCapabilities (tell the client application what the server can do) and GetMap (send back the requested map in the format solicited). WMS enables programmers to add interoperability to their geoprocessing systems. The specification defines a request-and-response protocol for the interaction between web-based clients and map servers.

Web client support for WMS means in simple terms that with just the URI of a map server that supports WMS, a desktop or browser-based application can tap into layers of data from any WMS-compliant servers, no matter what software is being run at the back end or front end. Technically, the WMS specification determines how the client and the server communicate about what data are available, how they are delivered, and how information about map features is delivered.

The publication of the WMS specification has caused traditional geographic information system (GIS) vendors to wrap their proprietary server products, adding Open GIS interface software that allows them to respond to native (proprietary) and WMS (open) queries.

The WFS Interface

Similarly, the Web Feature Server (WFS) specification focuses almost exclusively on data encoding and transmission. How the data sets that are being served should be displayed and used is left to the developers of web client applications.

WFS describes data manipulation operations on Open GIS simple features, such that servers and clients can communicate at the feature level. Therefore, a Web Feature Server request — like those supported in many GIS and relational database management system (RDBMS) packages — consists of a description of the query and data transformation operations that are to be applied to WFS-enabled spatial data warehouses on the web. The request is generated on the client and is posted to a WFS. The WFS reads and executes the request returned in a feature set as GML. A GML-enabled client then can use the feature set.

Whereas WMS delivers the data as an image, WFS when implemented in a client supports the dynamic exploitation and access to feature data and associated attributes on the web from any server that implements WFS. This capability makes it possible to use the servers implementing WFS for enhanced spatial analysis, modeling, and other operations based on the intelligence of the attributed data.

Beyond feature access, there is an additional set of interfaces in the WFS for supporting simple transactions on a feature: *create, delete*, and *update*.

The Open GIS Catalog Service

The Open GIS Catalog Service interface specification defines a common interface that enables applications that conform to it to perform discovery, browse, and query operations against distributed and potentially heterogeneous catalog servers.

Spatial catalog servers typically contain metadata about spatially referenced information such as maps, schematics, diagrams, or textual documents. The specification describes metadata and spatial location data to identify and select layers of interest, and provides for interoperability in catalog update, maintenance, and other librarian functions.

The Open GIS Catalog Service specification is designed for catalogs of imagery, geographic information, and mixtures of the two. It specifies a set of open APIs that provide discovery services, access services, and interfaces for catalog managers, including a complete Catalog Query Language.

Detailed implementation guidance is provided in the specifications for establishing and ending a stateful catalog session to query the catalog server properties, check the status of a request, cancel a request, issue a query, present the query results, and get the schema of a discovered collection.

Open GIS Filter Encoding

In a request–response system, the LBS needs to use a query filter to get the information it needs, since this is a subset of the total information set in the database. This query interface should preferably be XML based, such as OGC's Filter Encoding specification. This enables the integration with other XML documents, for instance, the web services formats defined by the OGC. In the query, the developer specifies the type of location content he wants the user to see displayed.

To achieve application-independent location data, and thus location-independent location-based applications, several conventions will have to be respected (all of which can easily be implemented using the OGC Filter Encoding standard):

- An application-independent interface for querying the data or service
- A location authority-independent representation of a location as a data object
- A universal federation in which all location models combine to create a virtual location model to cover the (indoor) world in one framework

In addition to GML, the Open GIS Filter Encoding serves as a common data API that provides the functionality needed for typical location-aware applications and hides the details of the underlying data management. Queries to this interface are formulated in an XML-based language, which contains the following elements:

1. A restriction, which is a Boolean expression made of relations between attributes of objects and fixed values
2. A filter, which allows an application to remove attributes it is not interested in from the result
3. A closest predicate, which allows narrowing of the result to the n objects closest to a given position

The OpenLS Platform

The OpenLS GeoMobility Server architecture is an open location services platform. It uses OpenLS interfaces to access network location capacity (provided in a mobile network through a GMLC, for instance) and provides a set of interfaces allowing applications hosted on this server, or in another server, to access the OpenLS Core Services.

The OpenLS APIs are interfaces (XML schemas), used for implementing interoperable location-based services applications, and are used for accessing directory services (such as Yellow Pages), route determination, location determination gateway, geocoding, reverse geocoding, and portrayal services using standard web protocols.

These APIs allow developers and deployers of services, for instance, telecommunications companies, traditional GIS technology companies, and location-based services application providers to efficiently implement interoperable location-based services applications that seamlessly access multiple content repositories and link with other location-based services. They also enable seamless communication handover among cellular wireless communication networks (and devices).

The OpenLS objective is to provide a standardized solution for carriers (operators) that allows them to choose and implement standard interfaces and components into their location-based systems, ensuring that application developers have standard tools and services to use when building location-based services. The OpenLS Navigation Service, for example, determines routes between two or more (outdoor) points having specific (x, y) coordinates.

Part of the OpenLS Platform is the location-enabling server, which is a middleware module. The software interfaces to the MPC/GMLC and all other carrier (operator) systems (e.g., WAP-GW, SMSC/MMSC, OAM&P, and billing systems). This centralized interface minimizes integration efforts each time a carrier/operator launches a new LBS application.

Perhaps more important, this location-enabling middleware also handles privacy, presence, and personalization, which are absolute must-haves for location-based services, especially in the vertical mass market. Finally, it interfaces to a GIS engine, commonly referred to as a GeoServer. In some cases, the GIS engine is considered a transparent layered function of the location-enabling middleware.

There are several commercial offerings for location-enabling middleware. An example is the ESRI's ArcLocation Solutions that consists of a spatial server (GeoServer) built on top of industry standard SOAP and the OGC's OpenLS XML API suites. This middleware handles services chaining, MPC/GMLC integration, and mobile application server integration. It also consists of tool kits for third-party developers.

Geography Modeling Language

Geography Modeling Language (GML) is an XML language for the encoding, transport, and storage of geographic information, including both their geometry and properties of geographic features. GML is also an XML representation of the OGC simple features, which is a specification for vector-based map content (geographic features) for GIS.

GML does not contain any styling information. How a feature is represented on the screen is up to the application designer. Maps can be created using SVG, Java, or generated as raster output. Other ways of viewing the information than maps are of course also possible.

The Open GIS GML specification defines a standard set of primitive data types and operations on those data types. It also describes how to provide a service, how to request a service, and how to determine whether a request is a request for data, a request for an operation on data, or both. It uses the W3C XLink and XPointer specifications to express relationships between geospatial entities. In other words, such relationships can be expressed between features in the same database or between features across the Internet. You can also construct relationships between GML feature elements in different databases without requiring any modification of the participating databases. No more than read access is required to establish a relationship between data elements.

GML is not perfect, however. It lacks formats for storing topology, which means that conversions from map formats using vector graphics may be problematic. There is no way of expressing coordinates in higher than three dimensions (you can have coordinates in four dimensions, if you have a point with elevation and time, for instance). You can store elevation and other aspects as attributes, but it becomes problematic if you want to use them as the main key. GML also can only handle lines, not curves, which means conversion from computer-aided design (CAD) formats can be difficult.

The GML geometry schemas support points, lines, and polygons as their basic geometric elements. They also support geometric collections, such as multipoints, multilines, and multipolygons. The language is built on a hierarchical vector model, with feature collections compromising features that can comprise more features.

GML can be expanded via application schemas to add elements like PointCircle, PointEllipse, and PointArc, which can be used to accommodate the CIRCLE, ELLIPSE, and ARC elements from the LIF Mobile Location Protocol (which in turn are derived from the 3GPP specifications for generating mobile positions). These elements are used to describe the error estimates (the probability error) of the mobile device location.

GML can also be used to model symbolic, hierarchical objects (based on a nongeometric location model). In a hierarchical nongeometric location model, it is assumed that an object is located relative to another object. This can always be done using GML properties, which can be associations, as in the example below:

```
<abc:Building gml:id = "building1">
<gml:name>Store ABC </gml:name>
<abc:frontsOn xlink:href = "#r1"/>
<abc:numStories>3</abc:numStories>
</abc:Building>
```

GML is used for the geometry description of common data formats (i.e., GML: polygon). Other kinds of relationships, including hierarchical containment, can easily be constructed using a dictionary, expressed as an XML schema (dictionary.xsd) of property elements (e.g., <abc:frontsOn>). One dictionary has to be created for each feature relationship, but it is possible to describe any relationship this way. This schema is then used to construct the desired relationships in the application schemas. In terms of the mapping infrastructure, there is a need to model and map geometric points such as user's locations. The locations of other objects, like printers, are symbolic (printer A is "in Room 101").

Note that GML does not define which coordinate system should be used. Instead, a spatial reference system (SRS) has to be given, relative to which the coordinates are defined.

There is also a GML component (in the GML namespace) called observation, which can be used to denote a measurement. This could also be used in conjunction with a geometry element, to define an observation regarding the object's location.

GML is also organized in three sets of profiles of different complexity. The first profile is defined in three DTDs: the Feature DTD, the Geometry DTD, and the SRS DTD. The Feature DTD defines the structure of features and feature collections. The Geometry DTD defines the structure of the geometric properties and geometric elements. The last DTD is the SRS DTD, which defines the spatial reference systems.

The second profile also consists of three DTDs: the Geometry DTD and the SRS DTD, and an application-specific third DTD that specifies structures of features, so that features can store their properties in other elements than Property and GeometricProperty. Depending on the application, this might make sense, for instance, if you want to highlight specific features in your data model.

In the third profile, an RDF schema is defined for each application.

The Open GIS specification is based on a lexicon of common geodata types defined in terms of primitive data types (such as those available in all programming languages). This model is the Open Geodata Model. It is limited to those vocabulary elements that are needed to communicate geospatial information.

The model is an object model, but it is not a pure object model defining the interfaces of the objects. It also defines an API for the manipulation of the objects. This API is abstract, not mapped to the individual systems — that is something to be done for each specific system, such as Common Object Request Broker Architecture (CORBA) or Component Object Model (COM), or indeed the World Wide Web. The object-oriented model provides the same capability but is usually presented to the application developer as an object API (in other words, a class library).

Every geoprocessing system has a geodata model that serves as a guide for digitally representing Earth features and phenomena. The Open Geodata Model is a universal geodata model that enables interoperability interfaces to be specified, referring to those parts of the Open GIS Essential Model that focus on data: geometry, spatial reference systems, transformations, shapes, locational geometry structures, topology, the well-known structures from which feature geometries are constructed, coverage extents, schema range functions, coverage generation functions, and so on. Simple features are features whose geometric properties are restricted to simple geometries for which coordinates are defined in two dimensions and the delineation of a curve is subject to linear interpolation. A geographic feature is an abstraction of a real world phenomenon; it is a geographic feature if it is associated with a location relative to the Earth.

In other words, a digital representation of the real world can be thought of as a set of features. The state of a feature is defined by a set of properties, where each property can be thought of as a {name, type, value} triple. The number of properties a feature can have, together with their names and types, are determined by its type definition. Geographic features are those with properties that can be geometry valued. A feature collection is a collection of features that can be regarded as a feature; as a consequence, a feature collection has a feature type and thus might have distinct properties of its own in addition to the features that it contains.

There are two fundamental geographic types recognized in the Open GIS specification: features and coverages. Both features and coverages can be used to map real-world entities or phenomena.

- A feature, as described previously, is a representation of a real-world entity or an abstraction of the real world. It has a spatial domain, a temporal domain, or a spatial/temporal domain as an attribute. Examples of features include almost anything that can be placed in time and space, including desks, buildings, cities, trees, forest stands, ecosystems, delivery vehicles, snow removal routes, oil wells, oil pipelines, oil spills, and so on. Features are usually managed in groups as feature collections. A GIS thematic map layer for a city that shows only roads, for example, is a collection of features — each of which is a feature of type road. Features usually represent entities.
- A coverage is an association of points within a spatial/temporal domain to a value (of a defined data type, possibly a complex type). That is, in a coverage, each point has a particular simple or complex value. A coverage is a function from a spatial/temporal domain to an attribute domain. That is, a coverage in the Open GIS specification is simply a function that can return its value at a geometric point. Scalar fields (such as temperature distribution), terrain models, population distributions, satellite images and digital aerial photographs, bathymetry data, gravitometric surveys, and soil maps can all be regarded as coverages. Coverages usually represent phenomena.

Coverages have all of the characteristics of features, so they are a subtype of a feature. Therefore, features and feature collections are the central Open GIS Geodata Model elements. A coverage has a data value associated with every location. For instance, a city defined as a feature does not return a value for each point. At a given point, it might contain another feature or it might contain a coverage, but by itself it does not return a value. A city defined as a coverage returns a value for each point, such as an elevation or an air quality index value. A coverage can be derived from a collection of features. A collection of features can be used as the starting point, and one or more attributes of these features can be used to define a coverage — the value of the coverage at a point being the value of the attribute of the feature located at that point.

On a map, there are certain features that you cannot help but to agree on being part of the representation of the Earth. A mountain is undeniably there, and its height can be unambiguously measured. But when you involve other information types, the composition of features will be a social phenomenon. Someone has to suggest a feature, and others have to accept it. The population density of a city will be meaningful to some users and irrelevant to others. The more a feature is grounded in society and less in the purely physical reality, the more it will be subject to interpretation and therefore not unambiguous. The Open GIS Consortium maintains a record of consensus about features, coverages, Open GIS services, and so on, in a number of bookshelves containing topic volumes that together form the Open GIS abstract specification.

A collection of features in GML is comprised of features, the basic unit of digital geospatial information. Features can be defined recursively, so there can be considerable variation in feature granularity. For example, depending on the application or interests of the information gatherer, any of the following items could be a feature:

- A segment of a road between consecutive intersections
- A numbered highway consisting of many road segments
- A georeferenced satellite image
- A single pixel from a georeferenced satellite image
- A temperature overlay on a weather map
- A triangulated irregular network
- Those segments of a dynamically segmented road that fall between two other roads
- A drainage network
- A single seismic event magnitude contour

Features can be recursive (for example, features can contain many subfeatures that contain, in turn, more subfeatures) and can contain a collection of subfeatures or coverages (the other type of element in the Open GIS specifications) that form a logically consistent grouping in terms of resolution, accuracy, content, and context. Because the design enables recursion, it might also lead to the creation of logically inconsistent collections. Such inconsistencies will be a serious problem for anyone integrating data from different sources. XML validation is one way to get rid of part of the problem, but the main task lies with the database designer, who has to make sure the collections are consistent. While the spatial reference system has to be declared in all data, there is no way of declaring other attributes that might affect consistency, such as accuracy, for instance. This may become a problem when using GML data from different sources in applications, which demand knowledge about accuracy, such as construction work applications. It is, of course, possible to define an attribute in the application-specific data model for storing the accuracy. But that attribute cannot be expected to be present in all GML data, and this will not solve the problem.

In the Feature DTD, every feature element has a type-name attribute describing the object type of the feature itself. A feature can also contain an arbitrary number of property and geometric property elements. The property element has the attributes typeName and type. The value of the type-name attribute describes what relation the property has to the feature, and the value of the type attribute specifies the type of contained data.

The DTD itself can only constrain the data type of the contained data to be character data, which is why information about data type must be specified in an attribute. The value of the type attribute must be Boolean, integer, real, or string.

Features consist of three basic elements: geometry, semantic properties, and metadata.

The first of these is *geometry*, with an associated spatial or temporal reference system, including a statement of the resolution and accuracy of the geometric model. They can be defined by using simple, primitive geometric shapes defined as instances of well-known types of objects in the Open GIS geometry, such as polygons, line strings, polyhedrons, and other Open GIS shapes. The rules for representing feature types with well-known types must be explicit. (For example, a rule can specify that a brick house is seen as a polyhedron). Given a description of the objects in a feature collection, this situation will mean that they can be translated to a graphic representation. Instances of well-known types are called well-known structures, and the model of the well-known structures must carry sufficient information to enable the reconstruction of the extents of the features to which they contribute. That is, the geometry components must know how they contribute to complex geometries. (For example, the highway segment geometries must know the sequence in which they concatenate to become an entire highway feature.)

A geometric property element has the attribute typeName and contains a geometry element that belongs to the entity GeometryClasses. The geometry elements belonging to this entity are:

Point
LineString
Polygon
Multipoint
MultiLineString
MultiPolygon
GeometryCollection

The Open GIS abstract specification manages geometry independently of the representation used by the specific geoprocessing applications. All geometries must be able to use a coordinate geometry representation (for example, (x, y)). Just like other properties, geometric properties must be named. So, the River feature type might have a geometric property called centerLineOf, and the Road feature type might have a geometric property called linearGeometry. It is possible to be more precise about the type of geometry that can be used as a property value. In the previous examples, the geometric property could be specialized to be a line string property. Just as it is common to have multiple simple properties defined on a single feature type, so too can a feature type have multiple geometric properties.

The second basic element is *semantic properties*, or the definition of the entity or phenomenon. Unlike geometries, which are the same between information communities, semantics can vary from group to group in the same way that the conceptual view does. Definitions and meanings might vary slightly or radically, as in the different ways that farmers and civil engineers might define roads.

The third, *metadata*, is other information that might be needed to position the phenomenon in the context of the application environment or user community. Metadata content requirements and standards are often defined through professional societies and are used to trace lineage and provide a measure of quality assurance to the user. Metadata is a subset of the properties of a feature (or, more typically, of a feature collection), but it is data that describe the data (or the instance), not data that contribute to the presentation or modeling (in the current application — what is metadata to one application might be metadata to another). For example, a property of a coverage representing an aerial photograph might contain simply a name, such as date flown and a value from a date type. The complexity of the metadata can be adjusted to meet the demands of the application. A metadata dictionary (schema or data mapping) listing metadata categories and their data type might look like the following:

```
element name
type
acquisition date
date
percent cloud cover
float
```

A feature is made up of geometry, semantics, and metadata. But there is no rigorous requirement that mandates a value for each of these three elements. For example, a developer or an application can create a feature that has no geometry, and therefore no spatial/temporal reference system (in effect, no location). Because the GML specification is designed for use in geographic applications, most features will have a location. There are also better ways of representing features that do not have a location.

Feature instances are identified in operational software by an object identity (often known as an object ID, or OID). The OID is, ideally, unique for each feature in all data sets everywhere through time. What the OID should be is implementation dependent. Because OID is a proper type (typically implemented as a formatted string or a long integer), it can be used as a value of a property. The use of OIDs corresponds to pointers in object systems and reference values in SQL3-based relational systems.

Geographic elements fall into two broad categories: entities and phenomena.

- Entities are recognizable, discreet objects that have relatively well-defined boundaries or spatial extent. Examples include trucks, buildings, streams, certain landforms, and measurement stations.
- Phenomena vary over space and have no specific extent. Examples include temperature, soil composition, and topography. A value or description of a phenomenon is only meaningful at a particular point in space (and possibly time). The phenomenon called temperature, for example, only takes on specific values at defined locations, whether measured or interpolated from other locations. Tourist information is another example of a phenomenon.

These are not mutually exclusive sets of information. In fact, there are many components of the landscape that are part entity and part phenomena, making their ultimate classification subjective and open to interpretation. For tourist information, the distinction might not be so difficult, but a highway can be thought of as a feature or a collection of observations measuring accidents, shoulder quality, asphalt composition, or other structural status items.

The basic unit of the geospatial information interchange between applications is the feature collection. Feature collections can be any size. They can contain only a single point, grid post, pixel, road segment, or several terabytes of data, depending on the context in which the feature collection will be used (the context of the interchange transaction in the language of the Open GIS specifications). A feature collection, for instance, might take the form of an online database with a front end to support on-the-fly map generation and retrieval. A database of geospatial information can be made available online for users to access and request tailored subsets of the feature collection (which might, in itself, become a feature collection).

GML provides information about which SRS is used in the srs-name attribute. This attribute can be specified for individual geographic elements. In a hierarchical structure, the top-level element contains the SRS declaration, and this can be overrridden by other SRS declarations further down in the structure (which is standard XML). When combining data, the srs-name attribute can be used to set up the proper coordinate transformations so all data in a document reference the same SRS. Another approach is to keep the original coordinate values and carry out the transformations when the data are displayed. If the reference systems do not match, one of the data sets can be sent to an automated coordinate transformation service. This service uses the information about the reference system to set up the appropriate transformation parameters.The measuring stick is to be capable of encoding all of the reference systems, which can be found at the European Petroleum Survey Group (EPSG) web site, and the specifiers' claim. In addition, the encoding scheme allows for user-defined units and reference system parameters.

Feature collections can employ a number of representation methods. They can contain vector, gridded, and raster data, or any combination of these. They are assumed to be owned by information communities, which can be a group of people who share a trade or who are concerned with a feature from different aspects (for instance, a highway might concern ecologists, engineers, and politicians, but from different perspectives).

The community groups its collection of features into a catalog, which is the way it shares the collection and its semantics with the world. Translating between different catalogs is done by using a set of Extended Stylesheet Language — Transformations (XSLT) transformation sheets, mapping the semantics of one catalog into the semantics of the other.

It would be nice to say that all these problems disappeared if you used XML, but that is unfortunately not true. XML only provides the mechanics; it does not define meanings. To understand why data sharing is complex, consider the analogy between human language and spatial/temporal computing. When we share a context (such as a culture, or even at a lower level, such as a workplace or an association), we use a common language to describe that context and to set up a similar frame of reference in regard to it. The members of the culture see the world through the same eyes and characterize it by using shared descriptors. Standardization of meanings facilitates unimpeded, accurate communication. But although the semantic intent of a feature might be consistent across two feature class definition schemes, the content of the supporting schemes might diverge from one another. Even when the attribute sets for two comparable feature classes in different schemes match to a large degree (but not completely), there are still opportunities for the loss of attribute information.

There are a number of cases in which information can be lost when communicating between different language groups and, by analogy, between information communities. Defining a vocabulary in a formal language avoids this problem, because there is a common reference. The Open GIS Forum gives a hydrological example of how misunderstandings can occur and what results they can have.

Hydrography, according to the *Department of Defense Glossary of Mapping, Charting, and Geodetic Terms*, is "the science which deals with the measurements and description of the physical features of the oceans, seas, and lakes, and their adjoining coastal areas, with particular reference to their use for navigational purposes." Hydrographic data have an important geospatial component. Two well-known systems of feature class definitions used to describe features within the hydrographic discipline are the S-57 Object Catalog and the Feature and Attribute Coding Catalog (FACC). The S-57 Object Catalog is part of the IHO Transfer Standard for Digital Hydrographic Data, developed by the International Hydrographic Organization. The FACC is part of the Digital Geographic Exchange Standard (DIGEST) developed through an international cooperative effort by the member nations of the Digital Geographic Information Working Group (DGIWG).

Both of these schemes have a robust ontology of hydrographic features and attributes to support the geospatial use of these data, although each is used for a slightly different purpose. The S-57 Object Catalog is primarily intended to support the visual display component of electronic charts onboard commercial seagoing ships and is used in the U.S. by the Department of Commerce in producing digital hydrographic charts for the Electronic Chart Display Information System (ECDIS). The FACC, as part of DIGEST, was developed to support broadly applicable geospatial analysis requirements and is used in the U.S. primarily by the Department of Defense in the generation of the Defense Mapping Agency's Vector Product Format (VPF) products, including the Digital Nautical Chart (DNC). Semantic translation between the two schemes might be necessary to support the exchange of hydrographic data under international exchange agreements aimed at updating and improving the hydrographic charts and safety of navigation information produced and maintained by both communities.

Therefore, there are two different feature class definition schemes for the same objects. When translating between two feature class definition schemes, there are at least six different results that can occur. It is possible to create an exact match of the meanings of the definitions between the two feature classes with no loss of information in the translation. For instance, an AQ070 Ferry Crossing in the FACC is the same as a FERYRT Ferry Route in the S-57 Object Catalog.

If there is no direct semantic match between definitions in the two feature class definition schemes, it is possible to do an exact translation by using information in the attributes of one or both feature definitions. For example, a LITFLT Light Float in the S-57 Object Catalog is not an exact match for the FACC feature class BC00

Light, but through the use of the FACC attribute BTC Beacon/Buoy Type Category, the information can be recovered without loss when an attribute value of BTC006 Light Float is used.

To translate between feature classes, you need to aggregate features. Essentially, you need to map the classes in one schema to the classes in another. For example, the FACC feature classes BJ040 Ice Cliff, BJ065 Ice Shelf, BJ070 Pack Ice, BJ080 Polar Ice, and BJ100 Snow Field/Ice Field can be aggregated to the feature class ICEARE Ice Area in the S-57 Object Catalog. If you aggregate features, however, you cannot determine whether ICEARE Ice Area was originally an ice shelf, an area of pack ice, or something else. The feature content is lost in the translation, but you need to have a description somewhere of how the aggregate is created and what translation process was used (for instance, an XML Schema and XSLT transformation sheet used to create the collection).

Not only do you need to create new aggregates, but you also need to deconstruct them to translate between feature classes. Using the feature classes from the previous example, a translation of an ICEARE Ice Area feature from a data set based on the S-57 Object Catalog to one based on the FACC is not possible. If you want to make such a translation, you would have to modify the FACC schema required to create a new Ice Area feature class. If it is not feasible to modify the schema, you need to verify the source data (or perform an image analysis or ground truth reconnaissance process) to classify the Ice Area as one of the existing snow- and ice-related feature classes supported in the FACC. That might take weeks or even months. The only feasible option is to classify ICEARE Ice Area in the FACC, incorrectly and misleadingly, as one of the snow- and ice-related feature classes. In this case, it would be necessary to include a caveat with the feature to capture the discontinuities in the translation process and to maintain an accurate lineage of the feature.

Matches between the meanings of two comparable features in the schemas of different information communities are possible, but you need further clarification on the definition of the supporting feature classes before the match can be verified, or the translation might be dependent on the representation of the feature in its respective information communities. For instance, a BA020 Foreshore in the FACC might be the same as an ITDARE Intertidal Area in the S-57 Object Catalog. Both definitions specifically refer to measurement of the shoreline, but in respect to different data. BA020 Freshore references Mean Low Water, while ITDARE Intertidal Area references Mean High Water. You need to define a conversion that takes these differences in the reference datum into account so that the information community you are targeting might use the same instance of a shoreline differently. But you need to document the nature of the conversion, and the translation process, to maintain an accurate lineage of the feature.

Matching two feature classes from different information communities might not be possible without losing significant information. An example would be the FACC feature class GB040 Launch Pad, which has no counterpart in the S-57 Object Catalog.

Geographic feature definitions become more specialized the more they are focused on narrow applications. A road might seem like a fairly simple object, but take four different GIS information communities and they will define four very different phenomena that have different sets of definition information. If other communities are involved, the definitions of objects can be defined in even more divergent terms.

Depending on what you want to represent in the real world, there are different ways you can use GML. If you want to represent a city for tourist purposes, you would represent its buildings, roads, and rivers as features and create a feature collection describing the city. If you wanted to represent the city as a climactic feature, you would probably represent it as a coverage of measurements.

```
<os:Road>
 <gml:description>Georgia Street</gml:description>
 <os:numberLanes>4</os:numberLanes>
 <gml:centerLineOf>
  <gml:LineString srsName="EPSG:4326">
       <gml:coordinates>0.0,100.0 100.0,0.0</gml:coordinates>
  </gml:LineString>
 </gml:centerLineOf>
</os:Road>
```

The geometry elements in GML consist of coordinate lists, which consist of coordinate tuples (value pairs). When working with simple features, there are only two dimensions, and therefore every tuple has two coordinates. In the structure of a GML document, every geometry element has to include a coordinate element. The coordinate element of a point element has one coordinate tuple. The corresponding coordinate element for a line string has more than one coordinate tuple, and a linear ring, used for defining the extent of polygons, has at least four tuples, where the last tuple duplicates the first. The reason for this is that both the starting point and the endpoint have to be declared, even if they are the same. The following is the GML for the linear ring in Figure 7.1.

When defining a polygon in GML, you define a linear ring containing the outer boundary, and then (if necessary) a linear ring(s) defining the inner boundary. If a

```
<LinearRing>
   <coordinates>
       0.0,0.0
       100.0,0.0
       50.0,100.0
       0.0,0.0
   </coordinates>
</LinearRing>
```

FIGURE 7.1 A linear ring in GML.

polygon declares the SRS attribute, the spatial reference system will apply to everything that is inside the polygon.

The linear ring is the base for polygons. There are also other possibilities, based on homogenous aggregates of geometry elements: MultiPoint, MultiLineString, and MultiPolygon. If different kinds of elements are to be aggregated, the element for heterogeneous aggregation, called GeometryCollection, is used.

Here is another GML example:

```
<MiddleSchool ID ="1451">

<extentOf

<Polygon srsName="epsg:27354">

  <outerBoundaryIs>

      <LinearRing>

   <coordinates>491888.999999459,5458045.99963358
491904.999999458,5458044.99963358

491908.999999462,5458064.99963358
491924.999999461,5458064.99963358

491925.999999462,5458079.99963359  491977.999999466,5458120.9996336

491953.999999466,5458017.99963357 </coordinates>

</LinearRing>

            </outerBoundaryIs>

        </Polygon>

   </extentOf>

</MiddleSchool>
```

Note that this coding has no properties (other than the geometry). If we add properties for the building, we can get something that looks like the following example:

```
<MiddleSchool ID ="1451">

  <description>Balmoral Middle School</description>

  <NumStudents>987</NumStudents>

  <NumFloors>3</NumFloors>

  <extentOf>

        <Polygon srsName="epsg:27354">

            <outerBoundaryIs>

                <LinearRing>

   <coordinates>491888.999999459,5458045.99963358

491904.999999458,5458044.99963358491908.999999462,5458064.99963358
491924.999999461,5458064.99963358 491925.999999462,5458079.99963359
491977.999999466,5458120.9996336
491953.999999466,5458017.99963357</coordinates>

                </LinearRing>

            </outerBoundaryIs>

        </Polygon>
```

```
    </extentOf>
  </MiddleSchool>
```

If you are working with a large number of geometry elements, or a number of geometry elements close together that have common properties (representing only one feature or feature collection), you can specify a grouping element to give the application a hint of which elements are present within the area. This element, a bounding box, is a box with the edges parallel to the axes, containing all elements in a defined set of elements. A feature collection must always contain an element called "bounded by," which in turn contains a box element. It is also possible for features to have this property, but it is not required.

Bounding boxes are defined in GML using the geometry element "Box." A box element is defined by only two coordinate tuples, both of which are depending on the coordinates of the elements in the set of geometry elements. The first represents the corner having the lowest coordinate values on all axes, and the second represents the corner having the highest coordinate values. When the box defines a bounding box for a set of elements, the highest coordinate values are the same as the highest coordinate value on each axis, and the lowest coordinate values are the lowest value on each axis, of the elements contained in the box.

For modeling mobile objects, GML has a schema called dynamicFeature.xsd, which provides support for time-dependent features. This schema includes a class called MovingObjectStatus, which provides information about the state of a moving rigid body (its location, speed, and direction). This was created to support the requirements of location-based services.

GML can also be used for data exchange, and is well suited for transmitting small to medium-size volumes of information. What in this context is "small" and "medium-size" will depend very much on the transmission link. What appears to be a small amount of data when transmitted over 2 Mbps is an enormous amount if transmitted over 300 kbps.

Since it is an XML language, fully compliant with the XML standards, GML is usable with all standard XML tools, including XSLT, which is a language to transform XML data into other representations, incuding text and SVG. This way a GML database containing full-feature descriptions can be published into more limited versions. If the data are stored in a database, other tools (mostly Java based) are available to handle the same extraction and publication more efficiently and cheaply. However, if the data are to be transformed at wire speed, systems to handle this mediation exist, some implemented as stand-alone boxes.

Since maps can be created as SVG, it is possible to transform the GML database directly into a map. This means that application development can largely be a matter of XML transformations, using either custom applications in Java or transformation scripts in XSLT.

NEXUS AUGMENTED WORLD MODELING LANGUAGE AND AUGMENTED WORLD QUERY LANGUAGE

The language is dependent on the world model. GML attempts to be agnostic of the world model, allowing developers to add the features and their relations using XML schemas. Other languages are more tightly coupled to the world models that they describe. The Nexus augmented world model is described using the Augmented World Modeling Language (AWML). Location-aware applications may query the current state of this model by using the Augmented World Querying Language (AWQL) and, as a response, receive information about the current state of the model described by the AWML. Both languages are defined using XML schemas. The responses to queries are serialized in XML (using the AWML schema). Other XML formats can easily be embedded in AWQL or AWML.

AWML itself uses GML for object descriptions. Objects in AWML have attributes that give their geometry relative to some coordinate system. The objects belong to classes that are structured in a hierarchical class schema; i.e., a church is a building, which in turn is a static object and a Nexus object. Nexus uses several different coordinate systems, e.g., World Geodetic System 1984 (WGS84) coordinates (used by GPS), Gauss–Krüger, and UTM coordinates.

In addition, AWML not only models geographic location and the geometry of objects, but also symbolic descriptors of the objects, such as room numbers and explicit relationships between objects, e.g., the "part of" relation. This is especially important when linking together different parts of the model that may be supplied from different providers.

In summary, Nexus AWML's features are:

- Object geometry (GML)
- Coordination system (absolute systems: WGS84, UTM; relative systems)
- Symbolic description (i.e., IDs, names, room numbers)
- Relationship between objects ("part of" relation)

NAVIGATION MODELING LANGUAGES

Location models can be connected to construct navigation models. These provide additional elements, for instance, routes or vectors, which can be used to describe paths that lead to objects, and objects along those paths. These languages define points, routes, information elements, and child elements under those that can hold information of various kinds. The points are intended to be along a route, which can be defined by an external entity. The route could also be prefiltered depending on the user's interest. The concept of routes is as difficult to define as that of areas. In the context of NvML, it implies a vector with duration in time.

8 Service Deployment

In this chapter, we show how to set up a location service using different positioning systems (including Ekahau, MIT Cricket, and Place Lab) and related technologies (e.g., WRAPI).

When deploying a Wi-Fi- based positioning system, the following steps take place:

Step 1: Performing the site survey (includes creating the signal strength model)
Step 2: Creating the positioning model
Step 3: Calibration
Step 4: Access point placement and configuration
Step 5: Tracking
Step 6: Maintenance

STEP 1: SITE SURVEY

Site survey comprises deploying access points (APs) in the defined space and performing a walk-through of accessible areas within the defined space. Moreover, site survey is a process that searches the best APs' locations to ensure that the AP placement provides the maximum coverage and data throughput before actually installing APs permanently. The ultimate goal of a radio frequency (RF) site survey is to supply enough information to determine the number and placement of access points that provide adequate coverage throughout the facility. In most implementations, adequate coverage means support of a minimum data rate.

Note that site surveying is recommended before recording data for the positioning model, which we discuss later.

Depending on which wireless local-area network (WLAN) attributes are to be used in the localization process, at this stage, the objective is to find existing APs and their network names (Service Set IDs, or SSIDs), channel assignments, signal strengths, and approximate locations. Take Place Lab, for example, with its map-based (database of APs' IDs and their corresponding physical locations) localization approach. It records the radio signal strengths from APs and stores the multiple-access control (MAC) addresses in a database directory that is then used to map Wi-Fi APs' MAC addresses to physical locations. Wi-Fi-enabled mobile devices detect APs and then look up the APs' MAC addresses in a local directory. They can identify the origin of a signal from a database giving the location of Wi-Fi base stations. Using the signal strength from at least three base stations, they can then triangulate the user's location. Similarly, Skyhook Wi-Fi Positioning System (WPS) includes a reference database of over 1.5 million private and public APs along with their locations' ID numbers of several Wi-Fi access points stored in Skyhook Wireless database, even if the signal is not strong enough to provide a connection. With these IDs, the system can map where the mobile device is.

Site surveying is needed due to the fact that it is very difficult to predict the propagation of radio waves in wireless systems and detect the presence of interfering signals without the use of test equipment. Before deploying any wireless LAN, it is very important to perform a radio frequency (RF) site survey with the goals of detecting the presence of RF interference and identifying the proper placement of access points. Even when using omnidirectional antennas, radio waves do not really travel the same distance in all directions. Instead, walls, doors, elevator shafts, people, and other obstacles offer varying degrees of attenuation, which cause the RF radiation pattern to be irregular and unpredictable. As a result, it is often necessary to perform an RF site survey to fully understand the behavior of radio waves within a facility before installing wireless network access points.

Due to the massive size and irregularly shaped facilities, the RF site survey can be somewhat challenging. The need and complexity of an RF site survey will vary depending on the building/environment type. For example, a small three-room office may not require a site survey. This case can probably get by with a single access point located anywhere within the office and still maintain adequate coverage. If this access point encounters RF interference from another nearby wireless LAN, you can likely choose a different channel and eliminate the problem. A larger facility, such as an office complex, apartment building, hospital, or warehouse, generally requires an extensive RF site survey. Without a survey, users will probably end up with inadequate coverage and suffer from low performance in some areas. Doing a site survey outweighs relocating and adding access points to the facility after installing and interconnecting many access points.

In proceeding with a site survey, simply walk through an office space, warehouse, or multifloor building — any interior space that needs to be surveyed — and take signal measurements. Next, place those measurements on top of any structural floor plan to get a view of any WLAN based upon MAC addresses, received signal strength indicator (RSSI), SSID, and more.

It is worth visually inspecting the facility before performing any tests to verify the accuracy of the facility diagram. This is a good time to note any potential barriers that may affect the propagation of RF signals. For example, a visual inspection will uncover obstacles to RF such as metal racks and partitions, items that blueprints generally do not show.

In terms of floor plan, locate or, if not available, prepare a floor plan drawing that depicts the location of walls, walkways, etc. See below for more on creating maps. On the floor plan diagram, mark the areas of fixed and mobile users. In addition to illustrating where mobile users may roam, indicate where they will not go. This might result in fewer access points if the roaming areas can be limited. Export AutoCAD files into floor plan site initiator software and scale the rooms and walls for measurement overlays.

It is possible to perform a site survey analysis at a small site with a simple utility (recording measurements on paper), but this approach quickly becomes tedious and

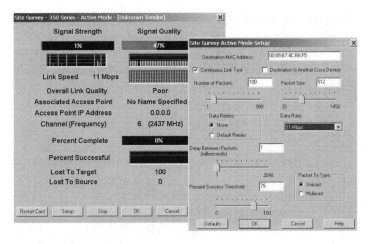

FIGURE 8.1   WLAN site survey utility.

time-consuming. As a result, site survey analyzers can help by gathering data before, during, and after that task.

Many WLAN adapters are supplied with site survey utilities, the Cisco Aironet Client Utility (Figure 8.1) being an example. These are handy for spot-checking signal strength, quality, and loss; however, a thorough site survey requires much more than just a client utility.

Overall, analyzers play an essential role in the site survey process. Most WLAN analyzers typically can discover existing APs and decommission unauthorized APs, but WLAN must live in harmony with neighbor APs. This means factoring those APs into the site survey so that it can avoid co-channel interference. Many wireless LAN vendors provide free RF site survey tools that identify the associated access point, data rate, signal strength, and signal quality. The site survey tools found in many WLAN analyzers can monitor or record detailed metrics associated with discovered devices. (To discover wireless devices in an area, a stumbler (e.g., see "PlaceLab Stumbler" section) can be used to walk inside the space, creating a file of discovered devices for later reference. Due to signal propagation limitations discussed in previous chapters, for best results, stumbling is to be repeated on different days, at different times, until the discovered device count appears stable.) This software can also be loaded on a laptop or PocketPC to test the coverage of each preliminary access point location.

For example, the Network Instruments Observer survey (Figure 8.2) provides minimum/maximum/average signal, quality, and data rate for different frame types (management, control, data), for each transmitter. This kind of information can be used both as input to planning and to validate results after plan implementation.

Or you could use a handheld site survey tool available from several different companies. For example, Berkeley Varitronics Systems offers a line of handheld devices, such as Grasshopper and Scorpion, that provide advanced site survey functions. For reasons tied to the utility of more capabilities, advanced wireless site survey systems are available from a variety of sources, including WLAN

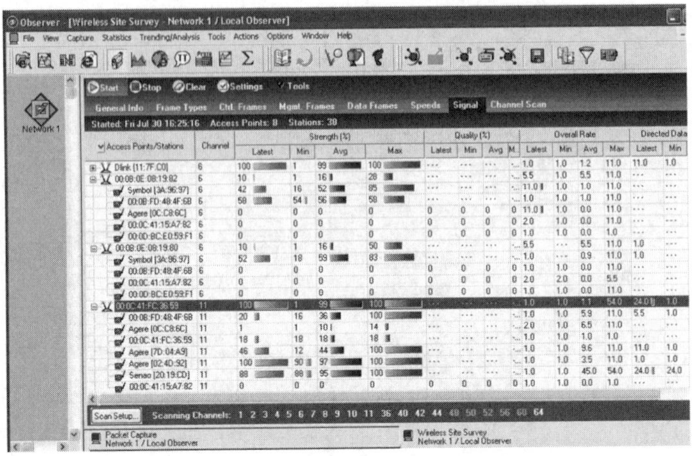

FIGURE 8.2 Network Instruments Observer survey.

switch vendors (e.g., Airespace, Nortel, Trapeze) and software suppliers (e.g., AirMagnet, BVS, Connect802, Ekahau, VisiWave). These systems help to design WLANs by using field measurements to plot radio coverage areas on floor plans, predicting signal, noise, data rate, and capacity. Obstructions, building materials, ceiling height, existing APs, and other sources of interference may all be factored in to recommend AP number, placement, power output, and channel assignments. Install an access point at each preliminary location, and monitor the site survey software readings by walking varying distances away from the access point. Different points should be observed for data rates and signal readings, especially in the outer bounds of the access point coverage. In cases of a multifloor building, perform tests on the floors above and below the access point. Keep in mind that a poor signal quality reading likely indicates that RF interference is affecting the wireless LAN. This would warrant the use of a spectrum analyzer to characterize the interference, especially if there are no other indications of its source. Based on the results of the testing, you might need to reconsider the location of some access points and redo the affected tests.

Using a site survey system, let us consider the Ekahau Manager to perform a site survey. The following positioning attributes are to be observed:

1. Number of detected APs at each location — accurate positioning requires at least 3 APs (preferably 5 to 10).
2. RSSI of each AP signal — stronger signal is better in general; however, adjusting APs or using directional antennas to provide different RSSIs for different areas may help differentiate the areas and improve accuracy.
3. Noise level — this is available separately in Ekahau Site Survey. Significant radio noise will decrease positioning accuracy. It is recommended to avoid noise with standard channel planning.

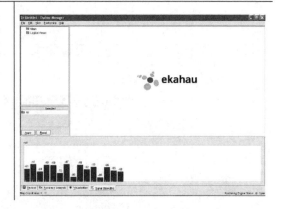

Ekahau Manager with Signal Strength (RSSI) display

FIGURE 8.3 Using Ekahau Manager for site survey.

Also, before using the system for a specific application, the Ekahau software offers an accuracy gauge to determine whether it is necessary to move or add more access points to optimize accuracy.

Site surveys are often conducted by positioning APs in probable locations, such as the center of a floor. Tools are then used to record signal, noise, speed, and loss at defined distances (i.e., taking measurements every 10 ft) from each AP. With the laptop, users can walk around the area while observing the number of detected signals and their RSSIs. Note that if necessary, "dummy" APs should be moved or added to have at least 3 (preferably 5 to 10) AP signals available in each location (Figure 8.3).

Some WLAN analyzers include tools to automate the measurement process. For example, AirMagnet (Figure 8.4) can record measurements to a file when specified events occur, like change in association state, signal strength, or data rate. Or a user can move at a consistent pace between two points, recording measurements every N seconds.

AirMagnet and BVS analyzers generate data that can be fed directly into related site survey products. For example, BVS Bird's Eye Site Supervisor runs on a Yellowjacket. As users move through a site with Yellowjacket, they tap their location on the floor plan to record data points. Those results are consumed by a Win32 program, Site Investigator (Figure 8.5), to plot RF coverage by AP, SSID, or channel. The more data points recorded, the more granular and accurate the coverage map.

In addition, most stumblers indicate whether APs use some kind of security (e.g., WEP, TKIP) and are currently active (e.g., first/last time seen). For example, scan output from KisMAC, a free stumbler for MacOS X, is shown in Figure 8.6.

Some stumblers also provide real-time traffic or signal graphs, like the NetStumbler received signal strength indicator plot shown in Figure 8.7. NetStumbler, a freely available tool, can be used to measure signal strengths. The software logs signal strengths to a file over a period of time. A Perl script can be used to go through the logs and parses the signal strengths to calculate the average. To improve the

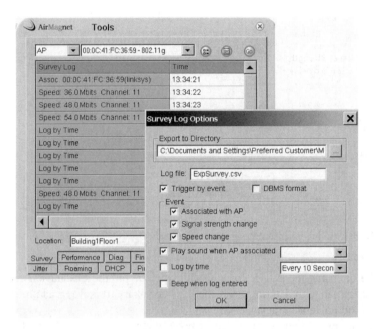

FIGURE 8.4 AirMagnet WLAN analyzer automation tool.

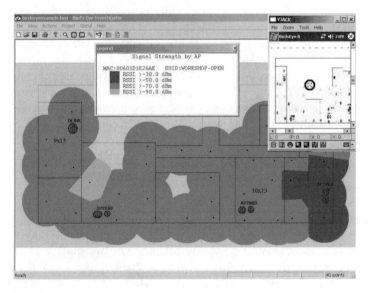

FIGURE 8.5 Site investigator.

accuracy of the measurements, directional antennas can be used. This provides more stable signal readings.

Site survey systems provide other advanced features, like active surveys, "what if" simulations, and automated AP (re)configuration.

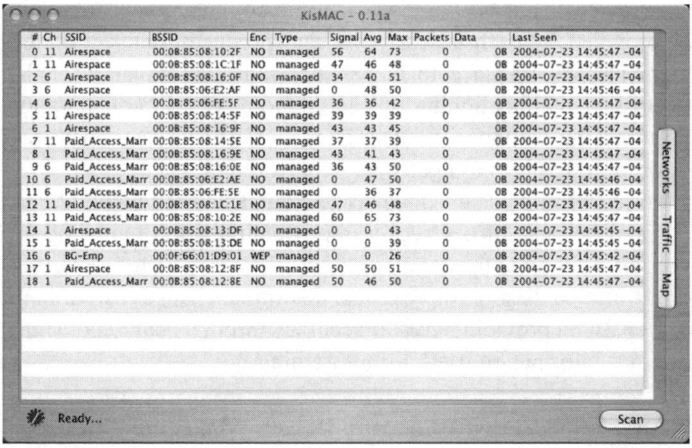

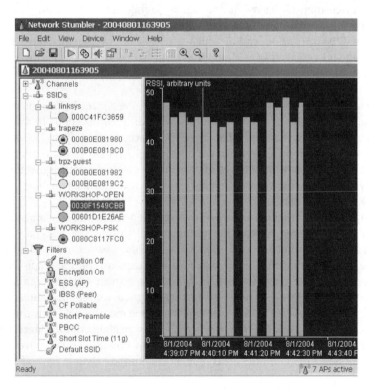

FIGURE 8.6   KisMAC stumbler.

FIGURE 8.7   NetStumbler.

A monitoring system may be in place that monitors the status of the APs (which may interface with access point manager). For instance, such a subsystem may be configured to determine if an access point is malfunctioning, turned off, or inoperable, if a new detector has been added, or some combination. The monitor obtains status data provided by each access point and produces status/error messages. A system might also be in place to adjust the statistical signal strength model in response to loss or malfunctioning of one or more APs. In collaboration with the location and tracking components, data from certain APs may be selectively suppressed, in order to reduce ambiguity in signal strength data.

Using the feedback mechanisms, network status and access point layout can be monitored. Feedback can also serve to improve simulation modeling and provide error correction estimates, by comparing actual vs. simulated data, for instance. Feedback may also be used to improve and determine changes useful in the training model (see step 3), by providing greater accuracy through analysis of signal strength and access point information used during training.

CREATING THE SIGNAL STRENGTH MODEL (RADIO MAP)

As explained in Chapter 4, the signal strength model defines, for each AP within the physical space, a pattern of signal strength reception that is anticipated from a mobile device transmitting within the space, taking into account the obstructions and placement of the APs. With a plurality of APs, a plurality of signal strength patterns will be defined, several of which will typically overlap to some extent. The APs can either be used as signal sources to be measured by the mobile device during the survey, or serve as listening posts measuring the signal strength from the mobile device.

There are two methods for creating a radio map. The first is the empirical method, which involves a mobile user walking to several different locations in the building and recording the physical coordinates of each location (e.g., using a floor layout map as spatial reference), together with the signal strength of the beacon packets from each of the APs within range. The second is the mathematical method, which involves computing the received signal strength using a mathematical model of indoor RF signal propagation

A signal strength model or radio/coverage map shows the fundamental signal propagation and related effects, like multipath and shading, for a specific environment. It is mainly useful for network planning and the analysis of signal attenuation by physical obstruction. Moreover, it serves as a database for the empirical verification of the signal propagation model. In order to convert received signal strength indicator (RSSI) values to a range estimate to be used with triangulation methods, it is necessary to use an appropriate signal propagation model. Constructing the radio map requires deciding the granularity of the readings (the distance between every reading), how many readings are needed to obtain a realistic average, and what is the best time to do the measurement (nighttime, daytime, or both).

SITE SURVEY AND THE RESULTING DATA

The resulting data collected from the site survey can be used by the signal strength module to develop a statistical model, using a manual, automated, or dummy approach.

With the manual approach, the model developer merely selects the areas of primary interest or locales on the digital map of the space for which to build the model and determine the positions of the APs.

With the automated approach, this limits the areas in which the APs can be located to where the planner designates. The automated approach instead involves using a statistical technique to deduce the number of highly recognizable locales with strongly distinctive signal profiles by either the user specifying the number of locales or a designated statistical confidence factor.

With the dummy approach, another manner of building the statistical model includes using simulated APs and simulated mobile device readings within the context of the digital map of the physical space. In such a case, the signal strength modeler assumes certain reception and transmission characteristics of the APs and of the mobile devices within the context of the space in the digital map. The statistical signal strength model is generated as a function of these assumptions. Preferably, the system allows for editing the assumptions (including the positioning of obstructions and APs) to yield different statistical models using the user interface of the system.

Overall, these approaches make use of a statistical mechanism to provide a correlation of the communication signal strengths obtained during the site survey with locales in the defined space. The actual signal strength data received from the APs are used to build a statistical signal strength model associated with the digital form of the physical space. Any one or more of a variety of known statistical modeling approaches may be used to build the signal strength model, such as a Markov model. According to this approach, building the statistical signal strength model includes performing a signal strength site survey of the defined space.

Either the mobile device or the network (APs) can be used as the signal source for the purposes of the site survey. There are two architectural choices for measuring signal strengths from different APs:

1. The local client measures signal strengths of multiple access points.
2. Multiple access points are queried by a third party for signal strength of client.

The ideal solution would be the network-based approach, to have APs queried for signal, but for fast-implementation projects, it is easier to implement the first option, where the client locally measures signal. During the online phase, the same configuration would be employed, with which to provide readings to the statistical model to determine the location and movement of the mobile device, by a location and tracking manager utility.

With the first approach, a mobile device can be used to capture location determination data, which can then be sent to a server from the mobile device. The signal strength model can be created by having APs installed in the physical space, and actual signal strength data are collected through migration of a transmitting mobile device through the space. The server calculates the distance between the AP and personal digital assistant (PDA), and determines the location of the PDA using signal strength-based location determination; the device gathers necessary data on the network structure of Windows CE. The location determination data might include signal strength, connected Basic Service Set ID (BSS ID), signal-receivable BSS ID list, and MAC address. The WLAN Native Software Development Kit (SDK), which is supplied by WLAN or mobile device manufacturers, can be used for this purpose. If the SDK is not available, the data can be captured by accessing network drivers directly using object ID (OID) that is supported over NDIS 5.1.

Different types of APs have different interfaces for the programmer to query them for signal information. Some types of APs, like the Axon AP, use multiple antennas connected to each AP. The mobile clients associate with an antenna and in turn with an AP. The problem is that the antennas could be quite far from the AP, and there is no easy way to determine with which antenna the client is communicating. One possible solution would be to try and use the unique MAC addresses of the antennas.

Signal strength values can be measured by both access points and the wireless network adaptors. If signal strengths of clients are measured by an AP, then the AP could theoretically be queried for the information. This would allow an administrator to monitor changing signal strengths of various clients, and from that calculate the changing locations of various clients. By querying the APs there would be no need to install any software on clients.

This AP-centric approach has two main problems. First, the programmatic interface to the access points is not uniform and different pieces of software would have to be written to query different AP models. Second, there would be issues with stale information. An AP maintains state about clients that are associated with it. Once a client roams away from that AP, then its state is marked as away, which would present stale signal strength information if that AP were queried for the client that has roamed away.

The alternative to AP-centric measurements would be client centric, where signal strengths of APs are measured from the point of view of the AP. If the application is designed with modularity in mind, then the portion that is responsible for signal strength measurements should be completely separate from the rest of the application. Since there are far fewer common operating systems than there are APs, writing modules to measure signal strengths on different operating systems would be simpler than writing modules to communicate with the different APs. Once signal strengths of visible APs are measured by the client, the location can be determined using two different positioning approaches.

Positioning systems such as Ekahau combine a digital definition of the physical space with a statistical signal strength model to provide a context within which mobile devices may be detected and tracked. With the digital form of the physical space defined, the signal strength model can be determined. A detailed RF site survey is essential for good accuracy in location determination using pattern-matching techniques, as well as for determining optimal locations of APs to provide the best possible coverage. Sniffers are used to measure signal strengths from the APs' locations and record them into a file. Usually, about 60 to 80 locations are calibrated at each floor and 20 RSS samples from all the available APs are recorded. Also, in order to consider the effect of all the orientations, the mobile device is rotated slowly

Example of how to place Tracking Rails™ in office environment

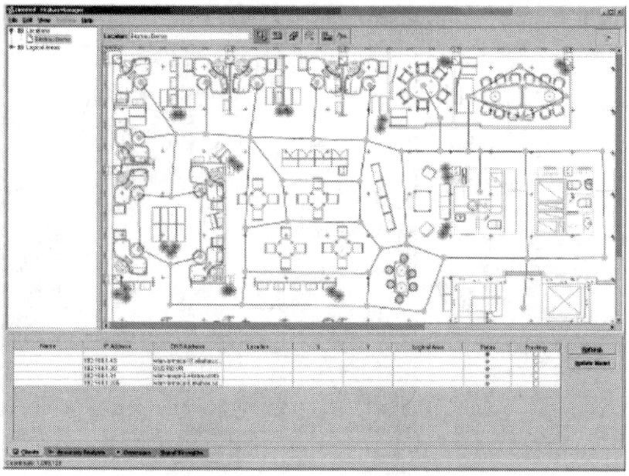

FIGURE 8.8   Using Ekahau Manager for creating a positioning model.

through all orientations. The resulting log file is then used to perform signal attenuation and coverage area analysis.

With the digital map of the physical space defined, the signal strength model can be generated. The system includes a signal strength modeler that can access the digital map in the database.

STEP 2: CREATE A POSITIONING MODEL

Creating a positioning model using, for example, Ekahau Manager, starts with opening any floor plan or map image (BMP, JPG, PNG). In the case of Ekahau, a map image is of an area with minimum accuracy of 10 dots/ft (30 dots/m) in normal viewing scale, 1:1. Note: To ensure that the Ekahau Positioning Engine functions accurately, a map scale must be defined for each map image. Also, you need to calculate how many map pixels equal the map's selected distance unit. This can be done by using the Measure tool within Ekahau Manager, where points on the map are selected and the distance between is measured (e.g., using a measuring tape).

Once three AP signals have been marked everywhere, tracking rails (Figure 8.8) can be drawn on the map to increase positioning accuracy and stability. Drawing the tracking rails on the map creates the positioning model. However, before you can use the positioning model for locating devices, it needs to be calibrated (i.e., you need to record network data from various map locations). Before recording data for the positioning model, tracking rails must be placed on the map to indicate possible travel paths between rooms, corridors, floors, and other locations. It is recommended to record a sample point at least every 3 to 5 m. The location estimates depend much on the rails that are placed on maps; hence, marking only the correct paths that are accessible and used by the tracked devices should be done with caution.

For better network planning, the Visualization Tool is used to display the five strongest AP signals at each location. Displaying the accuracy analysis error vectors

Signal strength displays the coverage area of all (enabled) access points in the positioning model

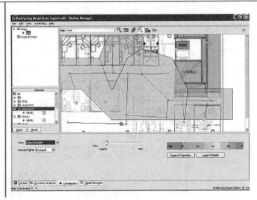

FIGURE 8.9   Using Ekahau Visualization Tool for better network planning.

Location probability displays the probability for any device to be located inside the grid square

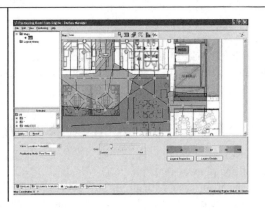

FIGURE 8.10   Using Ekahau Visualization Tool to display location probability.

on the coverage areas provides insight to network planning problems that might cause decreased accuracy. For example, poor network coverage typically correlates with decreased positioning accuracy in the area. To solve the problem, change access point locations or antenna directions, or add more access points in the problematic area. At least the three strongest signals should provide average or strong signal strength for good accuracy (Figure 8.9).

Figure 8.10 shows that the Visualization Tool can be used to display the probability for any device to be located inside the grid square, using a device history of around 5 sec (location update interval × 2.5 sec).

Figure 8.11 shows that the Visualization Tool can be used to display expected errors. Note: The Expected Error view is not using devices' history information, meaning that the accuracy is typically lower than in reality. The engine utilizes location history by default for higher accuracy.

Figure 8.12 shows how the Visualization Tool can be used to display accuracy analysis. The accuracy analysis process can record some additional test data. You

Location probability displays the probability for any device to be located inside the grid square

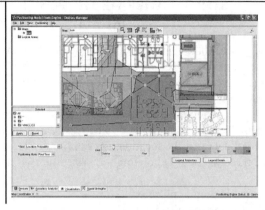

FIGURE 8.11 Using Ekahau Visualization Tool to display expected error.

Accuracy analysis displays the recorded average error in colors for each location

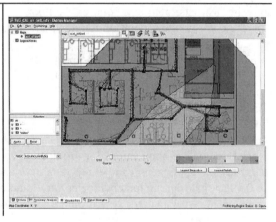

FIGURE 8.12 Using Ekahau Visualization Tool to display accuracy analysis.

can visually analyze the positioning error vectors and statistics to find areas where additional access points or calibration sample points are needed.

One potential challenge is the merging of positioning models. This is more of a case with an indoor–outdoor merge. The Ekahau Positioning Engine supports merging indoor models to share calibration work between multiple users. For example, to calibrate a 10-map building, you can first agree that three people will record the data and merge their changes directly in the Positioning Engine.

Once the positioning model is created, Ekahau Manager can save it either as Ekahau Database files (.edb) or directly in the Positioning Engine's internal database. Typically, the best positioning model is always saved in the engine, while test calibrations and exercises are saved as files. After saving a positioning model in the Positioning Engine, it takes a few seconds before the engine reads the new model into memory and starts using it for tracking. As the tracking history queue is emptied, positioning accuracy can slightly decrease for the next 5 to 20 sec.

STEP 3: CALIBRATE THE POSITIONING
MODEL/SIGNAL MAP (TRAINING)

The next step is calibrating the positioning model. The system is calibrated by a user walking around and recording measurements. These measurements will be stored in the database. Measurements made during run time are compared to database values to calculate position.

The process of calibrating a signal strength model is referred to as training the area or system. The calibration phase is necessary to train the system and provide it with recorded data that the location algorithm requires.

Usually, the process of calibrating the system is as follows:

Start the calibration application.
Enter floor name to be calibrated.
Start calibrating.
The floor map is displayed.
The software instructs to go to a point on the floor.
Click on the corresponding point on the map, and signal strength measurements for that point are recorded.
The software repeats the previous two steps as necessary.
When all the points necessary are recorded, the software indicates that calibration is complete.
Floor level list is displayed.

Calibration data are stored in a database, which can be implemented as a flat file or an object dump. The applications for installation, configuration, calibration, and the main run-time application can be integrated into one main application.

Despite the fact that training methodologies vary in number, source, and values of readings obtained, what matters is that data exists through the site survey to develop a profile of various locales within the defined space.

To account for APs that may have been deployed after the radio map was generated, and for lost beacons from APs, the following rules can be followed. First, during the positioning phase, if any AP that never appears in the radio map is discovered, that AP can be ignored. Second, when matching fingerprints to an observed scan, if the fingerprints with the same set of APs as heard in the scan cannot be found, the search to look for fingerprints that have supersets or subsets of the APs in the observed scan should be expanded. With these rules, it is possible to match fingerprints that have at most p different APs between the fingerprint in the radio map and the observed scan. According to Place Lab, these rules helped improve the matching rate for fingerprints significantly from 70% to 99%. Fingerprinting is based on the assumption that the Wi-Fi devices used for training and positioning measure signal strengths in the same way. If that is not the case (due to differences caused by manufacturing variations, antennas, orientation, etc.), one cannot directly compare the signal strengths. To account for this, a variation of fingerprinting (ranking) can be used with an algorithm proposed for the RightSpot system. Instead of comparing absolute signal strengths, this method compares lists of APs sorted by signal strength.

More specifically, training may be done by collecting labeled data for each location. The data are uniquely labeled and associated with their corresponding access points. The data are a set of samples, each of which has a measurement from one or more APs, averaged over a period of time. Many of these samples (typically about 25) compose what is called a signature. In some forms, unlabeled data could be used to augment or replace the existing data, but preferably the association with an access point is retained. Further, the signatures may be composed of fewer than about 25 samples, either by simply collecting fewer, by automatic decimation, or by algorithmic selection of which samples to retain.

In yet other forms, the signatures could be changed in representation from a set of samples to any number of other schemes, including using support vector machines (SVMs) or similar schemes to select critical samples, Gaussian clusters to estimate the densities, or any number of other density estimation schemes.

From these signatures, silhouettes are generated internally. Each signature yields a silhouette. However, in other forms, silhouettes composed of multiple signatures could be generated, if useful. A silhouette is generated by examining each source sample in a signature and identifying the other (i.e., target) samples (from all signatures combined) that are densest in the vicinity of that source sample, as discussed below with respect to terraced density estimation. A source sample is a sample from a sample set, associated with an access point, and selected for processing, and a target sample is a sample from the same sample set that is not the sample being processed, but is used for reference, comparison, or otherwise in relation to the processing of the source sample. In other forms, target samples could come from other sample sets. The signature that is the most heavily represented in these resultant target sample densities is counted. This is done for each source sample in a signature, and the resulting count of target signatures is tallied and becomes a silhouette. Most of the operations done in post processing (i.e., operation) are performed on silhouettes, though they are often presented in the user interface as signatures. This is done because there is a one-to-one mapping between them, and it avoids confusing the user if silhouettes are not mentioned at all.

A terraced density estimation scheme can be used to estimate signature sample densities, primarily for convenience of implementation. In the preferred form, a Parzen Window scheme with a series of stacked box kernels is used. However, any number of other known density estimation schemes could be utilized to good effect.

Other techniques include a variety of other kernel- based estimation schemes, as well as knn or Gaussian clustering.

Such a statistical model can be implemented by the signal strength modeler using a Markov model, with the state variable representing the locales within the defined space and transition probabilities representing the movement likelihood between them. The Markov model can be either continuous or discrete, affected by the desired tracking resolution, number of signal sources or APs, and their variations over the space. The signal strength modeler can generate the statistical signal strength model using the Markov model, or it can be the result of applying some other probabilistic fitting technique to represent the signal strength distribution in locations of interest or locales. Similarly, multiple distributions can be employed to represent the impact of different environmental profiles, such as, but not limited to, time of day, expected communications network load, transient environmental factors, and other physical or weather-related phenomena.

Using Ekahau Manager

Using the Ekahau calibration client (e.g., Ekahau's calibration software, Ekahau Manager, on a laptop), walk around the area, stopping on the nearest tracking rail every 5 to 10 m (15 to 30 ft), and use the Calibration Tool — click the map to record sample points containing received signal strength intensity (RSSI) samples. This will associate processed network data to the location. Note that no information about the AP locations is required, and there is no need to work on-site while drawing the rails, unless you want to observe the area when placing the rails. Overall, this step takes approximately 1 h/ 12,000 ft^2 (1200 m^2).

Using WRAPI and JWRAPI

The Wireless Research API (WRAPI) (http://ramp.ucsd.edu/pawn/wrapi/) provides access to signal strengths as observed by a client. WRAPI provides an easy-to-use, common API that enables applications (mobile and end stations) to monitor and control WLAN-specific parameters in an 802.11b wireless LAN. WRAPI is completely hardware agnostic; i.e., its API can be used independently of the underlying wireless network hardware. WRAPI 1.0 is implemented on the Windows XP operating system and is a hardware-independent tool that works with any IEEE 802.11b wireless network hardware vendor.

For example, WRAPI can be used to deploy a user location determination system that uses WRAPI to detect the strength of signals from multiple APs within range. Another use of WRAPI is to retrieve information about wireless network parameters and behavior, such as average load on the APs, the operating frequency, network identifier, etc.

The WRAPI documentation describes the tools and their implementation as such. The WRAPI software library (wrapi.dll) allows applications running in user space on mobile end stations to query information about the IEEE 802.11 network they are attached to. WRAPI works with any IEEE 802.11b wireless network hardware vendor. WRAPI functions obtain information about the wireless LAN using the NDIS User Mode I/O Protocol (NDISUIO). NDISUIO is a connectionless, NDIS 5.1-compliant protocol driver. It allows user-mode applications to establish and tear down bindings to network adapters (Ethernet, WLAN, etc.). Further, it also supports setting packet filters, sending and receiving data, and handling plug-and-play events. Therefore, as an NDIS-aware component, NDIS-UIO can directly open an NDIS miniport driver (i.e., network card driver) to send requests, set, and query information. NDISUIO provides an interface between a user-mode application and NDIS using DeviceIoControl (similar to the UNIX ioctl). The NDISUIO driver (ndisuio.sys) is already installed in your system under C:\WINDOWS\system32\drivers.

Accessing WRAPI functionality written in C++ in Java-based systems means having Java code, which calls a DLL written in C++. In Java this is accomplished by using the Java Native Interface (JNI). To use JNI, class methods that need to access native code are marked with the "native" keyword, and their implementation is provided inside of a DLL. Calling one of these methods will then result in a call to the DLL method, which could be written in any language, including C++. There are two choices when it comes to using JNI with WRAPI, with the difference being if we wanted to make any changes to the WRAPI DLL. In order for Java native methods to directly call WRAPI code, the WRAPI DLL source would have needed to be modified to conform to the JNI guidelines for native method implementation. The second option is to create a native implementation inside a new DLL; the implementations of the DLL would simply load and call the WRAPI DLL methods. The second approach acts as a wrapper for the WRAPI DLL and requires no changes to be made to the WRAPI source code.

You cannot simply include the WRAPI binaries provided on the web site because they may be built with a different version of the C Runtime Library than the version that might be available on your development environment. Any application that will use the WRAPI DLL has to be linked with the same version of the libraries as the DLL was linked against. The cleaner option might be to rebuild the DLL on your system and link it with the available library version, as opposed to attempting to replace the version of the library on your system.

In order to make systems available on other platforms, the equivalent of the system's specific WRAPI application class will need to be reimplemented for that system. The wrapper class exposes an interface to WRAPI. As long as that interface remains unchanged, WRAPI could be replaced or a similar API developed for different systems, without any impact on other portions of the application. In order to access the hardware, operating system-specific system calls need to be made, which makes the API platform dependent. These system calls, though similar in concept, are different in their usage on different operating systems.

The WRAPI DLL needs to be built by including the nuiouser.h file from the Windows XP Driver Development Kit (DDK). It is important that the latest version of the DDK, Windows XP DDK SP1, is used. Previous versions of the DDK might not have the right definitions in the header file that needs to be included; older versions will result in improper run-time behavior.

Before running the WRAPI test application, the user needs to verify that the NDISUIO Windows service has been started and the Wireless Zero Configuration Service has been stopped. Wireless Zero Configuration is a Windows service that is normally started at boot time. It provides auto configuration for the 802.11 wireless adapters (NICs) by scanning for available access points and associating with the strongest signal. This service automatically binds to NDISUIO and does not allow other apps/dlls like WRAPI to bind to it.

The test application first calls WRAPI methods to bind with the correct device and attach to the appropriate network. It then calls a WRAPI method that retrieves the list of visible APs, which contains each AP's MAC address and signal strength. All memory to store the list of network devices and the list of APs is allocated and freed inside of the application and not the DLL.

TABLE 8.1
List of WLAN Parameters Supported by WRAPI

Parameter	Query	Set
Service Set Identifier (SSID)	Yes	Yes
Basic Service Set Identifier (BSSID)	Yes	Yes
Network types supported	Yes	No
Network type in use	Yes	Yes
Transmit power level	Yes	Yes
Received signal strength	Yes	No
Received signal strength trigger	Yes	Yes
Infrastructure mode	Yes	Yes
Fragmentation threshold	Yes	Yes
RTS threshold	Yes	Yes
Number of antennas	Yes	No
Receive antenna selected	Yes	Yes
Transmit antenna selected	Yes	Yes
Supported rates	Yes	Yes
Desired rates	Yes	No
Configuration	Yes	Yes
Statistics	Yes	No
Add WEP	No	Yes
Remove WEP	No	Yes
Disassociate	No	Yes
Power mode	Yes	Yes
BSSID list	Yes	No
BSSID list scan	No	Yes
Authentication mode	Yes	Yes
Privacy filter	Yes	Yes

JWRAPI (http://www.cdt.luth.se/~johank/javawrapi/doc) is a software library in Java and provides methods for accessing 802.11 devices in Microsoft Windows XP and sending raw Ethernet packets directly to device drivers. JWRAPI class provides a Java interface, using Java Native Interface (JNI), to the Wireless Research API (WRAPI), written in C++.

WRAPI parameters that are supported are listed in Table 8.1. The entries in the "Query" and "Set" columns denote whether the parameter is read only, write only, or read–write.

The WRAPI DLL provides a method called WRAPIGetAPList(), which calls the method GetAPList() of the CWRAPIApp object (this object implements the WRAPI code). GetAPList() creates a buffer in which it stores the results returned from the DeviceIOControl() call.

The test application makes a call to WRAPIGetAPList(), which is supposed to return a list of APs and their signal strengths. However, different results are known to be returned on each system. For example, the GetAPList() returns the correct number of APs on one machine; however, it only displays the correct MAC address and signal strength value for the first AP in the list.

When using WRAPI as part of the calibration software, for example, the logic for the calibration part is as follows: The calibration software would contain a reference to an instance of WRAPI to enable it to obtain required information from APs. The calibration software might use a method that takes in x and y coordinates as integers and records tuples for that position. To begin calibration, the user needs to enter the floor name following a naming convention like <institution-name>-<building-name>-floor-<floor number>. For example, the name of the floor to be calibrated is also entered in the text field in the correct format (e.g., building-floor-1). Calibration requires the user to go to the different points on the floor and take measurements with the application. After the user walks to the first point, he can start the calibration.

This would load the map of the floor that the user specified in the text box. After the map is loaded, the user is instructed to click on the map and indicate where the first point is.

Then the entered floor name can be used to display the correct SVG map, for instance. The user is then instructed to go to a point to record measurements at and click on the map. The user would then take measurements per point facing at different directions. (One calibration for every point might not be adequate as studies/tests show that there is a substantial variance in the measured signal strengths for different directions; hence, about four measurements are recommended, for which the values then would be averaged.)

The process keeps a count of the number of points recorded and the number of measurements taken per point. Each time the user clicks the map, the main counter is incremented and the user is instructed to go to the next point or the next direction. The recording would use WRAPI to get AP information and store it in temporary variables. After all recordings are taken for a point, the AP data would be averaged and added to a tuple database object. The x and y coordinates would be recorded only the first time a user clicks on a point. This process would be repeated for the different points, which results in an object containing tuples corresponding to the points. The tuple database object would then be stored onto the hard disk by calling a store database method of the tuple database class. The data are stored onto a file by using an object output stream. This procedure can be done for different floors.

GENERAL RECOMMENDATIONS

It is recommended that for the best tracking accuracy, only the supported adapters for site calibration and tracking are to be used. Also, using similar devices for creating the positioning model and tracking gives better results. One issue is that different Wi-Fi adapter manufacturers use different radio hardware and software in Wi-Fi adapters, resulting in incompatible scales for observing the RSSI values, which in turn translates to positioning inaccuracies (1 to 5 m, or 3 to 15 ft). Ekahau's technology, for example, minimizes the positioning inaccuracy caused by RSSI measurement differences by automatically applying the correct adapter model to mitigate the differences.

In terms of calibration resolution, there are two categories of inaccuracy with the system: inherent and run time. The inherent inaccuracy is due to the calibration resolution. The calibration resolution is the number of points that the user goes to and records signal strength values at. This is a set of points for a given floor, which splits up the floor into a grid composed of blocks. Therefore, the best-case results would place the client in the correct grid, whose dimensions are dependent on the size of the floor. The more points a user goes to during calibration, the higher the resolution (i.e., more blocks), and the closer the center of a block will be to the client's actual location. Run-time inaccuracy causes the client's location to be displayed in the wrong block. Two examples of run-time inaccuracies might include incorrect location updating, or the location not updating even if the user is moving.

In the case where the Wi-Fi network changes, there is a need for recalibration. In order to recalibrate, you need to de-select from your positioning model any APs that have been physically moved or uninstalled, and save them back to the Positioning Engine. Note that data from all APs are processed and saved in the positioning model, but only the selected APs are used for positioning and accuracy analyses. This makes it possible to quickly test positioning and accuracy analyses with different AP combinations without a need for recalibration.

STEP 4: ACCESS POINT PLACEMENT AND CONFIGURATION

This section describes AP placement and configuration. This process usually involves, first, selecting discrete/strategic locations of the beacons, constructing connectivity graphs, determining sensor placements, and defining regions (locales). Preliminary access point locations can be determined by considering the location of wireless users and range estimations of the wireless LAN products being used, approximating the locations of access points that will provide adequate coverage throughout the user areas. Plan for some propagation overlap among adjacent access points, but keep in mind that channel assignments for access points will need to be far enough apart to avoid interaccess point interference. With these preliminary findings, it can be better planned how to test the access point locations. Typically, the placement of the APs is presumed fixed since most communications infrastructure is connected to some back-end network fixed in location. A mobile infrastructure could be used if there exists a predictable movement or periodicity to the position of the APs, or if a frame of reference can be established in conjunction with another positioning system/sensor (this ties back to spatial reference systems and coordinate transformations, discussed in Chapter 5).

The system may include a module (e.g., AP manager) for determining the placement of the APs within the defined space. In such a case, the space in the digital map is defined, including a definition of the obstructions. Obstructions may be assigned values relating to the amount of interference they tend to provide. For example, a brick wall typically provides a greater amount of interference than does a window. Analyzing the interference characteristics in light of a range of signal strengths from a foreseeable set of mobile devices, and in light of the detection and transmission characteristics of the APs, allows access point placement to be determined. If there are detectors having different detection and transmission characteristics identified in the system, the system may determine not only placement, but also selection of detectors. The system may also determine placement of the detectors with respect to the locales.

Given a sufficient number of APs, it would be possible to deduce the relative location of the mobile devices without necessarily knowing the positions of the APs by applying the appropriate geometric constraints. Assuming a fixed placement of APs, the site surveying process can be used to determine an optimal placement of the APs to maximize both communication signal coverage and tracking accuracy throughout a given defined space.

At minimum, it is recommended that at least three AP signals are provided at each location. Note that Ekahau's Positioning Engine, for example, can locate a device with only one AP, but a minimum of three signals must be detected by a client device for better than 5 m (15 ft) average accuracy. It is suggested that the placement of APs is such that any spatial symmetry is broken relative to the traversable paths that maximize the dynamic range variation where possible. Different heuristics can be applied for different spatial geometries and the number of APs to be deployed.

LOCALES

Locales may be defined either prior to or after generation of the signal strength model. However, typically, once the digital map of the space is formed, the locales are defined and the statistical signal strength model is then defined. In other words, a locale may be defined accordingly to a signal strength region. An iterative process of defining locales, generating the signal strength model, and (optionally) positioning the APs may be used. A user may have different privileges, access, or rights with respect to functionality or data, depending on the current locale of the user. Transitioning from one locale to another locale may cause loss of privileges, rights, and access, and in some cases, selective loss or delivery of data. Obstructions may be elevator shafts, for example. Locale A may be a conference room. Locales B1 and B2 may be offices. Locales C1 and C2 may be separate waiting areas. Locale D may be an area that includes exterior space D1 and interior space D2. Locale E may be a location, i.e., a very small area or spot. And space F may be a common area locale, or an area for which a locale is not defined.

RESOLUTION

Resolution (granularity) is the spacing between grid points and influences the accuracy of the position estimate. Granularity depends on the application's accuracy requirements. Note that a position accuracy is usually reported as the error distance deviated from the actual position, while a position precision is reported in percentages of position information that are within the distance of accuracy. In general, increasing the granularity (decreasing the spacing) of the grid only slightly improves the positioning accuracy. However, when the size of database entries is increased, as a result, the search time for matching the fingerprint is also significantly increased. On the other hand, if the spacing is very large, it may reduce the DB size (and search time), but drastically decrease the accuracy. Note that in addition to being spatial (to what level of detail the information is required), granularity could also be temporal (how often the position location information is needed). As a general rule, a grid spacing of 3.5 ft (1 m) is found to be a suitable value.

The resolution of such a system is based on a number of factors, comprising:

- The number and complexity of environmental obstructions
- The number of APs and their placements
- The scalability of the network itself

The primary factor that affects the resolution is the dynamic range of the access point signals themselves. Through the addition of APs with specific signal pattern profiles having significant variation in signal strength over the desired space, the location tracking can improve substantially. Naturally, despite the predicted or expected resolution of a system, an actual deployment may have to consider interference from other unanticipated sources, or that the orientation of the communications signal transducer on either the communications access point or the mobile device play, a factor in the accurate measurement of signal strength.

Note that choosing a smaller grid spacing may increase the accuracy, but not the *precision* or the probability of correctly matching the fingerprint. This is because the fingerprints of two points on the grid measured a foot apart will be more or less the same. A larger grid spacing may provide a better precision, but may render the location information very coarse. Also, the grid spacing is closely tied to the application requirements and must be selected accordingly. It cannot be too small to provide a fine granularity or accuracy, as the precision may be poor. A hybrid scheme provides better accuracy and also good performance by reducing the search time. A smaller grid spacing also results in a more laborious site survey. Certain sites may, however, be more tolerant to a more granular grid spacing because of the radio propagation conditions. And during the site survey, the grid spacing may not be strictly uniform due to the inaccessible parts of the environment, such as walls and office furniture. Note that the actual mobile device positions are not limited to the grids defined in the database. Therefore, the sampled RSS vector measured by the mobile devices may not have the same mean RSS as recorded in the database.

INTERFERENCE OF OTHER DEVICES

Bluetooth signal is one source of interference in Wi-Fi's RSS measurements. The signal strength- based location determination is subject to variation in terms of radio signal environments. Thus, triangulation or profiles including devices' features are necessary to correct the above problem. Triangulation can limit an approximate location of a mobile device by finding the intersectional regions for each AP's signal area. To do this, a mobile device perceives at least two AP signals and holds their BSS ID list.

TIPS FOR INCREASING THE NUMBER OF APs

Note that increasing the number of AP signals increases accuracy, but at a minimum. The following are tips for access point placement:

- Start from the corners of the intended coverage area.
- With three APs, the test area size can be, for example, 20 × 20 m (obstructed) or 40 × 40 m (open).
- If the area has low-height obstructions, such as office cubicles, attach the APs (using indoor antennas) to ceilings or walls above the height of the obstructions.
- Do not place the APs (or antennas) in straight lines or near one another (less than 5 m). Use adhesive tape to secure the antennas for a site survey and testing before installing them permanently.
- Depending on the desired coverage area, refer to the AP manual to adjust the output power levels.
- If you need to move an AP after recording the positioning model, disable the moved APs from the positioning model, or recalibrate the affected area.
- Install dummy APs for additional signals:
  - If the highest possible accuracy is needed, increase the number of AP signals per location by adding dummy APs. These APs can be installed with just power on, but not connected to the network. Just powering up the dummies is enough.
  - Note: Channel configuration is not very important, as the dummies are not sending data and will not interference with data networks.
- Disable automatic power management features:
  - Disable optional power management and self-healing features for reliable positioning results.
  - Enable ESSID broadcasting/probe responses to locate Wi-Fi adapters that do not support passive scanning.
- Minimized interference (noise):
  - High interference may decrease data throughput and positioning accuracy. Noise can be minimized by proper channel configuration, for example, using 802.11b channels 1, 3, 6, 9, and 11 (for five adjacent access points), to keep the channels as separated as possible. Also try to minimize radio interference from other Wi-Fi networks and radio sources.
  - Strong but diversified AP signals.
  - Strong AP signals typically vary less than weaker signals. This is why the best positioning accuracy is usually achieved closer to APs, whereas areas with fewer and weaker signals provide lower accuracy.
  - Stronger is better, but moving APs or using directional antennas to provide different RSSI for areas in close proximity will help differentiate the areas and improve accuracy.

A quick positioning test with Ekahau Manager to test positioning accuracy is to use dummy APs only. Record a positioning model with Ekahau Manager, and click the Ekahau icon in the upper right corner to display the laptop's location. Walk around to test the accuracy.

WLAN Management and Location-Based Security

Effective WLAN management requires the capability to determine the exact location and behavior of every wireless device in and around your facilities. In this way, IT staff can reinstate tried management and security policies that are inherently based on location. For example, WLAN management capabilities related to location-based provisioning include point-of-activity computing (e.g., remotely connecting employees and authorized personnel with front-desk, reservation, point-of-sale, accounting, inventory management, and maintenance systems). With these capabilities, network managers need to leverage 802.1q virtual LAN (VLAN) tagging, so they can control which network resources various defined groups of users are permitted to access based on where they are situated in the facility. With this capability, organizations can establish access policies based on the nature of the area, the user, and device. For example, network managers can define locations like public spaces and conference rooms for visitors to receive basic Internet access only, while wireless access to sensitive information such as human resources, patient records, or research would be shut off completely from employees not employed or located in specified departments.

Location tracking brings granular functionality to indoor enterprise wireless environments, enabling IT staff to establish access control policies that are based on geographic location, immediately identify the source of unauthorized WLAN activity, such as rogue APs, and adapt to network conditions in real time for dynamic capacity management.

Tools for position-reported latitude/longitude, relative signal strength, and location finding can be used to physically track down suspicious devices that warrant action. For example, this AirMagnet Find Tool (Figure 8.13) can be used to walk in the direction of increasing signal strength for any detected AP or station. The Geiger Counter panel in BVS Yellowjacket can also help you find a signal source.

As a commercial product example, the Wi-Fi Watchdog (Figure 8.14) from Newbury Networks identifies all 802.11 traffic, including authorized and unauthorized use, from legitimate clients to rogue access points. Wi-Fi Watchdog utilizes Newbury's LocalePoint sensors to continuously monitor all channels across the 802.11a, 802.11b, and 802.11g frequencies.

As another example, using Ekahau Manager, for example, IT staff can prevent unwanted or unreliable APs from being used for positioning by de-selecting them with the tool. All detected APs do not necessarily belong to the network under your control. For example, another company nearby may begin to use Wi-Fi APs that can be observed with Ekahau Manager. This causes problems if you use all detected APs for calibration, and your neighbors change their AP locations.

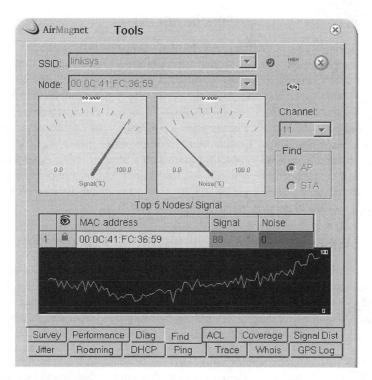

FIGURE 8.13 AigMagnet Find Tool.

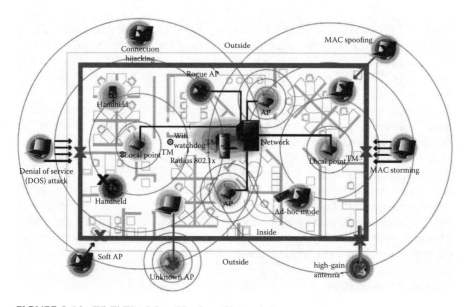

FIGURE 8.14 Wi-Fi Watchdog (Newbury Networks).

STEP 5: TRACKING

Tracking demonstrates where the user is located on a given floor. The client's current location is represented by a dot on the map. Ekahau Manager can be used to start tracking devices. In terms of serving location information to applications, after recording a positioning model and saving it in the Positioning Engine with Ekahau Manager, developers can use Ekahau SDK or YAX protocol to read client coordinates, logical areas, speed, heading, and other information from Positioning Engine into your application. Ekahau Client allows Ekahau Manager to retrieve network data for site calibration (typically from a Wi-Fi adapter installed on a local computer), and allows the Positioning Engine to retrieve network data from client devices for positioning. Note: This is why Ekahau Client must be installed and running on the Ekahau Manager laptop and on each client device that is to be tracked.

STEP 6: MAINTENANCE (PERIODIC ACCURACY TEST)

Positioning model inconsistency (the difference between recorded calibration data and actual radio environment) increases with changes in the environment, such as new walls, cubicle changes, large containers or furniture, and so on. Extensive tests in various dynamically changing environments have proved that during a continuous 30-day field testing the average positioning accuracy may decrease up to:

- 5 to 10% in areas with dynamic changes (manufacturing area)
- 10 to 20% in areas with significant dynamic changes (shipping and receiving area)

Periodic testing can be performed quickly by opening the positioning model from the Ekahau Positioning Engine, activating the tracking feature, and walking around while comparing the actual location to location estimate. A more accurate and recommended method is to record new sample points (5 to 10 % of the number of points in the positioning model) to be analyzed against the original positioning model in Ekahau Manager.

In case of decreased average positioning accuracy, the following positioning model maintenance methods can be used: (1) record more sample points or (2) recalibrate problematic areas. If adding new sample points did not help, or the average accuracy has decreased by less than 30%, delete the existing sample points. This method is very helpful if the area has changed only partially and neighboring areas do not need to be recalibrated. Recalibrate the entire positioning site. If adding new sample points did not help, or the average positioning accuracy has decreased by more than 50%, create a new positioning model and recalibrate the entire site. This method is required if positioning accuracy has decreased due to structural or other major changes in the environment.

CREATING FLOOR MAPS

Some systems, like Ekahau and MIT Cricket, include a digital mapper component to accept or facilitate a definition of the defined space in digital form. The digital mapper may receive tasking via the system manager and may interact with the graphical user interface (GUI) manager to facilitate generation and viewing of the digital map. The digital map may be formed by, for example, translating an architectural drawing into digital form or making use of an existing digital map of the defined space. In other forms, using typical computer-aided design (CAD) tools, a digital map may be formed. Preferably, digital mapper includes tools to accommodate any of the foregoing approaches to accepting or generating a digital map of the defined space.

Requirements for building a spatial/geographic information system (SIS/GIS) include the following: The process of generating spatial information maps should avoid manual surveying and be automated. Some systems, for example, MIT Cricket, include tools to convert CAD drawings into floor plans with annotated structural features (e.g., walls, doorways, etc.). Some companies, like NearSpace, specialize in generating indoor maps with off-the-shelf software to do it. Other components of an SIS/GIS include a space descriptor, a coordinate system, and a suitable data structure.

For buildings, maps are usually readily available as building floor plans. The graphical display portion of a GUI can use floor maps. Acquiring these floor plans in electronic format such as CAD drawings will significantly speed up their integration into whatever graphical display module is developed.

For example, MIT maintains inventories of its buildings and rooms, information that is managed by Space Accounting using the INSITE™ facilities management system. Included in these inventories is a database of floor plans for each floor of each building of MIT. The Department of Facilities (DOF) at MIT maintains a database consisting of architectural two-dimensional floor plans in AutoCAD format of every floor in every building at MIT. That is currently more than 50 buildings and more than 800 floor plans. These floor plans are updated frequently, and therefore mirror the actual floor plans at MIT. Also, DOF provides a two-dimensional base map of the whole MIT campus, showing building contours and their corresponding numbers.

A popular application used to create and view CAD files is AutoCAD made by Autodesk. AutoCAD is a large and expensive application and is not meant to be incorporated into other applications. Using such an application would also be overkill, since the final location-sensing solution would only need to display and not edit the maps.

The basic units of data in maps and CAD files are simple geometric shapes and line segments. These units are combined in various ways to form a map. Scalable Vector Graphics (SVG), used for many different purposes, including animations and maps, also uses these same basic units of data. SVG is an open standard, and there is open-source software for viewing of SVG files. Other key graphics technologies include Java 2D/3D, Flash, and OpenGL.

The process of making a map usually involves obtaining an existing or creating a new CAD file for the floor. This CAD file is then converted to an SVG file using Adobe Illustrator or another conversion utility. The map file is named according to the floor-naming convention described below. This SVG file is then copied to an appropriate application directory. Creating a naming convention involves preparing a list of building names and floor numbers. A floor name might be composed of the network name, building name, and floor number. A floor name might have the following example naming convention (without any spaces): [network_name]-[building_name]-floor-[floor_number]. The chosen naming convention should result in unique names for every floor.

In MIT Cricket, data representation of floor maps can be either a map space descriptor, space label table, or Quadedge. A map space descriptor contains general information about a floor map (version, unique map ID, coordinate value unit). It is used to identify the location of a space and consists of one or more attributes and value pairs, for example, [building=MIT LCS][floor=5][spaceid=06]. Note that in order for the assignment process to be automated, the spaceid field is assigned the same value as the one given on the CAD drawings (floor plans). Cricket allows defining arbitrary space boundaries, but space boundaries are used for the purpose of the automation. Note that since CAD drawings do not contain semantic information about space boundaries, it is difficult to produce perfect space boundaries. A space label table stores space labels and their coordinates. Sometimes a reference arrow is used to point a label into a target point's coordinates (target point's coordinates are inserted into the table).

MIT Cricket leveraged work from Drury (MIT LCS, 2001) in developing tools to convert CAD drawings into floor maps; see the following steps:

Step 1: AutoCAD DXF to UniGrafix conversion. UniGrafix is a convenient format for manipulating graphical elements programmatically.

Step 2: Feature recognition. Add semantics information to drawing elements (e.g., vertices, line segments) to indicate how they compose objects found inside a building (e.g., walls, doorways, staircases, elevators). Use feature recognition algorithms on the UniGraifx file to extract such objects (then used to annotate the floor map).

Step 3: Path extraction. Apply the constrained Delaunay triangulation (CDT) to produce a planar graph that binds the spatial relationship between different structural features of the floor plan. Applications can use this planar graph to extract accessible paths in a given floor plan.

Step 4: Labeling. Extraction of space descriptors from the original floor plan to be used in labeling each element in the floor map.

These steps are portrayed in Figure 8.15.

Step 1: AutoCAD DXF to UniGrafix Conversion

AutoCAD floor plan elements are represented by graphical primitives (e.g., vertices, segments, polylines, circles, arcs, and text). These primitives are organized into different layers, which name the group of primitives that represent load-bearing structures, ordinary walls, labels of rooms, staircases, and other objects. Examples are the following: A-WALL-CORE, A-WALL, A-AREA-IDEN. Unfortunately, the layering organization depends only on the convention used by the draft person, and hence is not useful for object extraction. For example, there is no separate layer to

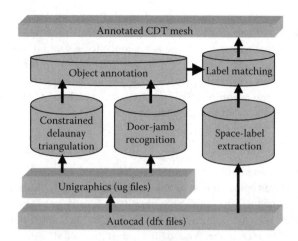

FIGURE 8.15 Steps for indoor map creation. (From Allen et al., CricketNav, MIT thesis, Cambridge, MA, 2002.)

identify things like staircase and doorway elements. DXF files are encoded in ASCII, grouped into code/value pairs. For example, a vertex object is represented by a specific group code followed by a coordinate value. This makes DXF files easy to parse. Conversation to UniGrafix takes place because a set of APIs has been developed to manipulate UniGrafix elements in C, useful in extracting features later in the map generation process. Conversation takes place using the acad2µg utility (originally part of the Building Model Generator (BMG) tool kit (Lewis, 1996) and later modified by Drury to convert MIT floor plans, due to different floor plan conventions).

Step 2: Feature Recognition

It is important to recognize doorways from graphical primitives so that applications can compute a customized shortest-path metric. Due to lack of semantics information, manual determination is needed. For example, MIT floor plans mark doorways as an arc. In addition, certain applications need to know the exact doorway boundary (to determine the space boundary of a room). Therefore, there might be a need to recognize the surrounding door jambs.

Step 3: Path Extraction

Accessible paths need to support navigation applications. For this reason, floor plans are divided into two types of spaces: (1) valid, an area that can be occupied by a person, and (2) invalid, an area occupied by a physical obstacle (e.g., wall). Hence, the desired path should not intersect any invalid space and should lie approximate in the center of valid space. Extracting accessible paths is done using the constrained Delaunay triangulation (CDT) to subdivide the floor plan into triangles. The CDT algorithm can be applied to a general set of input segments known as the constraining segments. The result is a triangulated planar graph (defines the boundary between valid and invalid spaces). One issue that might come up is the gaps/leakages between segments (need for a minimum tolerance level). Another potential issue is the gaps/leakages from ground floors of buildings (need to define a bounding box encompassing the outline of the building). Also, the acad2μg tool sometimes discards wall segments due to incorrect recognition (need for manual intervention). This results in a need for CAD tools that will embed standardized semantics information.

Step 4: Labeling

Separate tool based on acad2μg for extracting labels and their coordinates from DXF floor plans. (UniGrafix does not have an expression to represent textual info; hence, labels are not propagated to UniGrafix.) For step 4, it is helpful to annotate each planar graph element with a space label. A coordinate of a label defines the center reference point of a given space. A quadedge is a single data structure that statistically represents the CDT planar graph. A software package called IDT (incremental Delaunay triangulation) by Lischinski can be used to compute a CDT.

Dividing spatial information maps is also of some consideration. A floor map for each floor typically has a size of 300 to 400 kb. Elevators and staircases are linkage points between two floor maps. In MIT Cricket, applications are download-able from the SIS server onto a mobile device and a unique map identifier is provided (e.g., campus/building/floor number) for the download process.

In Chapter 3 we talked about locales and their relation to a signal strength model used for positioning. Typically, a defined space is comprised of a set of defined regions, areas, or locations, collectively referred to as locales. Each locale is defined within the system in relationship to the digital map of the physical space. A locale may be defined as an interior or exterior space or location, or a combination thereof. For instance, a conference room, office, or waiting area may each be defined as a single locale within a defined space.

THREE-DIMENSIONAL INDOOR MAPS

Three-dimensional representations of spaces are also possible. For instance, the graphics department of MIT is turning the two-dimensional floor plans of MIT into three-dimensional building models. The goal is to develop a process that converts all of these floor plans into a three-dimensional walk-through model of the MIT campus. A three-dimenisional model of each individual building is created by the Building Model Generator. This program takes each floor plan and extrudes all the elements of the floor plan, such as the doors, walls, stairs, etc. All of the extruded floor plans of a building are then stacked upon one another to generate a complete building. This method, however, is only capable of generating single walk-through models of buildings. In order to generate an entire campus walk-through model, the location and orientation of each building with respect to each other need to be known.

There is a cost associated with creating and maintaining a topographical model. However, in the future, three-dimensional models of cities, buildings, and malls will be created to support applications such as virtual tours, urban planning, and disaster management. The remaining issue would be whether these models would be available for free.

Some of the challenges linked with three-dimensional representations of spaces include the following: efficient construction of a topography model, storing and querying the topography model on the user's mobile device, and providing the user with a convenient interface to input information (e.g., landmarks) while also resolving ambiguities in the user's ID of, in this case, landmarks (e.g., relevance of tall, small).

In terms of creating a three-dimensional topographical model of a region of interest (e.g., a mall, city), there is a need to model any objects that either are likely to be identified as landmarks or are large enough to cause occlusion. The most promising approach is to use a large collection of photographs to deduce the physical structure of a scene in an automated fashion. Prior information, such as a CAD model of a building or a city, can facilitate construction of the model. Examples of relevant projects include the Virtual L.A. Project and the MIT City Scanning Project.

The following steps for constructing a three-dimensional model are taken from the MIT City Scanning Project:

1. Capturing images: Several overlapping photographs of the scene of interest are taken. Photographs are taken from several locations. The position and orientation of the camera are recorded each time a photograph is taken.
2. Constructing image mosaic: The individual images are aligned to form a seamless mosaic. Knowledge of the camera position corresponding to each image greatly facilitates this task. A mosaic offers a much larger field of view than any individual image, which results in better efficiency and robustness. Moreover, knowledge of camera position is very helpful because it reduces the number of unknowns that the reconstruction algorithm has to deal with.

Several algorithms for three-dimensional reconstruction are discussed in the MIT City Scanning Project. These include feature- and region-based algorithms as well as edge histogramming:

- Feature based: Attempt to infer structure based on proximity of features such as corners or edges.
- Region based: Attempt to infer structure by identifying a large number of pixels that appear to lie on the same surface. For example, the façade of a building may be identified in this manner. Another approach is to use voxels (volume pixels, the three-dimensional equivalent of pixels) and multiscale image processing to extract a volumetric representation of the scene.
- Edge histogramming: No feature correspondence is used, but edges are used instead to identify and localize prominent vertical façades under the assumption that such facades exhibit many horizontal edges. This technique produces a peak whenever many edges reinforce a single façade, yielding the orientations of the dominant façades in the scene. This technique yields façade geometry but not texture information. While the latter may be inconsistent across multiple images (each of

which may be partially obsured by scene clutter), geometry information correlates across images, so the multiple images reinforce each other to produce a consistent and high-confidence estimate of the shape of a structure such as a building.

USING THE EKAHAU POSITIONING SYSTEM

To be trackable, each mobile client (laptop or PDA) must be running the client software. Before the Positioning Engine can be used for tracking devices over the WLAN, it needs to be calibrated manually. A floor map is first uploaded, on which the Positioning Engine draws tracking rails to create a positioning model. The calibration of the system requires the user to physically move around the area with a mobile client equipped with the Ekahau Client. Approximately every 10 ft the current position has to be indicated by clicking on the uploaded map on the client to record sample points containing received signal strength samples. The software does not need the location coordinates of the access points. Tracking of mobile clients on the WLAN can be initiated after calibrating the system though the Ekahau Manager. Real-time location of users is shown on the provided floor plan. To locate clients, the software calculates the signal strengths from the access points and compares them to calibrated signal strength samples and to the map of the location. Since Ekahau's system requires the Ekahau Client to be installed on all mobile users, it can only be used over a private network where the company has some administrative rights over the wireless users.

As mentioned in Chapter 5, Ekahau Positioning Engine (EPE) 3.0 offers two ways for serving location information to local or remote applications: (1) Ekahau SDK Java API for Java developers and (2) character-based Transmission Control Protocol (TCP) Ekahau YAX for any programming language that supports TCP sockets.

Both options utilize TCP sockets to connect to the Positioning Engine. The SDK is just an easy-to-use Java front end for the YAX protocol. After recording a positioning model and saving it in the Positioning Engine with Ekahau Manager, you can use Ekahau SDK or the YAX protocol to read client coordinates, logical areas, speed, heading, and other information from Positioning Engine into your application.

EKAHAU JAVA SDK

The Ekahau Software Development Kit (SDK) Java package, com.ekahau.sdk, provides a quick and effective way for accessing location information, from either a local or remote computer. All the developer needs to do is import the Ekahau SDK package into his Java application and follow the provided examples. The SDK integrates the location information to external applications. The location information is displayed in the form of (x, y), floor, speed, heading, and logical area, such as "main conference room."

Using SDK requires a working knowledge of the Java programming language and Java 2 SDK (from version 1.4.2 up). The Positioning Engine's TrackedDevice objects represent wireless devices, so one object needs to be created for each physical device that is to be tracked.

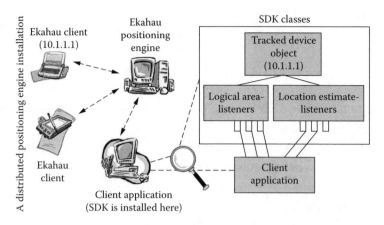

FIGURE 8.16   Ekahau-distributed Positioning Engine installation.

TrackedDevice objects are used to return the coordinates, time stamp, status, speed, map, and any logical area information. The client receives the information via three different kinds of Listener classes, which the programmer using the SDK should implement (Figure 8.16). The Listener interfaces are the following:

- LocationEstimateListener
- LogicalAreaListener
- StatusListener

YAX Protocol

YAX is a TCP-based ASCII protocol designed to be human readable for quick testing with Telnet, for example. YAX consists of request–response messages encoded/decoded using UTF-8 character set.

Developers can quickly test YAX with any Telnet program by connecting to Positioning Engine's Internet Protocol (IP) address using port 8548 (the default YAX protocol port).

Below is an example for finding a device by a certain device property, IP address in this case, from Positioning Engine that is running on a local computer.

YAX Telnet Example

Open Windows command prompt or any Telnet program and type in the italic text below to receive the responses (regular text below). Do not mind about the authentication and MD5 details at this point.

```
telnet localhost 8548
<HELLO 1 "cbda430474dc51d">
<HELLO 1 "" password=Llama>
<TALK "yax" 1 "yax1" "MD5" "">
<TALK "yax" 1 "yax1" "MD5"
```

```
"9ef4f785b26890941212da6792859c51">

<GET_DEVICE_LIST NETWORK.IP-ADDRESS=10.0.0.1>

<DEVICE_LIST

4

>
```

The number in the DEVICE_LIST response message is the device ID for the found device.

Parsing YAX Messages

The general YAX message format is:

```
<[#MSGID ]message

property1=value1

>
```

#MSGID (message ID) can be any string without white spaces. It can be used by the programmer to identify response messages. ID is optional — if it is omitted, the response will not contain it either. Line breaks (carriage return and line feed together, \r\n in C and Java) and the characters < and > will have to be escaped with a back slash if they appear inside of the message content of your messages, for example, \<. Responses will contain the escaped characters if they exist in the message body. Every message ends with character > and carriage return + line feed.

YAX Main Functionality

Getting the Device List

The device list can be received without any parameters, in which case all client devices that have been detected by the Positioning Engine are returned. The list can be filtered with specific parameters like NETWORK.DNS_NAME. The parameters can also be regular expressions instead of plain strings. Regular expression must use Java's regular expression syntax. For more information, see http://java.sun.com/docs/books/tutorial/extra/regex/.

Parameters may be omitted. There are two parameters that can be optionally used (below with example values):

```
NETWORK.MAC=00:E0:63:82:65:76
NETWORK.IP-ADDRESS=10.0.0.1
```

For reference of the device properties in different situations, see Chapter 5.

Request:

```
<[#MSGID ]GET_DEVICE_LIST

param1=value1

param2=value2

...

>
```

Response:

```
<[#MSGID ]DEVICE_LIST
deviceID1
deviceID2
deviceID3
...
>
```

The response contains device IDs as integers on separate lines; however, it is recommended to treat them as strings. The same device always gets the same device ID as long as the Positioning Engine is running, but do not count on the device getting the same ID after the Positioning Engine process has been restarted. Also, note that the response device list can be empty as well.

UNDERSTANDING LOCATION ESTIMATES IN EKAHAU

Before going into specifics, it is important first to understand location estimates when deploying a service using Ekahau. A location estimate for a device can be described as an arrow that may have its ends located on different maps. Accurate location is the most accurate location estimate with some delay (typically 5 sec), depending on positioning settings, and is more accurate than latest location because more history data can be used to calculate the estimate. Latest location is the most current location estimate. It is less accurate than accurate location but almost real time (typically 1 to 2 sec delay), and as a result, it is most useful to applications that require real-time locations.

Both location estimates contain rectangular x and y pixel coordinates that originate from the associated map's top left corner, growing from left to right and top to bottom.

Read the accurate and latest locations using YAX protocol's LOCATION_ ESTIMATE message:

- *accurateX* and *accurateY*
- *latestX* and *latestY*

Read the accurate and latest locations using Ekahau Java SDK:

- *LocationEstimate.getAccurateLocation()*
- *LocationEstimate.getLatestLocation()*

UNDERSTANDING LOCATION MAPS AND LOGICAL AREAS IN EKAHAU

Maps and location contexts have a one-to-one relationship. This is why both accurate and latest location estimates have their own location context references (Location.getLocationContext() in SDK and ContextID in YAX). Typically, both location estimates refer to the same location context, i.e., map.

Location context also contains the map scale (indicating how many map pixels equal one real-world meter or foot) and user-defined map properties (name–value pairs) if such information has been manually saved in the positioning model with Ekahau Manager.

In addition, the Ekahau Positioning Engine supports predefined logical areas to allow application developers to quickly display area names, find nearest printers, create virtual tours, add burglar alarms, etc. Before reading area information from Ekahau SDK or YAX, you need to draw and name logical areas and save them in your positioning model with Ekahau Manager.

USING AND REFRESHING LOCATIONCONTEXT

It was previously explained how to receive coordinates by using LocationEstimate-Listener. These are pixel coordinates of a map image, one of the maps that was added in the positioning model with Ekahau Manager. The same map can be retrieved through the SDK by using the LocationContext object, which is attached to both the latest and accurate location estimates.

Note that devices can move from a map to another. If you want to keep displaying the correct map that contains a certain device, first retrieve the map with the LocationContext.getMap() method, and detect the situation where the client has moved to another map, to retrieve and display the correct map. This functionality is implemented in TrackingView.java example.

If you want to implement the LocationContext check yourself, compare a new LocationContext object (from LocationEstimate or LogicalArea object) to the previous one: if the equals method returns false, use LocationContext.getMap() and redraw the map; otherwise, the tracked device is still located on the same map.

GETTING THE MAP IMAGE OF A LOCATION CONTEXT

The map image is returned as a binary file in PNG format. Note that the ID is the location context ID.

```
Request:
  <[#MSGID ]GET_MAP ID>
Response:
  <[#MSGID ]MAP ID
  size=X
  type=png
  data=binarydatabinarydata
  binarydatabinarydata
  binarydatabinarydata
  binarydatabinarydata
  binarydatabinarydata
  >
```

APPLICATION/SERVICE EXAMPLE: HAWKTOUR

A good example of deploying a location-based service (LBS) using Ekahau is the HawkTour project at IIT, an end-user application written in Java that automatically presents a scalable map of the local area and links points of interest in the vicinity.

When a user clicks on a link, the UI retrieves multimedia content of the point of interest from a content database.

Ekahau's Positioning Engine is used to determine the position of users. Each Tablet PC that the system tracks runs the Ekahau Client software. The Ekahau Manager was used as a platform for creating positioning models, tracking devices, and analyzing positioning accuracy.

Just like outdoor global positioning systems (GPS) guide tours, HawkTour's navigational capabilities tell students where to find classrooms, food service, restrooms, and specific administration areas. HawkTour does not have the capability to navigate to a particular location such as a classroom. Instead, it only displays text, image, and audio based on the user's current location. It does, however, show (as icons on the map) the location of nearby restrooms and food service areas. It does not give directions to getting there. It can even display text, pictures, and audio that describe the history of the immediate part of the facility. A moving map on the client device provides the user with situational awareness, making it easier to navigate to a particular destination. Hawk Tour can lead the user to a campus building or a specific room. Ekahau is the location technology that is currently used for HawkTour development. However, it is possible to use another location technology by simply exposing that technology as a web service, or making it a local service, such as the GPS service in HawkTour. A Cisco Wi-Fi wireless LAN enables Hawk Tour to function by offering communications links and supporting the location system. The Tablet PC communicates with the campus network and servers. Information regarding a particular part of the facility, for example, is sent to the Tablet PC over the wireless LAN. HawkTour is currently achieving 5 to 7 ft of accuracy in areas having very good signal strength and coverage. Coverage falls mostly within buildings, with some signals spilling outside. This causes signal strengths to vary across the campus, and some coverage holes persist. More on deployment issues in Chapter 9. The installation of Ekahau's positioning system is fairly straightforward, but calibration is necessary initially. Site survey is required throughout the relevant areas with a user device equipped with the Ekahau Client. On the device, a network administrator clicks a map to record sample points every 10 ft or so. At each sample point, a rotation of $360°$ is needed to capture access point signals from all directions. Once a site survey with a given map of the area is done, it serves as the spatial reference for the locations. Thus, some of the features, such as where is the nearest bathroom, can be provided visually (indicated by an icon). No real need for a GIS is needed for this purpose of mapping features on a map; however, a GIS is seen as beneficial for path finding.

HawkTour System Architecture

The following sections were adopted from the HawkTour application website literature. The following is a technical overview of the service, including the protocols and tools used by the application to provide the overall functionality. The HawkTour application runs on a Tablet PC. The HawkTour application is implemented in Java and is responsible for retrieving data from the content and location services and displaying these data. The graphical user interface is designed using Java's Abstract

Windowing Toolkit (AWT) and is currently tailored to fit Tablet PC displays. The AWT design will allow for simpler translation of the interface to smaller devices, such as PDAs or cell phones. Java was picked initially because it was the language that the first developer was most familiar with. Java provided all the basic functionality that was required for the HawkTour project, and also has excellent open-source development tools (Eclipse, Apache Tomcat) available. Ekahau also has Java classes available for use, which make integration with the location technology easier. This is part of the Ekahau SDK Java API. Java is the most likely language to run capably on mobile phones. Also, HawkTour is designed using standard communication methods (Simple Object Access Protocol (SOAP), for example), and modules are specifically designed so that they can be replaced by other modules that are implemented differently. Basically, if you want to implement a HawkTour content service in C#, Perl, or Python, you can (provided your language of choice can talk using SOAP). If you want to make a GUI in GTK# that connects to existing HawkTour services, you can.

There are two core services: location service, which is responsible for tracking devices in the HawkTour environment and providing that information to the application, and content service, which is responsible for hosting all of the HawkTour content and providing that content to the application when requested. Because these services are web services, they are available to be used by other web service-based applications. A third service, the hardware interface, provides an API that allows the HawkTour application to interface with hardware devices, such as wall displays, via Bluetooth.

The HawkTour infrastructure is based on web services, which allows for Hawk-Tour services to be contacted remotely and allows other client applications to make use of the HawkTour infrastructure. This is convenient to transfer data across components of the application and allows HawkTour modules to be virtually hot swappable, so that the application does not have to rely on any particular technology.

HawkTour Application Architecture

The application is based on a Model-View-Controller (MVC) architecture, and consists of five major components:

- Main Manager
- Location Manager
- Content Manager
- Map Manager
- GUI

The MVC is a software architecture concept that was first developed at the Xerox Palo Alto Research Center for its Smalltalk-80 system for the multiwindowed graphical interface — the first in the world. The Model section of the program is the function, algorithm, or state machine; the View is the output, usually a rectangular portion of the screen; and the Controller is the input handler/interpreter. Within this triad there is heavy communication between the View and the Controller, and com-

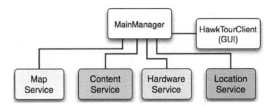

FIGURE 8.17   HawkTour hierarchy.

munication between the Model and each View and Controller. HawkTour is built on the MVC design pattern, with the HawkTourClient (GUI) corresponding to the View, the MainManager corresponding to the Controller, and each of the back- end services corresponding to the Model. Figure 8.17 shows the application hierarchy of Hawk-Tour. The sections and figures that follow illustrate how each of the component services works.

The main purpose, besides code readability, of HawkTour is a consistent interface for the application to communicate with the different services. The motivation for this was to make the different services completely independent of each other, so that the application would not be inherently tied to a technology used to implement one of the services. For example, HawkTour currently uses QuickTime for content delivery, and Ekahau is used for Wi-Fi positioning. Either of these components can and may be upgraded to newer or more favorable technologies in the future, or another development group may want to substitute a technology more appropriate for its purpose for its version of the HawkTour application.

Location Service

The location service is a web service implemented in Java and deployed on an Apache Tomcat server running Apache SOAP. The location service itself serves as a public gateway to an Ekahau Positioning Engine, which provides location data based on wireless signal strength from 802.11 access points. The location service gateway is multithreaded, which allows it to handle multiple client connections and handle the asynchronous transmissions of the Ekahau back end. The Ekahau back end is accessed by the application via the exposed methods of the Java web service. This communication is done with SOAP. At the time of writing, Ekahau's SDK was not used to design the service, but it is foreseen to be used in the future with the expectation that a more robust service can be designed.

The location service is implemented as a passive (request–response or pull) service, which means that the application desiring information from the service is responsible for connecting to it and requesting the appropriate information. The location service is responsible for reporting back to the application the current map information (e.g., content such as points of interest) for which the location information applies. This is in addition to the positioning information provided by the Ekahau Positioning Engine. This allows the application to provide maps of greater detail in appropriate areas. Figure 8.18 shows how the location service works.

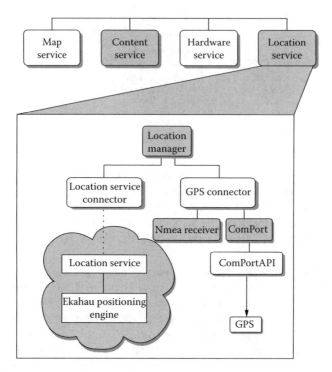

FIGURE 8.18   Location service.

Content Service

Similarly to the location service, the content service is implemented in Java and deployed as a web service on a Tomcat server running Apache SOAP, and is also a passive service. The content service itself serves as a public gateway to a MySQL database This is similar to the location service that serves as a gateway to a back-end Ekahau Positioning Engine. Communication between the service and the database is done via the MySQL Java Database Connectivity (JDBC) connector. There-fore, the content service and database can be located on totally different servers, since communication between the service and the database can be done remotely. The database makes use of the InnoDB database type, which allows for foreign key constraints and is searchable in a variety of different ways. The database contains universal resource locater (URL) references to content, which the content service returns to the application. The content service only provides references to the content; it does not deliver the content itself. This prevents the need to organize binary data, such as pictures and audio, into SOAP envelopes, and allows the actual content to be located in multiple locations. All the content is downloaded via a wireless Internet connection from a web server. The application simply gets a URL for all content it is supposed to display. A wireless network connection must be available to fetch or download the data from a URL. Outdoors on our campus there is no Wi-Fi available. Once the entire campus becomes wireless, then there will be no problem with

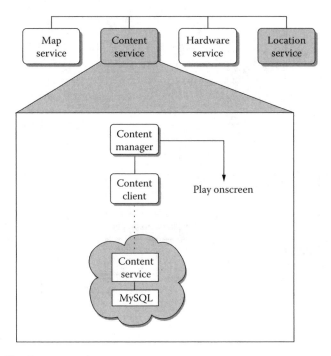

FIGURE 8.19 Content service.

downloading content using the same communication network. Figure 8.19 shows how the content service works.

Map Service (Map Manager)

The map service is responsible for managing all map-related functions of the client application. Location information is provided by the location service, which must then be displayed appropriately by the map service. There are two basic functionality points the map service must provide: (1) it must display the map on the application GUI, (2) it must perform the calculations necessary to translate the coordinates from multiple systems into display coordinates, and also provide some intelligent mechanism for determining submap hierarchy. The map service provides a simple mechanism to facilitate communication with the rest of the client application. Since the map is interactive, users are able to invoke commands by interacting in some way with the map. In addition, the map display is capable of accepting manipulation commands from elsewhere in the application. The MS must have a storage mechanism for map information. The map and subformats are probably suitable for this, so their redesign is probably not necessary. The map service also needs to strictly define what resources are necessary for its functionality, and where those resources should be stored. It would be useful to determine whether this information could or should be provided remotely (possibly as a separate mapping service) or stored locally. Figure 8.20 shows how the map service works.

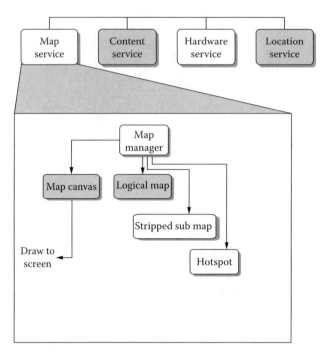

FIGURE 8.20 Map service.

Hardware Interface Service

The hardware interface utilizes hardware devices that might be available to the user. This feature is currently implemented for display devices, such as televisions and computer monitors. As the user equipped with a HawkTour device approaches a large display, the HawkTour application will prompt the user to interact with the display. Possible functions that the display can provide are as follows:

- Greet guests as they enter the building
- Display large campus and building maps
- Display videos

The hardware interface is built in Java and C++. It utilizes a layered structure that allows Java and C++ code to communicate with each other. This is made possible through the Java Native Interface (JNI). The Bluetooth hardware device is controlled using C++, which is then wrapped in JNI and Java and made available for the Java HawkTour application to utilize. Figure 8.21 shows how the hardware service works.

HawkTour Application Implementation

MainManager

MainManager is the working of all classes of HawkTour. It is a Singleton class and creates instances of the middle managers {LocationManager, MapManager, Con-

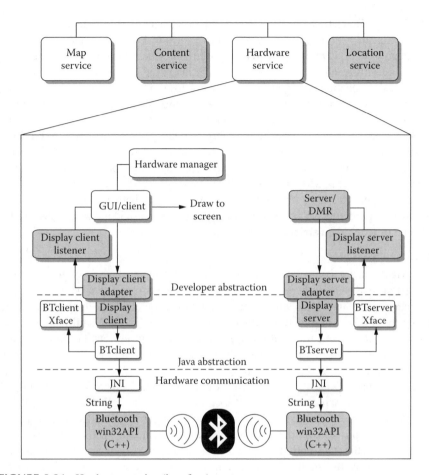

FIGURE 8.21 Hardware service (interface).

tentManager, and HawkTourClient}. It acts as the middleman between them. If, for
example, the application needs to find the current location, you just call the name
of the MainManager (mainMan for the sake of argument), as in mainMan.Location().

LocationManager

The LocationManager class is in charge of gathering location data from the LocationService. The LocationService in turn connects to the locator servers and puts
the information in a plain old vector, called LocationData. The format of Location-
Data is (in global coordinates) X (longitude), Y (latitude), speed, and heading. There
are currently two location services: Wi-Fi and GPS. The LocationServiceConnector
(package edu.iit.cs.scs.HawkTour.ServiceCommunication) retrieves the location data
from the Wi-Fi location server, which right now uses the Ekahau Positioning Engine.
If a different technology becomes more favorable in the future, a separate Connector
class would be written for it and the LocationServiceConnector would be deprecated.

GPSConnector, of course, gets the information from the GPS device, which
happens to be on the local host, through the ComPortAPI class (package

edu.iit.cs.scs.HawkTour.LocalServices). But it is cleaner to have a separate connector for each locator service, and they all return values to LocationService. In addition to gathering data, LocationManager also arbitrates which locator service the application is using, so that the source of information is hidden from the application. It simply asks LocationManager for the current location and returns an (hopefully) accurate value, from whichever service the LocationManager deems most suitable for the present conditions. Currently, automatic arbitration is not workable, so a manual mode switcher is used. An industry standard for handoff between GPS and Wi-Fi would be very much welcome and certainly needed. This is seen as immense help for future HawkTour projects instead of having HawkTour developers develop a proprietary system. Moreover, adding additional positioning technologies, such as cellular, requires having a standard. Otherwise, switching between these technologies becomes a problem, particularly when a positioning provider (be it services like Ekahau or a cellular carrier) changes.

MapManager

The MapManager controls the classes that are related to the mapping service in HawkTour. Right now there is one MapManager constructor that accepts a Main-Manager object, two floats, one string, and a dimension object as its starting values. Using the setLocation function, which will be called from the MainManager every 10 sec or so (a fixed time), the current PixelPoint and current GpsPoint are set together with the current PixelView, which will later be sent to MapCanvas to draw. The gpsToPixel class is used to convert GPS coordinates, which are obtained to pixel coordinates, which is used to display graphics on the screen. The MapManager keeps all the submaps for the current location in a vector called mapStack. When the current location changes, MapManager uses is_In functions to determine whether the current map is still valid. If it is not, and an adjacent map needs to be displayed, it traverses up the stack until the is_In function returns true, then uses the findSubmap function to find the smallest available map for the current location. MapCanvas handles the display of the map. MapCanvas takes in an image of a map, draws a portion of it, and displays it as our current viewable map. It also handles the size of the user location dot. LogicalMap's main function is to take in a map file, parse it, and load up the appropriate Submap (*.sub) and Hotspot (*.hot) files. The HotSpot class's job is to just return the hot spot ID and the coordinates of the hot spot to LogicalMap. The StrippedSubMap class is similar to the HotSpot class, but instead returns the coordinates of the submap and the name of the submap found within the coordinates of the current map. The coordinates mentioned are all global (GPS). In addition, LogicalMap provides is_In and findSubMap functions for MapManager to call when searching for the smallest available map to display.

ContentManager

ContentManager receives a list of URLs from ContentClient, which could be better named ContentServiceConnector. ContentClient uses SOAP to communicate with the ContentService. ContentService is basically a Java database accessor with a bunch of getter functions that are called by the getter functions of ContentClient.

HawkTourClient

The HawkTourClient handles the rendering of the main window, the user interface the user sees when running the application. It makes calls to MainManager and MapCanvas. It generates the map window (zooming/scrolling buttons and main map display) and menu options (buttons).

Hardware Service

The hardware section supports Bluetooth functionality. With Bluetooth, the application is able to communicate with digital media receivers (DMRs). A DMR is a Bluetooth-enabled PC whose video output goes to a large-screen TV or HDTV and digital media are stored on it. The DMR is called upon by the Bluetooth client (i.e., HawkTour running on a Tablet PC) to start displaying pictures or video. This is done when both devices, the tablet and the DMR, discover each other through Bluetooth when in near proximity.

In Figure 8.21, the Bluetooth functionality is divided in two: the client and the server. On the client side, the GUI/client is implemented in a SampleClient (a generic implementation written by a generic developer who wishes to use the Bluetooth function), which extends the DisplayClientAdapter. This setup allows the developer to implement just a subset of the methods in DisplayClient. The SampleClient receives events to display to the screen through the DisplayClientListener, which defines the events that are received from the DisplayClient and sends messages over to the DMR through the DisplayClientAdapter. This way, SampleClient and DisplayClientAdapter do not need to make constructors for each other, and message passing is simpler. The user (generic developer) never needs to know what is behind the DisplayClient. Similarly, the DisplayClient receives messages through the BTClientXface from BTClient and sends messages to BTClient. The BTClient, responsible for device detection, device information, and device connection parameters, is the lowest level at which the application transmits the messages before they actually enter the Bluetooth hardware. The Bluetooth Tx/Rx hardware is dealt with by the Java Native Interface and the Win32 API. On the right side is the DisplayServer class, which is called by SampleServer, or any other class that needs to use it, and DisplayServer will return data from the BTServer class. Again, the user will never need to know what is behind the DisplayServer class when writing an application like SampleServer. DisplayServer provides the functionality needed to create a server by encapsulating Bluetooth functionality and message passing. DisplayServer provides events through the interface DisplayServerListener. DisplayServerListener defines the events that are received from the DisplayServer. The Location, Map, and Content Managers are responsible for all functions relating to their names. For example, the Map Manager is responsible for all map-related functions, such as coordinate translation, map display, and managing the GUI map elements (Figure 8.22).

Each manager manages the GUI elements that belong to its functional area. For example, the Content Manager contains QuickTime GUI elements within itself that it passes to the GUI in order to create the GUI. The GUI itself is only responsible for displaying the interface to the user and passing user input back to the appropriate

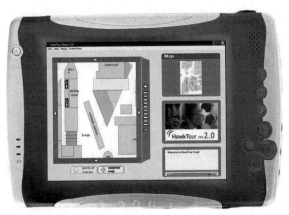

OR (depending which one is higher resolution)

FIGURE 8.22   HawkTour GUI. (Note: User trail feature does not exist.)

manager. The GUI itself is not responsible for executing anything other than simple method calls on the appropriate manager.

All communication is funneled through the Main Manager, which instantiates the client and facilitates communication between all the other managers. This architecture allows for a unified communication pathway between the managers and creates code that is more easily debugged. When the application initializes, it first communicates with the local Scarlet run time (this means that a Scarlet run time must be running on the client machine before HawkTour can be started) to determine where the content and location services are located. Once the services have been found, the application periodically queries the location service for updated location information. If the location has changed, the application updates the map display appropriately and queries the content service for updated references to appropriate content.

Once these references are received, the application initializes download threads to retrieve the information via a wireless network. The application makes use of Apple QuickTime technology to display the audio and image content. The QuickTime technology is threaded internally and provides a more robust content delivery solution than using Java's standard content delivery options. If for any reason the application is unable to communicate with the location service, there is an additional local GPS module that provides location information via GPS satellites. When the GPS unit is used, content delivery is disabled automatically, because network connections are assumed to be inaccessible. When the application regains a connection to the standard location service, content delivery is automatically reenabled and the GPS module becomes inactive. The GPS module also provides outdoor location awareness capabilities and allows the user to navigate in outdoor environments. Though content delivery is disabled in GPS mode, the mapping subsystem remains active and continues to update the location display appropriately.

USING MIT CRICKET

This section was adopted from the Cricket manual. It provides information to develop Cricket applications to run on handhelds, laptops, and desktops under Linux or Windows and maintain a Cricket installation. The Cricket manual can also come in handy when writing location-aware embedded wireless sensor-computing applications on the Mote platform. Writing such applications will not be difficult because the Cricket embedded software is written in TinyOS, the software platform for the Motes.

SYSTEM SETUP

The only configuration required in Cricket is setting the string for a space that is disseminated by a beacon. The specific string is a function of the resource discovery protocol being used, and Cricket allows any one of several possibilities (see below for information about the implementation platform and integration with an inertial navigation system (INS)). Cricket also provides a way by which the owner of a room can securely set and change the space identifier that is sent in the advertisements. This is done by sending a special message over the same RF channel that is used for the advertisements, after authenticating the user via a password. At this stage, it was decided that Cricket allows this change only from within physical proximity of the room or location where the beacon is located. This makes the system somewhat more secure than if it would be allowed to do this from afar.

The boundaries between adjacent spaces can either be real, as in a wall separating two rooms, or virtual, as in a nonphysical partition used to separate portions of a room. The precision of the system is determined by how well the listener can detect the boundary between two spaces, while the granularity of the system is the smallest possible size for a geographic space such that boundaries can be detected with a high degree of precision. A third metric, accuracy, is used to calibrate individual beacons and listeners. It is the degree to which the distance from a beacon, estimated by a listener, matches the true distance. While Cricket's experiments show that the distance accuracy of its hardware is smaller than a few inches, what matters is the precision and granularity of the system. These depend on the algorithms and the placement of beacons across boundaries. Cricket's goal is a system with a close to 100% precision, with a granularity of a few feet (a portion of a room).

SETTING UP THE BEACON COORDINATE SYSTEM

To enable the receivers to gather the distance information (ratio of height to distance), Cricket implemented a local coordinate system using four active beacons instrumented with known positions within the space. The compass determines its mean position as an (x, y, z) tuple by listening to beacon transmissions (Figure 8.23).

The beacons are configured with their $(x, y, 0)$ coordinates and broadcast this information on the RF channel, which is sensed by the receiver on the compass. At the same time, it also broadcasts an ultrasonic pulse. As the user moves around the building, the listener infers its location and asks the map server to provide the location on the map. Floorplan also learns about various services in the VSPACE and contacts

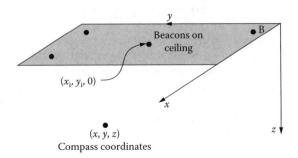

FIGURE 8.23 The coordinate system used in MIT's Cricket.

those services and downloads a small icon representing each service. These icons are displayed on the map; when the user clicks on an icon, Floorplan uses INS to download a control script or program for the application represented by that icon, and load the controls into a new window so the user can control the application. The goal in configuring the beacon coordinate system is to discover the coordinates of each active beacon being used in the application.

Step 1: Lay Out Beacons

Lay out three or more beacons on a flat surface (ceiling or floor, etc.). If four beacons are used, make sure they do not form a square or rectangle. The typical beacon operating range of a Cricket beacon is about 10 m (with the antenna and power levels that are set by default). Tip: Do not place a large number of beacons (> 8) within a given area, as that will cause excessive contention and could increase the latency of position tracking. Tip: If four or more beacons are used, the beacons cannot be placed such that any four of them are on the same circle (in particular, you cannot place them at the four corners of a rectangle or square). This is because when the beacons are all on the same circles, the resulting system of equations is degenerate and the listener will show up at a position on the imaginary plane. The positioning of a beacon within a room or space plays a nontrivial role in enabling listeners to make the correct choice of their location. For example, consider the positioning shown in Figure 8.24. Although the receiver is in Room A, the listener

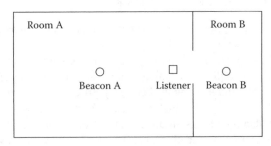

FIGURE 8.24 The nearest beacon to a listener may not be in the same geographic space.

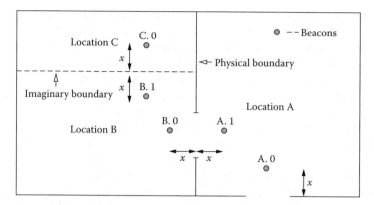

FIGURE 8.25. Real and imaginary boundaries (correct position of beacons).

finds the beacon in Room B to be closer and will end up using the space identifier advertised by the latter.

Cricket maintains a centralized repository of the physical locations of each beacon and provides these data to listeners. In order to preserve privacy and have a decentralized system, whenever a beacon is placed to distinguish a physical or virtual boundary corresponding to a different space, it must be placed at a fixed distance away from the boundary separating the two spaces.

Figure 8.25 shows an example of this in a setting with both real and virtual boundaries. Such placement ensures that a listener rarely makes a wrong choice, unless caught within a small distance (1 ft in Cricket's current implementation) from the boundary between two beacons advertising different spaces. In this case, it is often equally valid to pick either beacon as the closest.

Step 2: Select Three Beacons as References

Only three reference beacons need to be measured regardless of the number of active beacons used. In general, it does not matter which three among the set of installed beacons are selected for reference. However, coordinate accuracy should improve if the positions of the three selected beacons circumscribe the tracking area.

Step 3: Configure the Beacons (Calibrate the References)

Calibration works by steadily holding the listener directly above (or below) a reference beacon perpendicular to the plane on which the beacons are deployed, for a few seconds, until all the error bars in the Beacon Finder window fall under the indicated threshold. When a new dot appears in the Beacon Configuration window, move to the next step. The action of steadily holding a listener underneath a reference beacon lets the system measure the distances to all other beacons with respect to the reference beacon. SEPDIST CM defines the *maximum separation distance* (in cm) between the listener and a beacon during the beacon coordinate configuration phase. That is, the listener must be held within SEPDIST CM cm directly below a beacon during calibration. When SEPDIST CM is large relative to the distances between beacons (1.5 times or greater), the user should move swiftly from one beacon to another during the calibration process. Otherwise, BeaconConfig might start a measurement while the listener is not directly underneath a beacon and cause errors in the beacon coordinate computation. Upon estimating its distance to multiple beacons, a listener associates itself with the space advertised by the closest one, and infers its position coordinates using the known position coordinates of the beacons.

Originally, it was envisioned that each beacon would send its space and coordinate information on the RF channel. Over time, MIT researchers found that it was simpler in many (but not all) cases for a beacon to only send a unique ID, and for the mapping between the ID and the space/coordinate information to be maintained in a central database. Using this approach, the listener or host device downloads the database for each building of interest. The first reference beacon that is calibrated by the listener defines the origin of the coordinate system in the lower left corner. The distance measurements can be noisy and their user must hold the listener still (or tilt the listener slightly) until after the standard deviations of the distance measurements for all of the beacons fall below STDDEV TOLERANCE CM (default 10 cm). When this happens, the Beacon Configuration window will then show a dot (and sound a beep), representing the origin defined by this beacon. Note that a dot will *not* show the distance measurements if any one of the beacons is faulty (i.e., either the distances have high variations or the beacon is out of range or there are not enough samples to make a distance estimate).

Repeat the same process for the second reference beacon. When done, the Beacon Configuration window will display a second dot (and sound two beeps), representing the position of the second reference in the coordinate system. Thus, the line extended between the first and second beacons becomes the horizontal axis of the beacon coordinate system. Note that the order in which the reference beacons are calibrated defines the directions of the coordinate system axes. The coordinate system axes are defined by the order of listener placement underneath each of the three reference beacons. Repeat the process again for the third reference beacon. Note that the third reference beacon does *not* define the y axis. Rather, it defines the direction of the positive y axis. When done, the coordinate system configuration is complete and the Beacon Configuration window will display all the dots representing the positions of all the beacons that have been configured.

Step 4: Enter Tracking/Drawing Mode

Once all the references have been calibrated, the Beacon Configuration program automatically enters tracking mode, which displays the listener's position (represented by a red dot) with respect to the beacons. Use the listener device as a virtual mouse in free space to draw various objects, such as polylines, circles, and rectangles. The drawing interface is based on a "pen up" and "pen down" model, which behaves much like an old-fashioned plotter. This drawing program can be used to capture the general shapes and positions of various objects (e.g., furniture) in the environment by outlining the objects with the listener. If the listener can be set steadily for a length of time during the outlining process, one can produce an amazingly accurate capture of the environment. Due to the sliding window distance-filtering algorithm, there is a lag in tracking the listener's position. Depending on the beacon range, the level of contention, the noise in the distance readings, and various other factors, this lag may increase up to the size of the sliding window, which has a default value of 4 sec.

The "Aggressive" option bypasses the sliding window and enables CricketDaemon to compute position estimates based on the very last distance sample measured from each beacon. The tracking latency is reduced at the cost of increased errors from the occasional incorrect distance measurement. When enabled, the "Streamer" option draws the trail of the Cricket listener during tracking mode. Finally, there is an option to display the distance *annunus*, which graphically represents the distance measured by the listener with respect to each beacon. The thickness of the annunus represents one-half standard deviation of the measured distance. Thus, the intersection of the annuni graphically depicts the approximate position of the listener. The annunus display offers some intuition about how the position estimation works in real time.

SPACE CONFIGURATION AND SPATIAL INFORMATION MANAGEMENT

The only configuration required in Cricket is setting the string for a space that is disseminated by a beacon. (The specific string is a function of the resource discovery protocol used, and Cricket allows any one of several possibilities.) The spaces advertised by Cricket beacons may be separated by physical boundaries, such as walls, or virtual boundaries (e.g., different parts of a room may correspond to different spaces). Cricket is designed to accurately demarcate virtual spaces that do not have any walls between them. Because ultrasound does not travel through walls, Cricket can easily differentiate spaces separated by walls.

Metric 1: The *precision* of the system is determined by how well the listener can detect the boundary between two spaces.

Metric 2: The *granularity* of the system is the smallest possible size for a geographic space such that boundaries can be detected with a high degree of precision.

Metric 3: The *accuracy* is used to calibrate individual beacons and listeners; it is the degree to which the distance from a beacon, estimated by a listener, matches the true distance.

Maintaining spatial information for each beacon is straightforward, but coordinate information is considerably harder. The ideal resolution to the problem of configuring beacon coordinates is an autoconfiguration method. A common problem in the LPS community deals with *autolocalization* : Given a network of beacons and distances between beacons that are in range of each other, obtain a coordinate assignment for the beacons that satisfies the measured distances. While there are practical solutions to this problem, MIT researchers have found that those solutions do not apply well to a Cricket beacon network because obtaining interbeacon distances in a building-wide or floor-wide deployment of beacons is constrained by two factors. First, ultrasonic signals do not travel across walls, limiting the connectivity across rooms. Second, the directional nature of ultrasonic transmissions limit the interbeacon connectivity even when they are in the same room. These two constraints imply that a new solution is required.

The MIT Cricket approach is to use *mobile-assisted* localization, where a mobile roving Cricket transceiver is used to "patch together" disconnected portions of the beacon network to obtain a number of distance estimates.

WRITING AND RUNNING APPLICATIONS

Follow the following steps to use the Crickets to write and run applications. For troubleshooting, see Chapter 2 of the Cricket user manual.

Step 1: Set up a communication terminal. A serial communication terminal program such as HyperTerminal or minicom is needed.

Step 2: Communicate with the Cricket unit prerequisite: Connection exists with the Cricket unit over a serial communication link using a terminal program like HyperTerminal or minicom.

Step 3: Configure a Cricket unit to be a beacon. Cricket units are initially configured as listeners. Use the terminal program to set the beacon's space ID. The simple Cricket application, BeaconConfigDemo (part of the Cricket software release), can be used to configure beacon coordinates and track a mobile device.

Step 4: Test distance measurements. Use the terminal program to see (capture?) distance measurements. The terminal senses the measurements each time the listener senses them from the beacon. The output includes a DB field that assigns the distance (centimeters is the default) to the beacon.

Step 5: Install the Cricket embedded software image.

With TinyOS (http://www.tinyos.net/): At the time of writing, the version of the Cricket embedded software worked with TinyOS version 1.1.9. The embedded source code for the Cricket units has two parts and is available at the website (http://cricket.csail.mit.edu). The Cricket platform contains all the software that handles the differences between the Cricket hardware and the Mica2 hardware (Mica2 Motes are used for programming, using the MIB510CA model). The Cricket applies the software that incorporates the beacon and listener algorithms. If you get a Cricket kit from Crossbow, the embedded software will already be installed on it. If not, then you need to install the embedded software first. Note: The embedded software needs to be based on TinyOS 1.1.6. If you have an older version of TinyOS, you need to upgrade. Also, we have not tested Cricket with later versions of TinyOS; it might work, however. If you obtained a Cricket kit, verify that the CD that came with it has TinyOS 1.1.6 rather than an older version. Attach one of the Crickets to your computer's serial port. If your computer does not have a serial port, you can use a USB-to-serial converter.

Without TinyOS: You can program the software onto the Crickets using an already compiled version of the software.

An example of a parameter is duration (DR). The duration is reported by the listener under the same conditions as in DB, above. The duration represents the time of arrival (TOA) of the ultrasonic pulse, compensating for the various time offsets for accuracy. This value can be used to calculate a distance with more precision than the DB distance, because the latter uses a (temperature-compensated) value for the speed of sound, which may introduce some error. For example, to determine which of multiple beacons is closest to the listener, compare the DR values. The unit of the reported duration is microseconds. The uncorrected time-of-flight (TM) value is the same as the duration (DR) but without any compensation. It is also reported by the listener each time a distance measurement arrives.

Developing Cricket Applications in Java

The software package includes a library to help developers create Cricket applications in Java. This section describes the software architecture and the Java Cricket client library API, called `Cricketlib`. Figure 8.26 illustrates the Cricket software architecture. At the lowest layer, `cricketd` allows a Cricket host device to access the serial port API to configure low-level Cricket parameters and obtain raw measurements from the Cricket hardware device. The software package includes a `CricketDaemon` server application that connects to `cricketd` to filter and process raw Cricket measurements to infer the listener's spatial location and compute its position coordinates. Java applications may access the processed location information via the Java Cricket client library (Clientlib), which interfaces between the application and the `CricketDaemon`. At run time, one `CricketDaemon` processes location information for exactly one Cricket device.

Developers can use the sample application as a template to create their own Cricket applications in Java.

Clientlib API

The Java Cricket Client library (Clientlib) uses callbacks to feed location information to the application. As shown in Figure 8.27, a Java application instantiates one `ServerBroker` object. The `ServerBroker` object is an independent thread

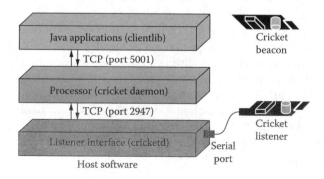

FIGURE 8.26 MIT Cricket software architecture.

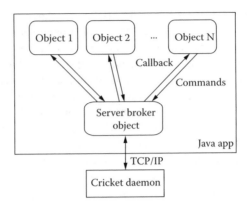

FIGURE 8.27 MIT Cricket Clientlib architecture.

that interfaces between a set of callback handlers in the Cricket application and the `CricketDaemon`. At least one object in the Java application implements a callback handler and registers with the `ServerBroker`. In the callback registration, the callback handler object specifies a *callback mask* that selects the type(s) of location updates that should trigger the callback handler being registered. The `ServerBroker` dispatches the callback handlers whenever it receives a location update from the `CricketDaemon` that matches the callback mask.

TROUBLESHOOTING AND DEPLOYMENT TIPS

Some of the problems that can occur while deploying MIT Cricket include the listener or the beacon does not respond to any command or that the listener does not report any events. Another problem is having the listener return erroneous distances. If the reported distances are wrong by more than a few centimeters, then something is not right with the setup. If the listener reports erroneous distances at certain locations but not others, this is probably caused by sources of interference on the path from beacon to listener. Some objects block ultrasonic signals and others reflect them; some even generate ultrasonic signals at the same frequency as Cricket. All these objects cause the Crickets to report wrong distances. Obstructions close to the beacon's ultrasonic transmitter or listener's ultrasonic receiver cause the biggest problems. These obstructions could be walls or doors (through which ultrasound does not pass) or could be people. Large metallic plates or cabinets on the path from beacon to listener can disrupt distance estimation by affecting both ultrasound and RF propagation. We have also found that some fluorescent lamps generate 40-kHZ ultrasonic waves that can interfere with the Crickets. It is recommended not to place beacons too close to large objects; if a large object is within 10 or 15 cm of a beacon, that beacon's transmissions may be blocked. Dealing with interfering fluorescent lamps could be harder, however. The interference usually comes from lamps that are close to dying.

Problem: The beacon blinks but does not respond to serial commands.
Problem: The listener associates itself with the wrong space identifier.

When using Cricket beacons to delimit spaces that are not separated by a wall, both RF and ultrasound from the beacons in the two different spaces may traverse the other space. Because the listener associates itself with the space advertised by the nearest beacon, it is suggested to place beacons with some care to achieve proper spatial demarcation. Figure 8.28 illustrates a bad configuration of beacons, where a listener on the left side of the partition could be closer to the beacon on the right, causing it to associate itself with the wrong space. Place the beacons corresponding to the different spaces at equal distances from the boundary between the spaces, as shown in Figure 8.29.

USING PLACE LAB

This section provides an overview of the Place Lab API and an introduction on using it to build location-enhanced applications. Place Lab currently runs on the platforms shown in Figure 8.30 and provides support for spotting the following beacon types.

Place Lab supports Global System for Mobile Communications (GSM) beacons on these platforms using a Bluetooth data connection to a paired Series 60 phone,

FIGURE 8.28   Incorrect placement of beacons for boundary detection between two spaces.

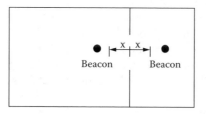

FIGURE 8.29   Correct beacon placement.

Operating systems	Architectures	802.11 abg beacons	GSM beacons	Bluetooth beacons
Windows XP	x86	•	•[3]	•
Linux	x86, ARM, X scale	•		
Os X	Power PC	•		
Pocket PC 2003	ARM, X scale	•	•[3]	•
Symbian	Series 60 cell phones		•	•

FIGURE 8.30   Platforms supported by Place Lab.

which actually receives the GSM beacons and forwards them to the master device. The Place Lab APIs for spotters, mappers, and trackers are consistent across all platforms assisting developers in porting their applications to different platforms, e.g., from a full-featured Windows laptop to a Nokia cell phone. Place Lab is written entirely in Java 2 Micro Edition (J2ME), except for some small amount of native code that is written for each spotter. It might be best to begin with the examples included in the sections that feature run able code from the org.placelab.examples package. The javadocs from each section offer more in-depth information. Once Place Lab is installed on the device of choice, the following are quick tips on how to verify the installation and how to get acquainted with Place Lab's basic features.

How to Create a Map-Based Database of Wi-Fi Locations

Capture signatures/trace of network availability of nearby RF beacons (Wi-Fi, Bluetooth, GSM, and other short-range radio sources) by walking around a neighborhood with a mobile computer equipped with a GPS device and a radio (typically an 802.11 card). This process of collecting traces of available beacons like 802.11 access points on a mobile device (which is also connected to a GPS unit) is referred to as stumbling. The resulting log has a time-stamped sequence of records (GPS readings) containing the latitude and longitude of where the record was taken, coupled with vectors of nearby beacons (a list of radio sources and associated signal strengths that could be heard at that time). PlacelabStumbler supports stumbling for Wi-Fi, Bluetooth, and GSM in various configurations. Mobile clients listen to nearby radio beacons. Most radio sources contain a signature (e.g., BSSID) but not location data in the signal itself.

Setting up Place Lab

Place Lab code is hosted on Sourceforge and can be accessed via CVS. Go to http://www.placelab.org/toolkit/ and download the proper distribution for your device. A hardware compatibility list is available (http://www.placelab.org/ toolkit/hcl.php). Each platform-specific distribution includes everything needed to run Place Lab, including a Java run-time environment, the source code for Place Lab, all precompiled jar files and platform-specific executables and scripts.

Directory Structure

The source for Place Lab is divided into two directories: src and srcx. Both are compiled to the bin directory by Eclipse. src (and package org.placelab inside it) is for release-quality code and is included in official releases. srcx (and package org.placelabx inside it) is for research, debug, or other not-ready-for-primetime code. No file in src should ever rely on a file in srcx.

The placelabdata directory also contains all the beacon database snapshots, maps, icons, and other data used by Place Lab. The directory is located in your home directory (e.g., C:\Documents and Settings\myusername\ on Windows XP) by default, although the distribution has a placelabdata directory which is used by the sample scripts in the run/ directory. The placelab.ini file contains many customization settings, including Wigle.net login information.

DEVELOPMENT TOOLS

Most Place Lab developers use the Eclipse IDE and the build is configured to support it. Builds can also be performed using Ant. If you are using Eclipse, make sure to copy the appropriate classpath.* file, to be named .classpath in your placelab directory after checking out the module from the CVS repository. If you make any global changes to your .classpath file, make sure you reflect those changes back to the classpath.* files.

Of course, if you are just trying to *use* Place Lab to build your location-aware applications, you can simply use placelab.jar (along with the necessary support files and native pieces such as spotter.dll) with whatever development environment you choose.

Building for the phones requires the Series 60 MIDP SDK available from Nokia, and you may want to be aware of Known Issues In The Nokia 6600 MIDP 2.0 Implementation v1.3. The "Phone Utilities How-To" section describes how to use the Place Lab helper tools, like error logging and persistent storage on the phones.

Eclipse Platform

The Eclipse platform (http://www.eclipse.org/platform/) subproject provides the core frameworks and services upon which all plug-in extensions are created. It also provides the run time in which plug-ins are loaded, integrated, and executed. The primary purpose of the platform subproject is to enable other tool developers to easily build and deliver integrated tools.

The Eclipse platform itself is a sort of universal tool platform — it is an IDE for anything and nothing in particular. It can deal with any type of resource (Java files, C files, Word files, HTML files, JSP files, etc.) in a generic manner, but does not know how to do anything that is specific to a particular file type. The Eclipse platform, by itself, does not provide a great deal of end-user functionality — it is what it enables that is interesting. The real value comes from tool plug-ins for eclipse that teach the platform how to work with these different kinds of resources. This pluggable architecture allows a more seamless experience for the end user when moving between different tools than ever before possible.

The Eclipse platform defines a set of frameworks and common services that collectively make up "integration-ware" required to support a comprehensive tool integration platform. These services and frameworks represent the common facilities required by most tool builders, including a standard workbench user interface and project model for managing resources, portable native widget and user interface libraries, automatic resource delta management for incremental compilers and builders, language-independent debug infrastructure, and infrastructure for distributed multiuser version resource management.

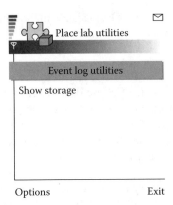

FIGURE 8.31   Place Lab utilities.

In addition, the Eclipse platform defines a workbench user interface and a set of common domain-independent user interaction paradigms that tool builders plug into to add new capabilities. The platform comes with a set of standard views that can be extended by tool builders. Tool builders can both add new views and plug new domain-specific capability into existing views.

PHONE UTILITIES HOW-TO

Developing for the J2ME phone platform presents many difficulties, especially when using functionality that can only be tested on the actual device. Since not all device functionality is implemented in an emulator, when prototyping software on the device, crucial debugging functionalities such as print lines or stack traces are lost. Place Lab is packaged with a Utilities application that assists developers in debugging J2ME applications for the Series 60 phone platform (Figure 8.31).

There are two utilities bundled together. The first is an Event Logger, which displays events a program has explicitly asked it to log. The second is a Storage utility for handling Place Lab preferences.

Using Event Logger

The following piece of code demonstrates how to use Event Logger. In this example, we are trying to make a local host connection to gain access to GSM information. If the attempt is unsuccessful, an IOException will be thrown. If this piece of code were running in the emulator, the line ioe.printStackTrace() would print the stack trace to standard error. However, if this piece of code were to run on the actual device, an error that occurs would be transparent, making it difficult to debug.

```
try {
    //native server runs on port 4040
    sc = (SocketConnection) Connector.open("socket://127.0.0.1:4040");

    ...

    } catch (IOException ioe) {
```

```
    EventLogger.logError("Cannot open stream to native GSM spotter");
    ioe.printStackTrace();
}
```

A static call is made to the EventLogger passing in the Exception as an argument. The EventLogger will log the class name of the exception, and also the message it gives by the call ioe.getMessage().

The Event Logger can also log String messages for debugging, such as the following:

```
if(map.isEmpty()) {
    EventLogger.logError("ERROR: map is empty");
} else {
    ...
}
```

The Event Logger is flexible enough to allow you to define your own types. The method call EventLogger.add(int,String) takes an EventType as the first argument, and the message to log as the second argument. This can be helpful if you do not want to log all messages as errors.

Viewing Events

The Place Lab Utilities MIDlet is the entry point for viewing events logged by the EventLogger. Since the EventLogger uses the RMS as persistent storage, creating the record store object to be publicly accessible, all events written to the logger are accessible from the Place Lab Utilities MIDlet.

Figure 8.31 shows the main screen to Place Lab Utilities. Select Event Log Utilities. On the next screen (Figure 8.32) select Show Event Log. This will list the events five at a time on the screen. To browse through all the events, there is an option on the left soft key menu for Next 5, which will list the next five events on the screen.

Viewing a large set of events on the phone can be time consuming and difficult considering the screen size constraints. There is an option to upload the event logs to a Place Lab server, via Bluetooth (Figure 8.33). This will allow you to view the

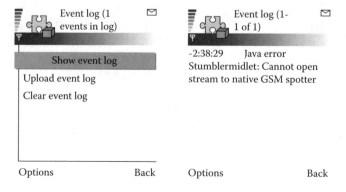

FIGURE 8.32 Place Lab event log.

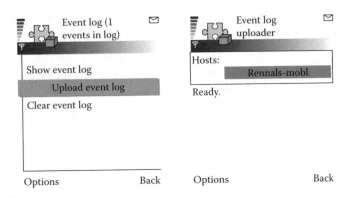

FIGURE 8.33 Place Lab upload event log.

events in a text file on your PC. The upload process is exactly the same as other Bluetooth transfers and is documented in the "Bluetooth Transfer How-To" section.

Storage Data

The Storage utility provides a persistent space for developers to set preferences. For instance, a section of code that should only run in the emulator could check the Storage to see if the value for the emulator preference is true or false.

The Storage class has two methods of interest:

```
Storage.add(String type, String key, String value)
Storage.get(String type, String key)
```

Both of these methods are statically accessible and can be a useful tool for developers. Screen shots on how to set preferences from a user interface are shown in Figure 8.34.

Bluetooth Transfer How-To

As the Place Lab team started development for Series 60 phones, it was clear that some sort of transfer utility needed be created to get important data onto and off of

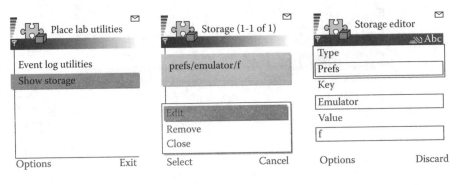

FIGURE 8.34 Place Lab preferences.

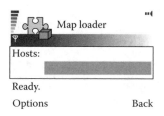

FIGURE 8.35   Place Lab map loader.

the phone. Along the path of developing application-specific transfer utilities we kept duplicating the same workarounds and handshaking. So we rolled these transfer procedures into a reusable standardized user interface.

Each end of the transfer has its own piece to run when establishing a connection. On the PC side, the BluetoothServer handles all incoming connections from phones. On the phone side, there is more than one entry point, but the user interface usually looks like that shown in Figure 8.35.

Starting a BluetoothServer

Note: The Place Lab distribution uses Blue Cove (http://sourceforge.net/projects/ bluecove) for javax.bluetooth support. The current Blue Cove release only works under Windows XP Service Pack 2 (SP2). Since Place Lab itself uses the standard Java Bluetooth interface (JSR-82: javax.bluetooth), it should work on any platform supporting this. Some of the instructions below are Blue Cove specific. Right now the Place Lab distribution only supports BluetoothServer on Windows XP Service Pack 2. If this is not the platform you are working with, then you might want to look into a javax.bluetooth for your particular platform.

The BluetoothServer can be started by either running org.placelab.midp. server.BluetoothServer from the placelab.jar or simply running `PlacelabPhone_ BluetoothServer.jar` in the phone distribution. After it starts you should be able to see output that looks something like this:

```
initialized bluetooth service.
Registering Log Upload Proxy
Registering Map Loader Proxy
Registering Activity Downloader
Registering GPS Time Downloader
Registering ESM Question Uploader (0)
Registering Bluetooth Console
Registering Place Downloader
listening…
```

If you have the Nokia PC Suite installed on your machine, make sure mRouter (the system tray item that looks like two disconnected plugs) is not configured to listen on any Bluetooth ports. This can interfere with the connection process between the PC and phone.

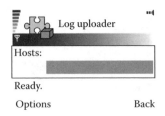

FIGURE 8.36   Place Lab stumble log upload (1).

The final step is to make sure that the computer running the BluetoothServer is discoverable. On Windows XP right-click on the Bluetooth icon in the system tray. Select "Open Bluetooth Settings." Go to the Options tab and select the "Turn discovery on" check box. If you get an error that looks like the following,

```
java.lang.UnsatisfiedLinkError: no intelbth in java.library.path
   at java.lang.ClassLoader.loadLibrary (ClassLoader.java:1517)
...
```

it usually means that you do not have BlueCove installed correctly. Make sure BlueCove.jar is in your classpath and intelbth.dll is in a place in which the JVM can find it (e.g., c:\windows\system32 or the directory the JVM is executed from).

Connecting to a BluetoothServer

Since there is no one utility for transfer data on the phone, I am going to use stumbler log uploading as an example. To access the log uploading, start the stumbler application on the phone. Select "Upload Stumble Logs" from the Options menu. The screen should look something like that shown in Figure 8.36.

The phone keeps a cache of previously seen servers to connect to. If the computer where your BluetoothServer is running is listed in the "Hosts:" list, then select it. If not, then select "More Devices" from the Options menu. This should start a search of the current environment for new servers to connect to (Figure 8.37).

If the computer running the BluetoothServer does not show up in the list of hosts after a new search, then it is probably the case that your computer is not set

FIGURE 8.37   Place Lab stumble log upload (2).

as discoverable in the Bluetooth preferences. Look back at the previous section as to how to do enable discoverable.

After selecting the correct server, select "Connect" from the Options menu. A bunch of statuses should start showing up where it used to say "Ready." If all goes right, then it should look something like what is shown in Figure 8.38.

Common Errors

The error shown in Figure 8.39 usually means the computer you were trying to connect to was not in range. It is possible you are trying to connect to a stale entry in your server cache. Try refreshing your device list by doing a "More Devices."

The error shown in Figure 8.40 usually means the computer you were trying to connect to was not running the BluetoothServer.

Make a New BluetoothService

It is pretty easy to create a new service that takes advantage of this preexisting BluetoothServer/BluetoothClient code base. All the following code references are in the org.placelab.midp and org.placelab.midp.server packages.

The first step to create a new service is adding a unique service type in the BluetoothService interface. Just add another constant that is different than the others listed there.

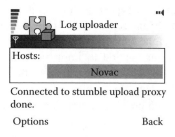

FIGURE 8.38 Place Lab stumble log upload (3).

FIGURE 8.39 Place Lab stumble log upload errors (1).

FIGURE 8.40 Place Lab stumble log upload errors (2).

```
public interface BluetoothService {

  ...

  public final static byte PLACE_UPLOAD_SERVICE = 7;

  public final static byte EXAMPLE_SERVICE = 8;

  ...

}
```

Next, a server component needs to be created by implementing the Bluetooth-Service interface. The getServiceType() method should return the constant that you just defined in the BluetoothService interface. The getName() method is just used as a human-readable name of the service and should return something that is sensible. The final and most important method to implement is the newClient(DataInput-Stream in, DataOuputStream out). In this method, the actual details of the data transfer should be worked out. Be sure to make use of the read<type>() and write<type>(). Java is on both ends, so this makes your life considerably easier.

```
public class ExampleService implements BluetoothService {
  public String getName() {
    return "Example Service";
  }
  public byte getServiceType() {
    return BluetoothService.EXAMPLE_SERVICE;
  }
  public void newClient(DataInputStream in, DataOutputStream out)
  {
    out.writeUTF("hello");
    out.flush();
    if (!in.readUTF().equals("bye"))
      System.err.println("Bad message!");
  }

}
```

After creating the new BluetoothService class, it needs to be registered with the BluetoothServer by calling addService() in the main() method.

```
public class BluetoothServer    {

  ...

  public static void main(String args[]) {

    ...

    bs.addService(new PlaceService());

    bs.addService(new ExampleService());

    ...

  }

}
```

A matching BluetoothClient needs to be created for the human-readable name. And the getServiceType() should return the same exact constant as the matching BluetoothService. Instead of a newClient() method, BluetoothClient has the handle-Connection(DataInputStream in, DataOutputStream out) method.

```
public class ExampleClient extends BluetoothClient {
  public String getName() {
    return "Example Client";
  }
  public byte getServiceType() {
    return BluetoothService.EXAMPLE_SERVICE;
  }
  public void handleConnection(DataInputStream in, DataOutputStream out) {
    if (in.readUTF.equals("hello"))
      out.writeUTF("bye");
    out.flush();
  }
}
```

The final step is to integrate the new ExampleClient into your existing MIDlet. To do this, you must use the UIComponent interface to define a pathway to show the ExampleClient and then return back to the calling screen. For example,

```
public class SomeForm implements UIComponent {
  public Display display;
  public Form someForm;
  ...
  public void showUI(UIComponent from) {
    display.setCurrent(someForm);
  }
  public void showExampleClient() {
    BluetoothClient client = new ExampleClient(display, this);
```

```
        client.showUI(this);
    }

        }
```

This code can obviously be changed around to produce more clever behavior, but it lays out the skeleton.

SPOTTING

In Place Lab, spotters are the objects that do the environmental sensing (i.e., detect radio beacons) for the system. The root class Spotter.java is an interface that all spotters implement. In many cases, spotters will have a native code component that understands how to reach the specific hardware that does the sensing.

Public Interface Spotter

A spotter is a Java object that generalizes the function of an environmental sensor like a GPS unit or a Wi-Fi card. A spotter produces Measurement objects. The types of the measurements produced depend on the type of spotter.

Regardless of the underlying implementation of the spotter, all spotters support the following three modes of operation:

- Synchronous: Call getMeasurement() at application-defined intervals and get back a single measurement for each call. This mode of operation is simple, but may result in excessive wait times while the method blocks to collect measurements, or it may result in missed measurements if it is not called often enough.
- Continuous scan: Call startScanning() to have the spotter begin scanning for measurements in the background, notifying its registered SpotterListener objects when new measurements are available. The continuous scan will continue providing measurements until stopScanning() is called.
- Single background scan: Call scanOnce() to have the spotter do a single background scan and return a measurement to the registered SpotterListeners.

WIFISPOTTER EXAMPLE

```
package org.placelab.example;

import org.placelab.core.BeaconMeasurement;
import org.placelab.core.WiFiReading;
import org.placelab.spotter.Spotter;
import org.placelab.spotter.SpotterException;
import org.placelab.spotter.WiFiSpotter;

/**
 * A sample that creates a WiFiSpotter and uses it to get measurements.
 * This will only return readings if a WiFi card is present.
 */
public class WiFiSpotterExample {
```

```
public static void main(String[] args) {
  Spotter s = new WiFiSpotter();
  try {
    s.open();
    BeaconMeasurement m = (BeaconMeasurement) s.getMeasurement();
    System.out.println(m.numberOfReadings() + " APs were seen\n");
    if (m.numberOfReadings() > 0) {
      System.out.println(pad("MAC Address", 20) + pad("SSID", 30)
        + pad("RSSI", 10));
      // Iterate through the Vector and print the readings
      for (int i = 0; i < m.numberOfReadings(); i++) {
        WiFiReading r = (WiFiReading) m.getReading(i);
        System.out.println(pad(r.getId(), 20)

          + pad(r.getSsid(), 30) + pad("" + r.getRssi(), 10));
      }
    }
  } catch (SpotterException ex) {
    ex.printStackTrace();
  }
}
// Pad out a string to the passed length
public static String pad(String str, int len) {
  StringBuffer sb = new StringBuffer(str);
  for (int i=str.length(); i < len; i++) {
    sb.append(" ");
  }
  return sb.toString();
}
}
```

RemoteGSMSpotter

This spotter talks via Bluetooth to read GSM readings from a Series 60 Bluetooth phone. The phone must be running the BTGSM midlet, which creates the GSM sharing service. Other than the remoteness, this spotter behaves as expected. The phone is capable of buffering up to 60 sec of GSM readings. This spotter can be created to pull over either all of the buffered readings or just the latest reading. In the case that all readings are pulled over, repeated calls to getMeasurement will empty the buffer.

If you have a Series 60 cell phone (e.g., Nokia 6600 or 6620), your laptop client can position itself using both GSM and Wi-Fi at the same time — Wi-Fi in the laptop and GSM by the phone relayed to the laptop over Bluetooth. To make this work, you need both a laptop running Place Lab with Bluetooth and a Series 60 cell phone running Place Lab with Bluetooth.

First, you must pair the phone and the laptop. This only needs to be done once, but it is a critical step. Run your phone's Bluetooth utility, discover your laptop, and create a pairing. (This usually involves typing a password key on both devices.) Then on the phone set this laptop as "authorized" to allow it to make connections without having to select "acceptî every time the laptop talks to the phone. Next, as is the case for any of the Place Lab applications on the phone, run the server's native portion of Place Lab on the phone and go back to the application chooser without exiting the server's application. Next run the midlet named BTGSM on the phone. This starts a Bluetooth server that will relay out GSM beacons to devices that ask for them. You will need to answer yes to two questions. (If, after you start your laptop's spotter, it asks you for permission more than twice, you did not do the first step correctly.) When BTGSM first starts up, it displays the URL used to access the server. Write down or copy the host address and port number of this URL. (The other parts are the same for all devices.) For my phone, for example, this address is 000e6d43ec17:4. After a few seconds, the phone will switch to a different display showing how many readings it has taken, what the GSM coverage has been for the life of the run, and what the current cell tower's ID is.

The laptop must have a Bluetooth interface and must also have the Microsoft Bluetooth software installed. This software is included as part of Windows XP SP2, but can also be installed separately from SP2. (Visit http://eben.phlegethon.org/blue-tooth.html to find out how.) RemoteGSMSpotter takes a constructor argument of the address (e.g., 000e6d43ec17:4) that was printed by the cell phone. (This spotter's constructor also takes a second Boolean argument that enables or disables buffering; see the Javadocs for more about buffering.) Call open() on the spotter as you would any other, and the spotter is ready to go. For several reasons, the phone is hard coded to take a GSM reading once a second, so asking for measurements more frequently than that will not result in more data being collected. Below is sample output for org.placelab.samples. RemoteGSMSpotterSample getting five GSM readings from the phone, 2 sec apart:

Tower ID	Name	Signal Strength
310:380:1301:52025	AT&T Wirel:1301:52025:1301:52025	78
310:380:1301:52025	AT&T Wirel:1301:52025:1301:52025	83
310:380:1301:52025	AT&T Wirel:1301:52025:1301:52025	78
310:380:1301:52025	AT&T Wirel:1301:52025:1301:52025	75
310:380:1301:52025	AT&T Wireel:1301:52025:1301:52025	75

LogSpotter

LogSpotters are spotters that replay stored measurements from log files. A LogSpotter takes a trace file created by NetStumbler or Place Lab Stumbler and replays it, creating BeaconMeasurements, to simulate a route traveled.

LogSpotterExample

```
package org.placelab.example;

import org.placelab.core.BeaconReading;
import org.placelab.core.StumblerMeasurement;
import org.placelab.spotter.LogSpotter;
import org.placelab.spotter.SpotterException;
import org.placelab.util.Cmdline;


/** This sample demonstrates the use of the LogSpotter class
 *   We show how to parse a large text-exported trace,
```

```
*    report the size of the trace and enlist some key values in
the log.
*/
public class LogSpotterExample {
  public static void main(String [] args)
  {
    Cmdline.parse(args);
    String inputFile = Cmdline.getArg("tracefile");
    if (inputFile == null) {
      System.err.println("Usage: java " +
LogSpotterExample.class.getName() + " --tracefile filename");
      System.exit(1);
    }
    LogSpotter log = LogSpotter.newSpotter(inputFile);
    try {
      log.open();
      // output as many chunks as are there in the inputfile
      if(!log.logIsFinished()) do {
        StumblerMeasurement m = (StumblerMeasurement)log
          .getMeasurement();
        if(m == null) {
          System.out.println("log is finished");
          break;
        }
        if (m.numberOfReadings() > 0) {
          System.out.println(pad("Timestamp", 20)
            + pad("Latitude", 20) + pad("Longitude", 20)
            + pad("BSSID", 30) + pad("RSSI", 10));
          //iterate through the Vector and print the readings
          for (int i = 0; i < m.numberOfReadings(); i++) {
            BeaconReading br = (BeaconReading) m
              .getReading(i);
            System.out.println(pad(""
              + (((long) m.getTimestamp() / 1000L) * 1000L),
              20)
+ pad(m.getPosition().getLatitudeAsString(), 20)
+ pad(m.getPosition().getLongitudeAsString(), 20)
+ pad(br.getId(), 30)
pad("" + br.getNormalizedSignalStrength(), 10));
          }
          System.out.println();
        }
      } while(!log.logIsFinished());
    } catch (SpotterException ex) {
      ex.printStackTrace();
    }
  }
  public static String pad(String str, int len) {
    StringBuffer sb = new StringBuffer(str);
    for (int i=str.length(); i < len; i++) {
      sb.append(" ");
    }
    return sb.toString();
  }
}
```

Other Spotters

Bluetooth

BluetoothSpotter scans for Bluetooth devices on systems with JSR-82 implementations.

NMEA GPS

NMEAGPSSpotter listens to an NMEA GPS device to get GPSMeasurements. GPS devices that use some form of NMEA should use this class. The NMEAGPSSpotter uses the NMEASentenceGatherer, which uses $GPGGA and $GPRMC sentences to gather location and fix data, and thereby construct its GPSMeasurements.

GSM

GSMSpotter scans for GSM cell towers on phones. It polls a local host server running on the phone. This is a synchronous spotter that makes a blocking read request to the server, returning when the read completes.

The GSMSpotter now supports operation in the emulator by using log files that were uploaded by the StumbleUploadProxy placed inside the jar to get the measurements. The build.xml helps here by only placing those log files in the jar when the "do emulate" target is selected.

Mapping

Coordinate

A coordinate represents a point on EarthModel. It does not expose any methods with doubles, because not all Place Lab platforms support floating point math.

Place Lab maps both access points and devices in two dimensions. That is to say, Place Lab estimates latitude and longitude, but we do not model or estimate altitude at this point (there is preliminary support for three- dimensional coordinates). Coordinate is the class Place Lab uses to represent all of its locations. Coordinates are like regular two-dimensional points in that they have an x and y. x and y refer to the distance in meters from the origin of the Earth. This origin can be pretty much wherever you want; you can change it in CoordinateTranslator. Making the origin close to you makes the meters look reasonable (like 875), rather than huge (like 789,674), making debugging easier.

Behind the scenes (in org.placelab.core.Types) Place Lab chooses to use TwoD-Coordinate or FixedTwoDCoordinate, depending on whether it is running on a CLDC 1.0 phone or not. Applications that intend to run on both phones and larger devices will use the string-based interface to coordinate, but applications not targeted at phones can cast any coordinate to a TwoDCoordinate and use the double-based interface.

Coordinate Example

```
package org.placelab.example;
import org.placelab.core.TwoDCoordinate;
/**
 * A sample program that shows how to use 2DCoordinates
```

```
*/
public class CoordinateExample {
   public static void main(String[] args) {
      TwoDCoordinate c1,c2,c3;
      // The coordinates of Intel Research Seattle
      c1 = new TwoDCoordinate(47.656,-122.318);
      // The coordinates of Intel Research Berkeley
      c2 = new TwoDCoordinate(37.87042,-122.26780);
      double distance = c1.distanceFrom(c2);
      System.out.println("IRS is at
("+c1.getLatitude()+","+c1.getLongitude()+")");
      System.out.println("IRB is at
("+c2.getLatitude()+","+c2.getLongitude()+")");
      System.out.println("IRS and IRB are " + (int)distance/1000 +
            km or " + (int)distance/1609 +
            " miles apart");
      // Move c2 5000 meters north
      c2.moveBy(0, 5000);
      System.out.println("The point 5 km north of IRB has lat=" +
         c2.getLatitude() + " lon=" + c2.getLongitude());
   }
}
```

Mapper

The mapper manages the persistent cache of known access points. The mapper can take BeaconReadings and will return Beacon objects for the BeaconReadings if it knows about them. Mappers find the Beacon object associated with a given ID. If a beacon is not known, the findBeacon() method will return null. Mappers typically represent the on-disk cache of Beacons in keeping with the Place Lab philosophy of offline location computation.

MapLoader

MapLoader is a GUI application to load all worthwhile mappers from various sources of beacon data on the web.

Before running any applications that use the mapper, you must first load your local cache of beacons from databases on the web. The MapLoaderGUI utility allows you to select your region and download all the known beacons for it.

MAPPER EXAMPLE

```
package org.placelab.example;
import org.placelab.core.BeaconMeasurement;
import org.placelab.core.BeaconReading;
import org.placelab.mapper.Beacon;
import org.placelab.mapper.CompoundMapper;
import org.placelab.mapper.Mapper;
import org.placelab.spotter.LogSpotter;
import org.placelab.spotter.Spotter;
import org.placelab.spotter.SpotterException;
import org.placelab.spotter.WiFiSpotter;
```

```
/**
 * This sample is very similar to CoordinateSample with the addition
 * of a lookup in the persistent AP cache
 */
public class MapperExample {
    public static void main(String[] args) {
        Spotter s;
        if(args.length >= 1) {
            s = LogSpotter.newSpotter(args[0]);
        } else {
            s = new WiFiSpotter();
        }
        try {
            s.open();

            BeaconMeasurement m = (BeaconMeasurement) s.getMeasurement();
            Mapper mapper;
            // This Mapper can tell us where APs are
            // The default Mapper (set in PlacelabProperties) will be
selected
            // here. The first argument says to exit on error, and the second
            // says to cache Beacons in memory as they are accessed.
            mapper = CompoundMapper.createDefaultMapper(true, true);
            int knownAPs = 0;
            for (int i = 0; i < m.numberOfReadings(); i++) {
                BeaconReading r = (BeaconReading) m.getReading(i);
                // Lets lookup this AP in the map
                Beacon b = (Beacon) mapper.findBeacon(r.getId());
                if (b == null) {
                    System.out.println(r.getId() + " is an unknown AP");
                } else {
                    System.out.println(r.getId() + " (" +
r.getHumanReadableName()
                        + ") is thought to be at " + b.getPosition());
                    knownAPs++;
                }
            }
            System.out.println("\nOf the " + m.numberOfReadings() + " APs "
                + knownAPs + " were known.");
        } catch (SpotterException ex) {
            ex.printStackTrace();
        }
    }
}
```

Place Lab Stumbler

Place Lab Stumbler is an application that is used to collect and determine the locations of Wi-Fi, Bluetooth, or GSM beacons in conjunction with a GPS unit. It also can be used to create trace files that can be played back by a LogSpotter and that can be submitted to placelab.org for incorporation into the site database of beacons.

This GUI is only intended for laptops or other devices with a standard-size display and a full swt and jface implementation. The Pocket PC will have to use the text-based version or a scaled-down ui.

STUMBLING

Laptop Stumbling How-To

Stumbling is the process of collecting traces of available beacons like 802.11 access points on a mobile device that is also connected to a GPS unit. The resulting log has time-stamped GPS readings coupled with vectors of nearby beacons. PlacelabStumbler supports stumbling for Wi-Fi, Bluetooth, and GSM in various configurations. You will find PlacelabStumblerGUI in the package org.placelab.stumbler.gui.

INSTALLATION

Required Libraries

- rxtxSerial 2.1-7pre17 (you will find the jar in /placelab/lib and the jnilib in /placelab/native//rxtx).
- swt 3.0 release (you will find the jar in /placelab/lib and the jnilib(s) in /placelab/lib/swt//).
- jface 3.0 release (you will find the jar in /placelab/lib).
- eclipse runtime.jar (you will find the jar in /placelab/li.b)
- In addition, to do Wi-Fi spotting, you will need the Place Lab WiFiS spotter library (found in /placelab/native//spotter).
- To do Bluetooth stumbling, you will need Windows XP SP2 and Bluecove (you will find BlueCove in /placelab/lib/bluetooth).
- GSM stumbling has all the Bluetooth requirements, plus the requirement of a Series 60 phone with the Place Lab phone distribution installed on it.

OS-Specific Instructions[1]

Permissions

Make sure /var/lock is group writable and your ID is also in the same group of the serial device (e.g., /dev/ttyS0) and /var/lock. In RedHat/Fedora, add "uucp" and "lock" into your group in /etc/groups. Also, you have to log out or exit the X window to make the permission change effective. If the permissions are not set correctly, you might get "lock file denied" messages from rxtx.

USB–Serial Port Converter

If you are using a USB–serial converter cable for the GPS, you need the usbserial kernel module. Ideally, the module will auto load, but if not, you can use modprobe usbserial to install it manually.

---

[1] Thanks to Yu-Chung for instructions for Linux.

- Run dmesg to see which device file is associated with the GPS; typically it is /dev/ttyUSB0.
- For some reason the rxtx library only accepts device file /dev/ttyS*. You can run "ln -s /dev/ttyUSB0 /dev/ttyS100." Later run the stumbler with "placelab.gps_device=/dev/ttyS100. "

Scanning Mode Support

If you run the stumbler but no beacons show up, this is because some cards may not support scanning mode with the default kernel/pcmcia module (e.g., the orinoco card). Try "iwlist ethX scan" and see if it works. If not, you can download a scanning patch from http://www.ozlabs.org/people/dgibson/dldwd/. On Fedora, you just need to download the tar ball, compile it, and replace all related kernel modules (ex *.ko) with current modules in /lib/modules. No need to reboot. For a prism2-based card, the host AP driver supports scanning and monitoring. You can find more information by Googling hostap.

Instructions for Mac OS

Add yourself to the uucp group (create uucp if it does not exist). Note that you will have to log out and back in to make the group change effective. Create the directory /var/spool/uucp and make sure that it is writable by members of uucp.

Hooking up Your GPS

Use org.placelab.util.GPSEcho and pass to it a single argument, the name for the serial device that you are using (like COM1 or /dev/serial0 and so on). The output from GPSEcho should be the raw NMEA data sent, and it should look like this:

```
using /dev/tty.USA19H1b1P1.1
Devel Library
========================================================
Native lib Version = RXTX-2.1-7pre17
Java lib Version = RXTX-2.1-7pre17
$GPRMC,,V,,,,,,120704,17.1,E,N*0F
$GPRMB,V,,,,,,,,,,,A,N*13
$GPGGA,,,,,,0.00,,,M,,M,,*66
$GPSSA,A,1,,,,,,,,,,,,,*1E
$GPGSV,3,1,09,01,72,002,00,04,21,303,00,11,04,200,00,13,31,290,00*7E
$GPGSV,3,2,09,16,42,108,00,20,73,187,00,24,02,334,00,25,24,053,00*76
$GPGSV,3,3,09,27,09,233,00*4E
```

and so on …

It is also possible to use Bluetooth GPS devices. If you have Windows XP SP2, set the system property placelab.gps_device=bluetooth and placelab.stumble_ bluetooth=true (this can be done either with the –D switch on the command line or by placing it in your placelab.ini). PlacelabStumblerGUI will automatically detect and begin using your GPS device. If it fails to connect to your device, try removing the device from your computer in the Bluetooth control panel. PlacelabStumblerGUI will automatically re-pair with the device.

If you do not have Windows XP SP2, you will have to use a Bluetooth virtual serial port utility. For a Mac OS, you will find that at /Applications/Utilities/Bluetooth Serial Utility.app.

Some GPS devices must be put into NMEA mode before they will work. I just fiddle with them until I find the menu for it, but you may have more luck referring to the documentation for your GPS device.

Stumbling for Bluetooth Beacons (Windows XP SP2 Only)

If you are running Windows XP SP2 and have BlueCove properly installed, Placelab-StumblerGUI will also search for Bluetooth devices. You will see two tables, one for Bluetooth and one for Wi-Fi, if everything is set up right. The text Placelab-Stumbler is not set up to stumble for Bluetooth at this time. To enable Bluetooth stumbling, set placelab.stumble_bluetooth=true (default is false for stability), and it will stumble for Bluetooth. Note that automatic discovery of Bluetooth GPS devices requires that you also stumble for Bluetooth.

Stumbling for 802.11, Bluetooth, and GSM (Windows XP SP2 + Nokia Series 60 Phone Only)

If you have the Place Lab phone distribution successfully installed on your Series 60 Nokia phone, PlacelabStumblerGUI can use the phone to spot for GSM towers, and it will collate these measurements into its current log and display a table for the GSM towers seen.

Note that if you are trying to stumble for GSM alone using only a phone (i.e., no laptop involved), you should read the Place Lab "Phone Stumbling How-To" section.

Before doing anything, make sure you pair your phone with the computer you will be running PlacelabStumblerGUI on. To do this, press the menu button on the phone (blue button on the far left) and go to connect->bluetooth and click over to the "Paired Devices" tab. Click "Options," choose "New Paired Device," and select your computer (entering in the passcode on both the phone and computer if necessary). Once you have the pairing set up, select the computer in the pairing tab on the phone, and in the Options menu choose "Set as authorized."

Now navigate back to the phone's launcher menu and start the Place Lab application (if it is not already running) to start the GSMS potter on the phone. Leave that running and start GSMBT (answer yes to both questions it asks).

Once GSMBT starts up, it will display some statistics on the number of GSM towers seen, and it will also display the Bluetooth ID and channel the program is listening on. Launch PlacelabStumblerGUI with the system property placelab.cellphone set to that ID and channel (it will be of the form xxxxxxxxxxxx:x, where the last x is the channel) and GSM stumbling will be enabled.

Transferring the data to a PC for uploading to placelab.org is the last step in GSM stumbling. The LogUploadProxy tool transfers stumbler logs from the phone to the PC, ready for uploading to placelab.org. The bytes that are received are in compressed format, which the proxy decompresses before uploading.

More information on transferring data from the phone can be found in the "Bluetooth Transfer How-To" section.

COMMAND LINE OPTIONS AND SYSTEM PROPERTIES

System Properties

These may be set either by editing your placelab.ini or on the command line with -D=value.

- **placelab.StumblerFunnel.timeout=time_in_ms** : This controls how long the stumbler will wait for an update from the GPS device before it will go out and collect beacon measurements without GPS. I set mine to 3000 because all the GPS devices I have used produce usable measurements once every 1 or 2 sec, and I like to give a little extra time just in case.
- **placelab.gps_device=gps_device** : This must be set to the serial device where the GPS is connected. If you do not set it, or set it incorrectly, the stumbler will still work, but you will not get GPS readings. An error message will be printed to stdout. With Windows XP SP2, you may also specify Bluetooth here (see GPS section above).
- **placelab.stumble_bluetooth=true|false** : Default is false. See Bluetooth section above.
- **placelab.cellphone=phone_id:phone_channel** (e.g., 000F6D8368FF:4): Used for GSM stumbling with a paired Nokia Series 60 phone. See GSM section above.
- **placelab.gps_speed=baudrate**: The NMEA specification says that a baud rate of 4800 is to be used to communicate with the GPS device. If you do not set this property, it will be set to the default. If your GPS device can support a higher speed, taking advantage of it may allow you to get more frequent updates from the device.
- **placelab.stumbler.newapcmd**: If you are using the AudioNotifier (see below) this will let you specify a command line to be executed when a new AP is seen.
- **placelab.stumbler.lostgpscmd**: Same as above but for lost GPS signal.
- **placelab.stumbler.lockedgpscmd**: Same as above but for acquiring a GPS fix.

Command Line Options

- **--log /path/to/log** (text stumbler only; the GUI stumbler has a GUI save dialog): This will save everything seen to the log file specified. Note that the stumbler will choose a unique name for the log if the one you pass already exists.
- **--a** (text stumbler only; the GUI stumbler uses --audio): Enables the AudioNotifier, which will play sounds when new APs are heard, either by using defaults for the OS you are using (Windows and Mac) or by

executing the commands corresponding to the placelab.stumbler.* system properties above.

- **--audio** (GUI stumbler only; the text stumbler uses --a)

Phone Stumbling How-To

Place Lab determines position estimates based upon visibility of known beacons. The process of finding beacons and assigning them coordinates is called stumbling. This how-to outlines the use of the phone stumbler that runs on Nokia Series 60 2.0 phones. The phone stumbler is designed to be used with a Bluetooth GPS device to log GSM cell information and GPS data, correlating them together. It does not require a laptop to be carried around; thus, stumbling can be performed for many hours without requiring constant attention.

If you are trying to stumble using a laptop (i.e., for 802.11, Bluetooth, or through a slave GSM phone), you should read the Place Lab "Laptop Stumbling How-To" section above. The entire process involves stumbling for GSM beacons, uploading the stumble data, and downloading the most recent data to your phone. Once you have map data on your device, Place Lab is ready to be used.

Using the Phone Stumbler

In order for the stumbler to work properly, it must be able to connect to a local host native application called Servers. This is a native Symbian application that provides the Java portion with GSM cell information. Make sure this is running before starting the stumbler. If you are not running this application, no GSM data will be available for stumbling. You can make the Servers portion of Place Lab start automatically by installing the placelab-autostart.sis file instead of the placelab.sis file. The autostart file will always keep the Servers application running unless you explicitly exit the application (Figure 8.41).

From the main menu choose the option Start Stumbling. You will be asked to allow placelab-s60 to make a network connection. This is establishing a local host network connection to the native Symbian server described above. There is no

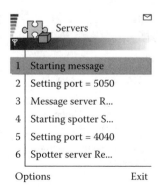

FIGURE 8.41 Place Lab connection to server.

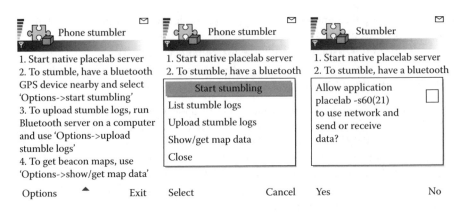

FIGURE 8.42 Place Lab phone stumbler (1).

outbound connection being made from your phone, and no charges will occur. Select yes to continue (Figure 8.42).

The following screen shows two items of interest. The first is the GSM readings. Every second a reading is taken for the current GSM cell information. The second item is the GPS readings. When the stumbler is started, a Bluetooth inquiry is done to find any Bluetooth devices advertising a string with the name GPS. If a GPS device is found, you will be prompted to connect to the device. Select yes to connect. In this example, our phone has connected to the TomTom Wireless GPS. As valid GPS information becomes available, it will be shown on the screen. However, if a GPS device is not found, the application will sleep for 5 min before trying another inquiry. Assuming a GPS device is found, you are now able to stumble everywhere you go (Figure 8.43).

Note, that although a Stop option exists while the stumbler is running, this is not guaranteed to work every time. We suggest closing the stumbler application, then starting it again if you need to stop stumbling.

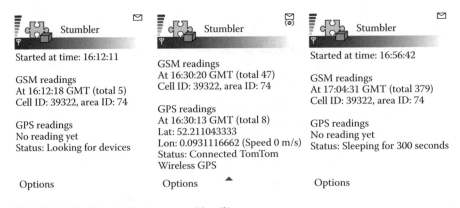

FIGURE 8.43 Place Lab phone stumbler (2).

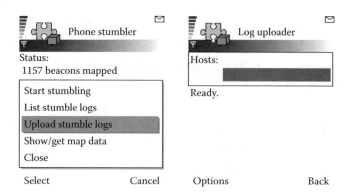

FIGURE 8.44 Place Lab phone stumbler logs.

Uploading Stumble Logs

After stumbling, the logs obtained can be uploaded to the main Place Lab database through the Bluetooth server. Select the "Upload Stumble Logs" option from the main menu. This will bring up a log uploader screen to choose a Bluetooth server to connect to. Refer to the "Bluetooth Transfer How-To" section above for instructions on uploading the logs (Figure 8.44).

All uploads are currently done with a default account phone with password "p\ hone." In order to use your own Place Lab username and password, set up your Place Lab properties file under the directory

```
C:\Documents and Settings\johndoe\placelab.ini

Here is a sample content for the placelab.ini file:

placelab.dir=C:\\Documents and Setting\\johndoe\\placelabdata

placelab.uploadLogs_username=johndoe

placelab.uploadLogs_password=placelab
```

Downloading Map Data

The stumbler application is the access point for downloading map data from the main Place Lab database. Choose the Show/Get Map Data option from the main menu. There are several options for loading map data (Figure 8.45).

1. The first method is using the Bluetooth server for downloading data from the main Place Lab database. Select the Update (Bluetooth) option, then follow the steps outlined in the "Bluetooth Transfer How-To" section, above, for connecting to the Bluetooth server. Using this method will always give you the most up-to-date map data for the phone platform.
2. The second method is to update from a local file packaged with this distribution. We have taken the most current map data at the time and packaged it with the application to provide a fairly recent, quick way of getting map data.

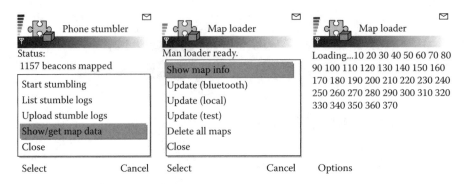

FIGURE 8.45   Place Lab downloading map data.

3. The final method for obtaining data is updating from a test file. This is a
    file packaged with the distribution that contains test beacons that are not
    in the main Place Lab database. This method is the way to load your own
    database of beacons that are separate from the Place Lab beacons. If you
    decide to do this, you must edit the TestBeacons.txt file in the midp-
    build/resources directory of the source and repackage a jad/jar distribution
    for yourself.

Place Lab is now ready for use.

TRACKING

Tracker

Tracker is the abstract superclass for the objects in Place Lab that aggregate measurements from the spotters into a single estimate of device position. Trackers are
used in the following way: measurements are fed to the tracker, and the tracker
produces estimated locations. Interested objects can also register with trackers to
get callbacks when the tracker's estimate changes.

To implement a tracker, you must furnish three functions. The first is the set of
measurement types the tracker understands. It is not required that a tracker understand all kinds of measurements. A tracker that can triangulate 802.11 signals, for
example, may not understand GPSMeasurements. The second is a method that
consumes the measurements coming from the spotter. Third, the system may call a
spotter to let it know that some time has elapsed without a measurement occurring.
This provides a sophisticated tracker the opportunity to update an estimate based
on its prediction of the velocity or path of the device.

Utility Classes

PlacelabWithProxy

The PlacelabWithProxy class is a wrapping of spotter and tracker creation and
maintenance. If the no-arg constructor is called, the system creates a WiFiSpotter,

a CentroidTracker, and the system default Mapper. The spotter is asked for measurements once a second, and the measurements are mapped and fed into the tracker, much as our last sample program does. Interested clients can register with the object to receive estimates from the tracker. This object also allows a spotter, tracker, and pulse frequency to be passed in on creation time. This allows a log spotter or an alternate tracker to be used instead. Passing in a pulse frequency of –1 will cause the PlacelabWithProxy object to not automatically ask the spotter for readings.

Servlets and Proxies

Place Lab includes a Hypertext Transfer Protocol (HTTP) proxy and servlet engine that allows applications to both create simple web front ends and include Place Lab services on remote web servers.

Place Lab communicates location information to web services via a web proxy that runs on port 2080. This proxy is started when a PlacelabWithProxy object is created and the createProxy() method is called on it. The proxy adds the location information by appending an HTTP request header called X-PlaceLab.Location that has lat,long as its value.

Place Lab also includes a simple servlet engine to allow the application to trap and answer HTTP requests based on their URLs. This trapping and answering allows a program to add simple behaviors with HTML UIs with little effort. The servlet behavior is encapsulated in objects of type Servlet, and they are installed by calling methods in ProxyServletEngine.

To use the proxy server or the servlet engine, set your browser configuration to use http://localhost:2080 as its HTTP proxy.

Maps and Visualization

MapWads

A MapWad is a special file that includes maps and sets of places. Mapwads can contain both raster maps and information in U.S. census Tiger/line format to make vector maps. A full specification of the MapWad format can be found in the javadoc for WadData.

MapView

MapViews consume MapWads and display interactive maps with pan and zoom functionality as an SWT control that is usable on all platforms supported by SWT (though memory constraints still apply for large maps, of course). MapViews may be extended by use of MapViewOverlays to show extra information. TrackedMap-View is a MapView that includes by default the ReticleOverlay for tracking position, APOverlay for viewing nearby access points and their activity, and the ParticleFilterOverlay for visualizing the particle filter (when a ParticleFilter-Tracker is used).

REFERENCES

Lewis, R., *Three-Dimensional Building Models from Two-Dimensional Architecture Plans*, master's thesis, UC Berkeley, 1996.

MIT LCS, *Generating a Three-Dimensional Campus Model*, MIT LCS, Cambridge, MA, 2001.

9 Privacy Concerns and Methods for Safeguarding Privacy

According to Intel's Place Lab and many others, privacy is the primary challenge faced in the development of location-based services (LBS). Furthermore, privacy is the most often cited criticism of LBS resulting in negative media coverage, and may be the greatest barrier to its long-term success.

Different access control policies might exist for different settings. Consider a hospital vs. a university. In a hospital, only doctors should be able to locate their patients. In general, an organization may want to allow only a certain group of people to discover who is in a specific room or building. For example, a floor of a building might be open to everyone who works there, but not to people from other floors and buildings. Also, a person's location should probably only be visible to a restricted group. For example, the managing director might be visible all the time to his or her secretary, while other people can see him only when he is in his office. As another example, it would be undesirable to allow just anyone (e.g., junk mail senders) to have access to all one's movement within a shopping mall. Yet, one might wish to be visible to, or reachable by, a select set of friends and family, or service ad providers, inside the mall.

LPS such as Place Lab have built-in advantages in that they are, in principle, passive systems. In a passive monitoring environment, mobile devices listen to the beacons' signals and are able to compute location (handset-based location computation). This principle gives the user control over when her location is disclosed. Unfortunately, current devices are not as passive as they would be desired in this case. For example, some 802.11 device drivers broadcast their existence to the infrastructure regularly.

Nonetheless, it is theoretically possible to construct a device that senses 802.11 and Bluetooth passively. Although the Bluetooth standard does not require that a device transmit its multiple-access control (MAC) address to neighboring devices when scanning, many of today's Bluetooth devices do so anyway.

Apart from passive scanning issues, another privacy trade-off is the manner in which mapping data are downloaded to clients from the mapping databases. Due to capacity issues, impoverished devices may only be able to load a small portion of the mapping database. However, if mapping data are downloaded for a small region, say a neighborhood, the operator of a mapping database has a reasonably fine-grained estimate of a user's location or potential location. Loading (and possibly discarding most of) a country or continent's worth of beacon locations gives the mapping servers less information about a user's location.

It may be possible to ensure privacy by hiding the identity of the user from the infrastructure. However, in some cases, like E911, the user's identity may be exposed to the infrastructure for reasons such as billing. Furthermore, by keeping cached copies of these directories, computers can calculate their location locally without transmitting information to any other computer. At this basic level, if one's personal device does not use the same wireless hot spot for communication, then this technique yields a totally private positioning technology. In these cases, one can still interact with cached offline content. This way, many services can be provided using cached information on the user's device, such as maps and schedules. However, more dynamic data, such as arrival and departure times, will still require a connection to a network. Other real-time applications include mobile commerce techniques that allow interaction between potential merchants and consumers.

In centralized architectures such as Active Badge, Active Bat, and PARCTab, the infrastructure consists of receivers deployed in places of interest, with end users beaconing out data. One's location data are initially determined on computers outside of one's personal control. In contrast, in decentralized architectures such as Cricket and radar, the infrastructure consists of beacons deployed in places of interest, signaling to end users that "You are here." Here, user's location is initially determined by the mobile device, giving users greater choice over whether to disclose their location to others and what information is disclosed. Client-computed location is the foundation for the most flexible privacy mechanisms and policies. In decentralized architectures, one's location is initially determined on a personal device, giving end users greater choice over whether to disclose their location to others and what information is disclosed. This means that users can select their location's resolution level (granularity) to be a room, building, block, or (at a larger scale) city. For example, a mobile user could choose to reveal her location at a room or store level to retrieve current sales going on in stores inside a shopping mall. This does not mean, however, that location information flows must be manually configured. Permissions could also be managed on automated criteria. For example, when a user is in the office, perhaps he wants his location to be generally disclosed. Some systems, like Place Lab, avoid the need for a central place where everyone's location information is collected.

Simple mobile transmitters, combined with active LPS, are well suited for inventory management and programs to monitor and secure objects and assets. People, however, may not prefer to be tracked like objects or assets while roaming the halls and rooms of indoor buildings and campus structures. A compromised active LPS could actually facilitate a data or even physical personal attack, as an intruder could watch — on screen — from a distance and wait for a target to move to a vulnerable location. Plus, any residual tracking records from the software could be analyzed for patterns or other variables that infringe upon privacy norms and regulations. Active systems must be highly secured and precisely engineered in order to guarantee the safety and security of users. Further, from a practical perspective, pure active systems alone do not satisfactorily address the problem of disconnected operation, although they can provide high levels of positioning accuracy.

Privacy concerns may be particularly sensitive as services allow colleagues, family members, or others to have real-time information on the location of individuals. Both privacy and security concerns do create resistance to LBS adoption. At the same time, a positioning capability is often used to increase security (e.g., tracking of children). In these applications, location information provides a compelling value proposition. In both of these examples, privacy and security are only maintained if access to the location information is restricted to authorized users.

How location information will be managed when the positioning capability becomes ubiquitous or seamless across all world (networks) is still uncertain.

For some applications, location positioning is a minor value add-on to the core service. In these cases, the positioning capability may be so marginal that it could be excluded or made optional to counter privacy concerns.

In short, applying these diverse perspectives on emerging indoor LBS applications provides insights into the types of users that are likely to adopt LBS, and major adoption barriers.

CONSTRAINTS

In any system deployment, there are economic constraints on the technology. In addition, there are constraints that do not come from the business model or customer behavior. These include legal liability for service failure and privacy requirements from the users of the technology. Such requirements can directly control the deployment of a technology. Current technology acceptance models, such as TAM, have many deficiencies for the range of contexts in which location-based services are deployed. For example, current acceptance models cannot explain the stark differences in the uptakes of WAP and i-mode service in Europe/U.S. and Japan, respectively. In contrast, it is important to integrate the perspectives of the user as a consumer, a network member, and a technology user. This approach highlights the need to use specific technology adoption models for LBS at the individual level, but also raises the need to develop multiperspective frameworks for understanding adoption behaviors at the group and organization levels.

In addition, more research is needed to explore how the factors influencing location-based services' adoption differ by level of analysis in the consumer market (e.g., families and nonwork organizations) and how location-based services' adoption in the business segment impacts use in individuals' private lives and vice versa. Services designers need to be able to understand the complex needs of all stakeholders in the overall social system, of which the location-based services are only a part.

There may be mobility implications that location-based services are not particularly well suited to address with their current capabilities. For example, while users frequently ask the OnStar service for directions to a particular store in a mall, the service can currently only provide the mall's street address. Carriers and others may be constrained by national or local laws regarding how they handle information. For example, in some relevant situations within the U.S., Customer Proprietary Network Information (CPNI) rules require that telecommunications carriers obtain customer approval before using, disclosing, or permitting access to individually identifiable

CPNI. See 47 U.S. Code Section 222 (http://www4.law.cornell.edu/uscode/47/222 html).

One of the main concerns of the users is privacy. Privacy is a well-recognized fundamental human right, defined as "the claim of individuals, groups, and institutions to determine for themselves when, how, and to what extent information about them is communicated to others." This is the stance taken by the legislators in the European Union, Canada, and other countries that have a specific privacy legislation (something that the U.S. does not). Note that according to this legislation, corporations and organizations do not have this right; it is a human right that can only be excercised by persons.

The European Commission has, in two directives, which have been implemented as laws by the member states, created directions for how privacy within the EU should be managed. The individual national laws sometimes add privacy enhancements to the directives, which in this respect can be said to represent a least common denominator for privacy. The extract from one of the directives describes the issues concerned in data protection:

ARTICLE 6

1. Member States shall provide that personal data must be:

(a) processed fairly and lawfully;

(b) collected for specified, explicit and legitimate purposes and not further processed in a way incompatible with those purposes. Further processing of data for historical, statistical or scientific purposes shall not be considered as incompatible provided that Member States provide appropriate safeguards;

(c) adequate, relevant and not excessive in relation to the purposes for which they are collected and/or further processed;

(d) accurate and, where necessary, kept up to date; every reasonable step must be taken to ensure that data which are inaccurate or incomplete, having regard to the purposes for which they were collected or for which they are further processed, are erased or rectified;

(e) kept in a form which permits identification of data subjects for no longer than is necessary for the purposes for which the data were collected or for which they are further processed. Member States shall lay down appropriate safeguards for personal data stored for longer periods for historical, statistical or scientific use.

ARTICLE 12

RIGHT OF ACCESS

Member States shall guarantee every data subject the right to obtain from the controller:

(a) without constraint at reasonable intervals and without excessive delay or expense:

- confirmation as to whether or not data relating to him are being processed and information at least as to the purposes of the processing, the categories of data concerned, and the recipients or categories of recipients to whom the data are disclosed,

- communication to him in an intelligible form of the data undergoing processing and of any available information as to their source,

- knowledge of the logic involved in any automatic processing of data concerning him at least in the case of the automated decisions referred to in Article 15 (1);

(*Controller* here implies the entity that is in control of the data, not necessarily its owner or the end user.)

According to many sources, contextual and location-based services are the primary concern for privacy advocates. The use of user profiles is also of concern. This consists of two issues: whether an identity can be bound to a location, context, or other set of parameters, and what the attributes of that identity (i.e., which information is accessible) may be.

The methods are agnostic to the attributes; the protection needs to be the same, however this is determined. The issue is the binding of the identity to some of these attributes, such as location, which could be used to reveal the identity of the individual.

The authentication of an endpoint is based on the trust relationship between an identifier and the corresponding identity. Once an endpoint wants to obtain identity privacy, it must be able to prove that it owns the related identifier without disclosing its identity. On the other hand, the location information often indirectly reveals the identity of an endpoint. Thus, the identity privacy and location privacy are implicitly bound together in the Internet Protocol (IP)-based communication.

Also, authentication can be applied when privacy issues that affect the owners of the beacons are of concern. Wi-Fi access points (APs) owners or providers (individuals or corporations) may be concerned if the existence and location of their APs are known and to be listed in a public database. There are a variety of potential ways to mitigate these concerns. 802.11 and Bluetooth beaconing can be manually turned off, making them invisible. As stronger authentication becomes available for wireless networks, some concern about beacon visibility may simply disappear.

Roaming subscribers may encounter service providers that they have not met before. And, vice versa, service providers may encounter unknown subscribers in their area. There could also be competing service providers in the same area, further complicating matters. The problem here is mainly one of cross-domain authentication and user profile management, which are outside the scope of this chapter.

In most cases, service providers will be obliged to implement certain generic security policies, such as nondisclosure to unauthorized third parties. Additionally, subscribers would specify an acceptable security policy for themselves.

In the following, we discuss in short the reasoning behind the conceptual separation between identifiers and location names, i.e., locators. The so-called identifier–locator split solves several mobility scenarios in a sound way. However, obtaining trust and sustaining privacy simultaneously is a challenging task in an identifier–locator split-based group communication.

The privacy issue can be addressed at three different levels: the connectivity level (where the IP address is bound to the location), the service and application level (regulating the terms where the service or application uses the location), and the application level (where the privacy policy in place is displayed to the end user).

Privacy-enhancing technologies fulfill one or several of the following purposes:

- Unlinkability
- Unobservability
- Anonymization
- Pseudonymization

Privacy is most frequently expressed as policies, e.g., the Platform for Privacy Preferences (P3P) Project, which is a language and an associated rules system to express privacy policies. However, alternatives exist, such as the PAW licensing system. Notable, however, is that both the service provider and user policies can be expressed in the same language. In fact, a generic language can be created for the expression of all types of privacy- and security-related policies. This has been done in REI, which is an OWL language that enables reasoning about policies.

CONNECTION-LEVEL PRIVACY: LOCATION–IDENTIFIER SPLIT

To keep the size of routing tables small enough, the Internet addresses are distributed hierarchically. That is, address prefixes and network topology are kept in rough synchrony, thereby allowing the routers to store less information than they otherwise would be forced to. At the same time, this practice binds the IP addresses to the topological locations in the network. This is true for any active location system, since there needs to be a dialog between the user terminal and the positioning system. However, if the location system is passive, i.e., broadcasting its position and allowing the terminal to draw its own conclusions about its location, there does not have to be any binding between the address and the location — unless the same network is used for location determination as is used for communication.

While the primary purpose of IP addresses is to make packet delivery possible, they are also used directly by the transport layer protocols, including Transmission Control Protocol (TCP) and User Datagram Protocol (UDP). In all of these, IP addresses are used for naming the transport layer sockets. That is, each communication context is named by IP addresses together with protocol and port numbers.

In Chapter 5 we talked about symbolic data models. In this chapter we will follow up with how symbolic (hierarchical) models can facilitate authentication and service delivery. A scale of privacy levels of the hierarchy can be defined as follows: Each level is defined to include more or less specific information about the location of a particular device. A user is able to assign a privacy level to entities that might request location information. Additionally, each node can have a privacy level associated with it. When a query from an application is received, the privacy software first determines who the query is from and the privacy level associated with the application or entity. The privacy software then evaluates one or more of the (sub) models to find a node that has a corresponding privacy level. When a corresponding node is found, information at that particular granularity is provided to the requesting application or entity. The next step would assign various privacy levels to the individual nodes in one or more hierarchical tree structures. Following, the system determines the privacy level associated with the application(s). The privacy software then traverses one or more hierarchical tree structures to find a node with a corresponding privacy level so that it can select the information that is associated with that node.

STATIC IP ADDRESSES

As long as a user is using a static IP address (or as long as she is using the same session with a dynamic address), it is possible to link her actions together and form a profile about her. With a little work, it is often even possible to link this profile to her real-life identity.

The address-tracking and profiling problem is slightly mitigated by the current practice of using dynamic IP addresses, especially Network Address Translation (NAT). However, with IPv6, the privacy situation is likely to detoriate since NAT is less likely to be used. Also, the privacy problem is simply moved from the receiving service to the service provider, since the mapping between the temporary address and the address used in the network is maintained by the service provider. In the European Union, legislation is now being introduced under the label of homeland security, which requires service providers to store such information for 3 years. Furthermore, even if NAT is used in IPv6, the translation will typically be one to one, without multiplexing several hosts behind a single address.

RANDOM IP ADDRESSES

If it becomes possible to use a random IP and link layer addresses, the problems related to IP address tracking more or less disappear. To be more precise, if an IP address no longer acts as an endpoint identifier, it is possible to take advantage of address translation at endpoints and middle boxes. As a consequence, the fact that it remains possible to keep track of IP addresses and find out their geographical location does not matter that much any more. The focus is moved to the endpoint identifiers, public keys, and fingerprints that must be protected.

The identifier–locator binding happens at a logical protocol layer between the transport and internetworking (IP) layers. The location names are bound dynamically to identifiers. The IP addresses become pure topological labels, naming locations in the Internet, while the EIDs identify an endpoint.

LAYER SOCKETS: SECURE SOCKETS LAYER (SSL)

It is good to notice that in the current Internet the processes are bound to transport layer sockets, and the sockets are identified using IP addresses and ports. In the identifier–locator split the transport layer sockets are no longer named with IP addresses but with endpoint identifiers. The endpoint identifiers (or their representations) are used in the socket application programming interface (API) and at the transport layer. Therefore, transport layer connections are bound to EIDs, instead of location names. As a result, an endpoint may change its IP address without breaking connections. The binding between EIDs and IP addresses is simultaneously dynamic and one to many, providing for mobility and multihoming, respectively. It is possible that horizontal and vertical handoffs are addressed to single connections that move independently of each other. The set of associated addresses can contain both IPv4 and IPv6 addresses.

However, the identifier–locator split still has some privacy-related problems. IP addresses together with public keys reveal directly the location and identity of an endpoint.

SERVICE-LEVEL PRIVACY: IDENTIFIERS AS A PRIVACY PROBLEM

If the connection level is de-identified, the user can freely connect to any information service without privacy concerns. However, if the service provider requires that the user identifies himself, and binds this identity to his behavior (current, past, or future), this becomes a privacy problem.

This is not solved using encryption. Instead, using public keys as endpoint identifiers is a source of privacy problems. First, a public key directly and strongly identifies an endpoint. Second, if an endpoint has just a single public key and uses it repeatedly, it is fairly easy to link together all the transactions made by the endpoint. However, an endpoint may have several public keys instead of just one. Some of the public keys can be used as more permanent identifiers, allowing the endpoint to be recognized. At the same time, some other keys can be anonymous, being temporary and periodically replaced.

Consider a Bluetooth and WAP push-based location-aware mobile advertising system. Encryption should be considered when device address data are sent from a Bluetooth sensor to the ad server. The ad server and the push sender send push SI messages as plain Hypertext Transfer Protocol (HTTP) requests where the user mobile phone number is associated with location information, which can be considered sensitive information that must not be disclosed to third parties. Ideally, mapping device addresses to location information should be done locally in the client device.

Recognizing the security implications of a location service (Spreitzer and Theimer, 1993), researchers at Xerox PARC argue that different environments need different levels of protection for people's privacy (Spreitzer and Theimer, 1994). They also advocate user control over the disclosure of location information. The approach allows for protection of anonymity via "secret" groups. They argue that in a large, heterogeneous system, only the user agent approach (as opposed to the location service approach) can deliver a meaningful protection of privacy. Within the Xerox PARC's user-centric architecture, the user agent implements the access control decisions as specified by the corresponding user. Access control can also be delegated to a central location broker to increase efficiency. Locations are not treated as first-class objects in this model; that is, no explicit policy regarding access to a specific location can be specified.

It is important to make a difference between anonymity and identity protection. If an endpoint uses an unencrypted identifier, it deliberately reveals its identity to outsiders, breaking identity protection. On the other hand, one can openly use an anonymous public key and remain anonymous.

In identity protection, one of the goals is to prevent malicious nodes from tracing any identity. Therefore, if we are able to offer complete identity protection for any type of identities, public or anonymous, the role of anonymous identities is changed. They are no longer needed to protect from man-in-the-middle or eavesdropping attackers, but from legitimate peers. Other techniques for user tracking, such as HTTP cookies, can be used in the same way.

This is an issue for the service design, and the user will need to verify that the service provider does not misuse the information gathered. To verify this, the user and service provider need to agree which policies apply for the profile management, and the user needs to verify that the privacy policy is maintained by the service provider (and that the service provider is not, in essence, lying). This is, needless to say, extremely hard to do and does not easily lend itself to computerization (it would carry too far to discuss in this book).

Overall, the use of anonymity can keep track of how many people are at each location over time, but nevertheless, tracking can reveal some strong hints about who has been where. Multitargeted access control (policies) can come in three levels of protection:

- Access protection
- Location anonymity
- Personal anonymity

But as a developer, the first step in developing a method is to provide a mechanism. The Platform for Privacy Preferences (P3P) Project, developed by the World Wide Web Consortium, is emerging as an industry standard providing a simple, automated way for users to gain more control over the use of personal information on websites they visit. It was released as a W3C recommendation in 2002.

At its most basic level, P3P is a standardized set of multiple-choice questions, covering all the major aspects of a website's privacy policies. Taken together, they present a clear snapshot of how a site handles personal information about its users. P3P-enabled websites make this information available in a standard, machine-readable format. P3P-enabled browsers can read this snapshot automatically and compare it to the consumer's own set of privacy preferences. P3P enhances user control by putting privacy policies where users can find them, in a form users can understand, and, most importantly, enables users to act on what they see.

There is also an associated rules language: A P3P Preference Exchange Language 1.0 (APPEL1.0). It complements the P3P specification by specifying a language for describing collections of preferences regarding P3P policies between P3P agents. Using this language, a user can express her preferences in a set of preference rules (called a rule set), which can then be used by her user agent to make automated or semiautomated decisions regarding the acceptability of machine-readable privacy policies from P3P-enabled websites.

IETF GeoPriv

Another similar idea comes from the Internet Engineering Task Force (IETF) Geographic Privacy group, which focuses on the location privacy aspect. The target is to provide an API for access and management of link-based location information. This includes an API for handling location information specific to 3G wireless link technology. One of the latest drafts (draft-ietf-geopriv-policy-03, October 2004) describes a document format for expressing privacy preferences for location information. It describes a protocol-independent model for access to geographic information.

Consider an example of having an ad location-based system (AdLoc), which allows for permission-based advertising based on geolocation information. It allows personal digital assistant (PDA) users to discover their geolocation and send it to a central database, where it can be accessed only with a digital license.

This example holds true if the handset cannot compute the location on its own for technical reasons (low computing power). Then, in order to protect his privacy, the user would need to form a trusted relationship with his carrier or a third party, obligating them to compute the location and either provide it back to the user's device for serving or abide by the user's rules about privacy.

Figure 9.1 to Figure 9.4 demonstrate the following phases: discovery of location, collection of data, collecting of a license, and sending of an ad.

In Figure 9.3, a licensing agreement (contract) between the system and the end user, Alice, is portrayed. The system requires control of access to position information, and that a user can see not only who has tried to locate her, but also how often that has occurred. The questions are how to restrict access to positioning information in general, and how the positioning system should be implemented in order to allow different levels of access.

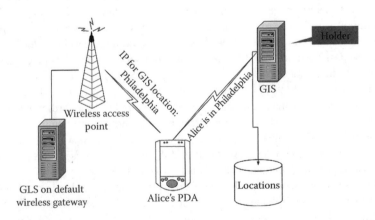

FIGURE 9.1 Discovering location.

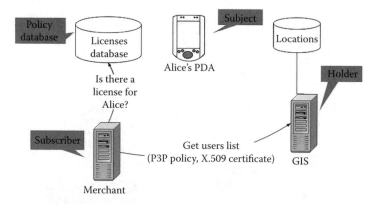

FIGURE 9.2 Collecting data.

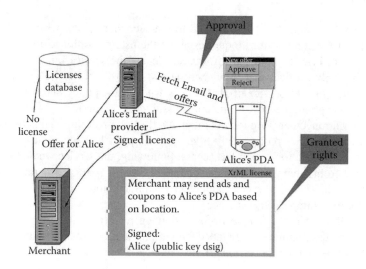

FIGURE 9.3 Collecting a license.

Hence, the user's application owns its position, and any external service has to request permission from the user's application in order to obtain the position. It is therefore possible to query the platform for a position, but it will automatically result in a request to the application.

The application is then free to accept or deny the request or to choose to ask the user if he wishes to permit the query. Queries can be accepted once or for a longer period, depending on a set of criteria defined by a contract between the user or application and the peer conducting the query. Criteria may include the ability for a service to query the platform a fixed number of times, unlimited queries in a specific amount of time, or any combination of the above two.

In addition, the user delegates the right to grant access to the user's position to a secondary party, such as a Wi-Fi positioning center, by constructing a contract. The contract will be limited by the user's selection of criteria. All queries to the

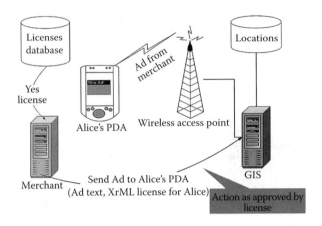

FIGURE 9.4 Sending an ad.

mobile device would then be redirected to the secondary party, which will follow the scope of the contract for all queries about the user's position. Service trust is a big deal here. In order to avoid information leaks, the service infrastructure must ensure that only services that implement access control checks are given location information. Plus, the service needs to check whether the policy maker (i.e., central authority) has granted access to the entity issuing the request.

Privacy arrangements are incorporated into the system design so that the services respond to a location request only after performing a location policy check that verifies that the location seeker has access. Following, the service would verify that the service from which it receives a forwarded request is trusted before returning an answer to it. Then the service can delegate both location policy and service trust checks to, for example, other services.

The following are a sample license allowing retention, limited redistribution, and sending ads. Comments are embedded within the eXtensible Markup Language (XML) code indicating the various rights.

```
<?xml version="1.0" encoding="utf-8"?>
<core:licenseGroup
    xmlns:core="http://www.xrml.org/schema/2001/11/xrml2core"
    xmlns:cx="http://www.xrml.org/schema/2001/11/xrml2cx"
    xmlns:dsig="http://www.w3.org/2000/09/xmldsig#"
    xmlns:sx="http://www.xrml.org/schema/2001/11/xrml2sx"
    xmlns:xsi="http://www.w3.org/2001/XMLSchema-instance"
    xmlns:priv="http://www.pdrm.org/XrMLPrivacy"
    xmlns:p3p="http://www.w3.org/2002/01/P3Pv1"
    xmlns:xs="http://www.w3.org/2001/XMLSchema"
    xsi:schemaLocation=
    "http://www.xrml.org/schema/2001/11/xrml2cx.../schemas/xrml2cx.xsd">
    <core:license
```

```
        licenseId="http://www.pdrm.org/examples/2003/SendAnyAd">
        <core:inventory>
        <!-- Device with ad -->
        <priv:mobile licensePartId="mobiledevice">
        <priv:locator>
        <priv:id>2155555050@MobileISP.com</priv:id>
        </priv:locator>
        </priv:mobile>
    </core:inventory>
    <core:grantGroup>
        <!--The company that is tracking us' specific key.-->
        <core:keyHolder>
        <core:info>
        <dsig:KeyValue>
            <dsig:RSAKeyValue>
                <dsig:Modulus>...</dsig:Modulus>
                <dsig:Exponent>...</dsig:Exponent>
            </dsig:RSAKeyValue>
            </dsig:KeyValue>
        </core:info>
        </core:keyHolder>
    <sx:x509SubjectName>
    The person allowing the company to track him/her-->
    <core:issuer>
        <sx:commonName>John Doe</sx:commonName>
    </core:issuer>
    <!--The period for which the company may track the user. -->
    <core:validityInterval licensePartId="trackingPeriod">
        <core:notBefore>2004-05-20T19:28:00</notBefore>
        <core:notAfter>2004-07-29T19:28:00</notAfter>
    </core:validityInterval>
    <!--Grants Company the right to track the user through the
    permission period. -->
    <core:grant>
        <priv:PrivacyPolicy>
        <!-- Disclosure-->
        <p3p:ACCESS>
            <p3p:all/>
        </p3p:ACCESS>
    <!-- Disputes -->
```

```
<p3p:DISPUTES-GROUP>

  <p3p:DISPUTES

    resolution-type="service"

    short-description="Customer service will

      remedy your complaints.">

    <p3p:REMEDIES>

    <p3p:correct/>

    </p3p:REMEDIES>

    </p3p:DISPUTES>

  </p3p:DISPUTES-GROUP>

<p3p:STATEMENT>

  <p3p:CONSEQUENCE>

We collect your location information for development purposes and
for tracking your individual movement habits.

  </p3p:CONSEQUENCE>

  <!-- Why we use it -->

  <p3p:PURPOSE>

    <p3p:develop/>

    <p3p:individual-analysis/>

    <p3p:individual-decision/>

    <p3p:current/>

  </p3p:PURPOSE>

  <!-- Who else can get this data -->

  <p3p:RECIPIENT>

    <p3p:ours/>

  </p3p:RECIPIENT>

  <!-- How long do we hold onto the data for -->

  <p3p:RETENTION>

        <p3p:legal-requirement/>

        </p3p:RETENTION>

      </p3p:STATEMENT>

    </priv:PrivacyPolicy>

    <!--The mobile device from the inventory-->

    <priv:mobile licensePartIdRef="mobiledevice"/>

    <!--The rights that we are giving-->

    <priv:sendanyad/>

    </core:grant>

    </core:grantGroup>

  </core:license>

</core:licenseGroup>
```

The user should also be able to deny all access to his position in a simple way, independent of any issued contracts to secondary parties. This could be done by revoking all contracts or adding additional criteria that override all others (i.e., deny criteria should take precedence over grant criteria).

There are, however, also other possible threats. The system could easily be compromised by false positioning sources, such as a Bluetooth beacon with a false position. Therefore, the sources should be separated into two groups, trusted and nontrusted positioning sources — authentication methods being in place for trusted sources and prioritorized by the service platform.

Access control should be distributed because of different services and interactions, and it should incorporate different positioning/location technology (global positioning systems (GPS), wireless, cellular). It should also be implemented with different administrative entities.

Moreover, policies should be able to limit information flow in other ways (these are privacy-preserving aggregation techniques/varying types of location disclosures):

Spatial granularity (scale/resolution): A policy can restrict the resolution level of the returned location. For example, a user can state that the building-level resolution be returned instead of the actual room. A policy mechanism can also be implemented that returns the number of people in a room as opposed to their identities.

Temporal granularity (history): For example, monitoring that the user was at the shopping mall sometime last month rather than "user was at the shopping mall Dec 1st."

Locations/users (one does not want a friend to secretly look him up without knowing first who it is): A policy can contain a set of locations that will allow a user to find out about another user's location only in that specified location (e.g., office). A policy can also include a set of users such that the answer to a query will include only users listed in the policy (provided they are in the room). This allows the user to easily limit the rate at which that buddy can query (by clicking a limit spot on the bar and thereby locking that level).

Time intervals (temporal freshness): Location policies can limit time intervals during which access should be granted, setting how long "guest" (other users, e.g., buddies) can access location information after the user becomes disconnected (walk out of Wi-Fi coverage).

Content/context priority level: A message submitted for delivery is delivered at the soonest acceptable time. Acceptable delivery time depends on the context of the recipient. For example, the recipient's profile may specify that messages below a certain priority level should not be delivered when the recipient is in a meeting.

REI

A third method of managing profiles is to integrate them into the general information management, for instance, using ontologies. REI is a policy language based in OWL-

Lite that allows policies to be specified as constraints over allowable and obligated actions on resources in the environment. It is intended for the management of both privacy and trust policies, and several other types of policies as well.

REI also includes logic-like variables, giving it the flexibility to specify relations like role value maps that are not directly possible in OWL. REI includes metapolicy specifications for conflict resolution, speech acts for remote policy management, and policy analysis specifications like "what if" analysis and use case management, making it a suitable candidate for adaptable security in the environments under consideration. The REI engine, developed in XSB, reasons over REI policies and domain knowledge in RDF and OWL to provide answers about the current permissions and obligations of an entity, which are used to guide the entity's behavior.

Companies do not have a right to privacy in the same way as an end user does. However, they can acquire certain rights to data, and control rights to data on behalf of the end user. This implies that mechanisms must be in place to handle this. Examples of these include Digital Rights Management (DRM).

OMA DRM

The Open Mobile Alliance (OMA) DRM specifications focus on the control of the distribution of the information provided to the mobile phone user. For privacy, the user takes the role of the entity, which does want to control its personal information. OMA DRM is implemented in many mobile devices and designed to protect content downloaded to the device according to the rights specified in the Rights Object. OMA DRM is founded on another markup language for rights management, Open Digital Rights Language (ODRL).

Another markup language for rights management is ISO/IEC MPEG21-5 Rights Expression Language (REL) and its associated Rights Data Dictionary (RDD). REL and RDD together enable management of the consumption rights of all forms of content. REL can also be used independently of RDD. REL provides the means to create expressions that can then be associated with audio/video and other content to express what the consumer can do with that content. REL expressions can be as simple as "play" and can also express the complex consumption models that can be devised.

The Open Digital Rights Language (ODRL) provides the semantics for a Digital Rights Management expression language and data dictionary pertaining to all forms of digital content. ODRL is a vocabulary for the expression of terms and conditions over digital content, including permissions, constraints, obligations, conditions, and agreements with rights holders. The ODRL can be extended by different industry sectors (e.g., ebooks, music, audio, mobile, software). ODRL is available in the spirit of open-source software.

In addition, the access to resources (such as services) needs to be managed. One method of doing this is XACML, an XML-based language, or schema, designed specifically for creating policies and automating their use to control access to disparate devices and applications on a network. The purpose of the XACML specification is to define a core schema and corresponding namespace for the expression of authorization policies in XML against objects that are themselves identified in

XML. The schema is capable of representing the functionality of most policy representation mechanisms. It is also intended that the schema be extensible in order to address that functionality not included, custom application requirements, or future features. XACML handles fine-grained access control, the nature of the requestor, the protocol over which the request is made, content introspection, and the types of activities authorized. Sometimes XACML is also described as the syntax to create trust.

Ideally, there should also be more generic ways to specify access authorization. For example, when attending a conference, a user should have the means to be visible to all the other attendees without actually knowing them. Similarly, I might wish to be anonymous in locations matching a given constraint, such as a motorway. Generic authorization constraints are especially important since the service is partitioned among many providers. These providers must rely on local knowledge to make access control decisions. Constraints that require frequent access to nonlocal information cannot be considered a scalable solution to this problem.

APPLICATION-LEVEL PRIVACY: VISUALIZATION OF PRIVACY

In a user-centric project, the visualization of issues pertaining to the user becomes paramount. This is especially true for issues where the user has been given control over some set of information by legislation, and service providers may not use the information without the user's permission. One such area is, of course, privacy.

There are two ways of visualizing privacy today:

- A number of graphic tools, which purport to indicate privacy, but actually show whether there is a risk of privacy violations
- Text representation of the privacy policy applied by the user or service provider

There are two ways of expressing trust policies in text form: as structured policies (following an ontology, or using a policy definition language, e.g., P3P) or free text. Since there is an infinite number of possible expressions of trust in free text form, and it has not been possible to identify any research directed at this method, we will disregard the free text representation as visualization; however, it should be noted that the reason for its prevalence is probably a response to the perceived difficulty in reading policies in structured text.

There have been studies where researchers investigated how non-English laymen understand privacy vocabularies, and proposed a phone display. Terms were selected based on the Privacy Bird interface (i.e., the subset selected for analysis of the textual representation was based on the subset selected for the graphical representation). The results were somewhat dismal (even using the suggested simplified language proposed by the P3P working group), since despite the familiarity of Swedes with English, the fact that many of the terms used in privacy vocabularies do not have an equivalent in Swedish seem to affect understanding quite drastically. It is noteworthy, however, that many users did not have an understanding of the Swedish terms used in the Swedish data protection act, which regulates privacy (terms that, as it happens, do not have a precise equivalent in English).

The relation between the textual presentation and the user interface presentation is also a crucial issue. A WAP was tested on several users, using a Swedish test of the vocabulary suggested by the Independent Center for Privacy Protection (ICPP) in Kiel, Germany. The results of this were significantly better than the vocabulary tests using English vocabulary.

If the users do not understand the terminology used to describe the actions of the system, it will become very difficult for them to exercise control over it in a meaningful way. The study recommends that the privacy policies be described in a vocabulary in the native language of the users, based on the ICPP vocabulary; a further conclusion is that this, too, may have to be simplified.

It is notable that even using a simplified, user-friendly vocabulary, it becomes difficult for the user to understand and visualize the control of all information that he legally has control over.

The most frequently used user interface to privacy information, apart from the purely textual display, is the AT&T Privacy Bird, which automatically searches for privacy policies at every website you visit. You can tell the software about your privacy concerns, and it will tell you whether each site's policies match your personal privacy preferences. The software displays a green bird icon at websites that match, and a red bird icon at sites that do not. The bird icon alerts you about website privacy policies with a visual symbol and optional sounds.

The AT&T Privacy Bird reads privacy policies written in the standard format specified by the World Wide Web Consortium's Platform for Privacy Preferences (P3P). The AT&T Privacy Bird is intended for use with Microsoft Internet Explorer web browsers (version 5.01/5.5/6.0 on Microsoft Windows platforms).

The only conclusion that can be drawn from the current systems is that this issue is in no way settled. There has to be more work, and which privacy policies may apply in an indoor location information service is by no means clear, nor is how they should be visualized.

REFERENCES

Spreitzer, M. and Theimer, M., Scalable, secure, mobile computing with location information, *Commun. ACM*, 36, 27, 1993.

Spreitzer, M. and Theimer, M., Architectural considerations for scalable, secure, mobile computing with location information, in *Proceedings of the 14th International Conference on Distributed Computing Systems*, Poznan, Poland, June 1994, pp. 29–38.

10 Development and Deployment of Indoor Location-Based Services

There are a number of components that have to work together to create an indoor location-based system. Location-based services can be classified into four types, which are in an increasing level of magnitude in terms of implementation:

- Simple or basic: Users manually enter their location (i.e., room/building reference, place name, landmark, address, phone number)
- Location aware: Location determined automatically (i.e., triggered)
- Navigational
- Context aware (or ubiquitous): Adaptivity to user's activities and events

The first type simply provides the user with a map and the ability to search the map in a variety of ways (e.g., search for an address, search for a landmark, scroll and pan). The second type provides both a map and the user's current location/position. The third type provides a map, the user's location, and directions of some kind. The fourth type takes into consideration what the user is currently doing (e.g., user is in a meeting) based on, for example, the user's schedule and calendar. The first two types are distinguished because a system that provides the user's current location is much more complicated than one that does not, and generally requires both additional hardware and software. In second-type systems, the location that is provided to the user need not coincide with the map (e.g., hallways, etc.) system. However, in order to provide directions in the third type, the user's location must coincide with a map of the indoor venue.

Moreover, with the first type, a list of restaurants may be maintained in a web-accessible database where each restaurant is associated with a zip code in which the restaurant is located. When a user desires to locate a particular restaurant, he might simply enter the zip code where he is located to see a list of corresponding restaurants in that zip code. From the list of restaurants, he might be able to select one or two restaurants of interest.

This approach is undesirable for a number of reasons. First, the operation of the system is dependent upon a central server that is responsible for receiving user queries and executing the queries to return the information to the user. In the event the server fails, so too does the service. In addition, this particular service might be suited to finding restaurants, but possibly not other businesses. In addition, the granularity with which the results are returned to the user may foist some of the search burden on the user (i.e., the user gets a list of restaurants in a nearby zip code, but has to further explore the list to select which ones are of interest).

For discussion purposes, we have previously discussed the Buddy Finder and Product Finder applications. Both of these applications can be achieved using the simple or basic, location-aware, or context-aware form. It is up to the developer to decide what form is best, which will depend on what the customers want and how much they are willing to pay for the extra functionality.

It might be logical to start with the most basic form of the service and offer it to customers for free, which will, in turn, lock in a large enough customer base. With good marketing, this approach will attract new customers that would be required to pay for this service. The original customers (as well as new ones) would have the option to upgrade to the next form of the service (location aware) for a charge. Alternatively, the service is developed as a bespoke application, under a consultancy agreement. In that case, the business issues are not the concern of the developer.

In this chapter, we look at the choices a developer has to make when building a system, and the choices an IT manager has to make when deploying it.

As an example, a wireless local-area network (WLAN) can be used as a positioning system, as we have described in earlier chapters. From the previous chapters, we know there are, in general, three different ways to determine the user's location:

1. Use of some form of ground-based beacon that broadcasts its location to nearby users
2. Use of some form of radio positioning system that transmits information that the positioning system/application can use to determine the user's location
3. Use of some form of dead-reckoning and map-matching algorithms in which the user's pace of movement, direction of movement, etc., are continuously used to update his location

Given a positioning receiver, it is almost trivial to convert a first-type location-based service (LBS) application into a second type (i.e., one that provides both a map and the user's location). However, converting a second-type into a third-type LBS application (reconciling the user's location with the underlying map) can be much more complicated.

Specifically when talking about Wi-Fi-based positioning systems, there are three main ways of doing this: using the closest access point (AP) based on proximity, APs for trilateration, and fingerprinting. Using WLAN (Wi-Fi) for trilateration means that each AP has to be given a position, either a relative position in relation to the other access points (such as "5 m away to the northeast") or an absolute position (in a coordinate system). This depends on the application and how the position measured should be used, as discussed in previous chapters. Whichever positioning method is used, care has to be taken during the deployment of the access points.

Here, three important issues have to be considered: complete coverage, minimal overlap between access point coverages, and three-dimensional coverage.

Other potential challenges include a Wi-Fi AP not being registered in the positioning database or the Wi-Fi signal is weak. One possible solution to the possibility of having a Wi-Fi AP not registered in the positioning database is to have this database updated by users contributing more information into the database. For example, assume a user goes to a particular room and receives beacons from three APs, but only two are in the database. The third AP can then be added to the database with some high confidence that it is near the location of the other two APs. Data can also be added when an unknown AP is detected temporarily between two known APs.

Also, a geographic statistical solution can be used to refine the details of the Wi-Fi positioning database as a side effect of people using their mobile devices. Clearly, the data being collected by the geographic statistical technique would be much more useful if it were sent back into the infrastructure and then redistributed to all users as part of the Wi-Fi positioning database.

TRILATERATION

Absolute positioning is possible only if the exact (x, y, z) coordinates of the APs are known. In addition, the mobile device must be in a straight line of sight for trilateration to work. Otherwise, the signal will be distorted by the intermediate objects, which may mean that the trilateration becomes unreliable. Trilateration can be used to determine both absolute and relative position (i.e., intraroom positioning, such as "the SW corner of Room 101"). Hence, a person would first be symbolically located and then additional processes could geometrically refine the position — if needed.

If the service is using a geometric map, the user may optionally display distances along segments of the route. If the service is based on a symbolic model (spatial relationships based on either associations, containment, or proximity), the directions will not be given precisely as to how many meters before making a right turn.

LOCATION FINGERPRINTING

Yet another solution is Wi-Fi location fingerprinting (pattern matching), which is the matching of one set of measurements with another reference set contained in a database. In other words, a mobile device takes a snapshot of signals from visible APs, and compares this with the snapshot of reference points stored in the database. The following describes how radio frequency (RF) fingerprinting works. Fingerprinting approaches the problem of location by determining how a signal will be received at any point in a building. To do that, RF fingerprinting creates a grid that identifies how every single part of a building floor plan looks to all access points. A grid point can be as granular as a half foot. In order to determine how the RF will sound at each grid point within an enterprise, a system must first predict how the RF will interact with the building. This is done as part of a site survey, which we discuss later in this chapter. Furthermore, it traces rays from every access point in the network and accounts for reflection and multipath to a destination. The RF prediction involved in RF fingerprinting takes into account the attenuation of the walls and other objects (e.g., elevator shafts) in a building. Reflection of the RF off of walls and multipath are also calculated. At the conclusion of prediction, a database is populated inside of the management system. That database contains each coordinate, and how each AP views that coordinate from a signal strength perspective — creating an RF fingerprint for each coordinate. The fingerprints of different locations are stored in a database and matched to measured fingerprints at the current location of a mobile user. A common signal modeling approach is to record samples of wireless signals from points in a large grid drawn to encompass either the entire floor or occupied areas of a building. Then it gathers real-world information APs to compare to these fingerprints to determine precise location, constantly updating the system for changes in the RF environment.

Note that signal strengths of the manufactured APs by the various manufacturers vary from each other. To work around the differences, profiles for signal strength for each AP can be constructed. These profiles are necessary to revise the variation between APs and mobile devices.

In spite of the additional load that databases present to a computing system, fingerprinting (a database-centric approach) has good applicability for indoor positioning, given the complexity of triangulation, considering indoor wireless propagation patterns over time. For any database-oriented approach, simplicity of the reference schema is a perpetual goal. However, it is not important to know the location of APs, as long as someone has visited each location (i.e., room) once, measured the RF signals, and saved it as a signal strength profile in a database.

The data for this local fingerprinting database can be broadcasted by having APs advertise or announce themselves, in a way similar to global positioning system (GPS) signals. Each AP would repeatedly broadcast all room numbers within its signal range, as well as profile information for fingerprinting. The alternative is that a user could figure out where she is going and download fingerprinting reference information from the Internet. In this case, an algorithm could determine that a user is probably closer to (or inside) Room 110. The advantage of this approach is that it is symbolic or contextual and no complicated geometry is required.

POTENTIAL ISSUES WHEN DEPLOYING A WI-FI-BASED POSITIONING SYSTEM

In this section, we talk about deploying a Wi-Fi-based positioning system. We start by pointing out some of the main considerations when deploying an indoor positioning system and recommend a few tips. Some of the most important questions to consider include:

- How many APs are required for a given level of accuracy (what coverage in terms of accuracy does an AP provide)? What is the best layout that should be applied when placing APs? What about floor mismatching in relation to the user's actual position?
- Granularity: What should be the spacing of the radio map grid where location fingerprints are taken to result in higher positioning accuracy?
- Is it feasible to provide the required accuracy or performance by performing some trade-offs?

- What is the impact of the architecture of a building, and thus the radio propagation characteristics?
- What is the significance of static data vs. dynamic data position data collection?

LAYOUT OF APS AND COVERAGE

The number and layout of the APs seen by a mobile device can vary depending on the received signal strength (RSS), path loss, interference, and multipath fading at different locations. As a general rule, for each position in the area covered by the Wi-Fi signal, there should be at least three to five APs (five APs giving the best accuracy). Still, with a minimum number of viewable APs for a given location, a positioning system can use the RSS metric, often aided by a database or the network topology.

Non-line-of-sight (NLOS) and other signal propagation issues discussed in previous chapters cause inconsistency of the signal strength reception. As a result, the positioning accuracy is reduced. In most cases, the environmental factors always have the most impact on the accuracy and maximum detectable range. Even NLOS does not prohibit RF transmission as that of infrared; it does create the multipath problem, though, meaning the signal can possibly take different paths to reach the receiver and result in interference among the received signals.

There is significant interference indoors, which will cause inaccuracy with the triangulation method. For example, Bluetooth signal is one potential source of interference in Wi-Fi's RSS measurements. The signal strength-based location determination is subject to variation in terms of radio signal environments. Thus, triangulation or profiles including devices' features are necessary to correct the above problem. Triangulation can limit an approximate location of a mobile device by finding the intersectional regions for each AP's signal area. To do this, a mobile device perceives at least two AP signals and holds their BSS ID list.

The alternative approach is a proximity-based one, which involves recording the signal strengths of several visible APs for a given number of known coordinates. Then, during run time, the observed signal strengths of APs are compared against the stored values and the closest match results in a location approximation. With this approach the sources of interference become part of the recorded and observed measurements. The downsides of this approach are that the resolution of the system depends on the number of recordings made and significant changes in the surroundings could render the recorded measurements invalid.

To better deal with the interference problems, more RF readers can be used to improve the accuracy. With more RF readers, better positioning measurements can be made because more data can be gathered by having extra readers to do the positioning. However, the RF readers are usually quite expensive, so placing more readers means extra costs for the whole system.

Also, positioning reliability (fault tolerance) is important. In many cases, some APs may be disabled because of local power failures, management, upgrades, etc. In such cases, the mobile device will still be able to obtain some position location service, e.g., the ID of the AP.

In addition, sometimes a subset of APs may be present in the mobile device's view at a particular time. However, at other times, some APs might never be seen for a significant duration of time at a location because of the dynamic changes, especially in multifloor environments. It is best to collect RSS samples at two very different times of day, when the presence or absence of people in the indoor area significantly affects the RSS values. Especially in a harsh environment, where the RSS has large variations, the positioning system may request multiple samples of the RSS vector. APs may be placed so that locations where higher accuracy and precision are desired can see at least four APs. It is recommended that as part of the design stage, a site survey is done to know the best placement locations for the APs.

Also, it is worth keeping in mind that Wi-Fi adapter manufacturers use different radio hardware and software in Wi-Fi adapters. Even individual adapters from the same manufacturer typically have slightly incompatible scales for observing the received signal strength indicator (RSSI) values. This can cause positioning inaccuracy if the calibration and tracking steps are performed using different Wi-Fi adapters. The positioning inaccuracy caused by different adapter scales is typically 1 to 5 m (3 to 15 ft), depending on the adapter, without using any mitigation technology.

Coverage is also an issue outdoors. For example, with the HawkTour project we presented in Chapter 8, the IIT campus has limited Wi-Fi coverage in the outdoor areas, and it is not feasible to install wireless LAN access points in all areas for the sole purpose of the tour system. As a result, a possible solution HawkTour is considering is to integrate GPS technology into the client device for obtaining position information when the user is outdoors and out of range of the wireless LAN. An issue with this, however, is that the lack of the wireless LAN eliminates the two-way communications link necessary for the tour system to operate. Because the student may roam to areas where wireless LAN coverage does not exist, even the current HawkTour application software must cache information. This allows the user to have the latest position and information content of a particular area. If wireless communications are lost, the user can at least make use of the most current information.

Floor Mismatching

Given that APs can have coverage areas with a radius well over 100 ft, including up and down, the nearest access point level of granularity will point toward the likely area within 30,000 ft² on multiple floors. This does not provide sufficient information to determine if devices are inside or outside a space. Moreover, sometimes mobile devices may link to the network through a base station that is directly below the room where its user was standing. The map would show the user on the wrong floor.

Determining the floor is important because usually it is a prerequisite to being able to calculate the coordinates.

Some attempts find the correct floor using the AP that the client was associated with. Using WRAPI, which we talked about in Chapter 8, it can be specified for a network to attach to WRAPI, which associates with an AP and provides a method to return the multiple-access control (MAC) address of the associated AP. However,

WRAPI does not necessarily always associate with an AP that is physically located on the same floor as the client, even given that there is less interference between the client and APs on the same floor, as opposed to APs on another floor of the same building. This also considers a list that can be created before calibration that contains the relationship between each AP and the floor that it is located on. The MAC address of the associated AP can then be retrieved for a reverse lookup to determine the floor. Unfortunately, this logic and approach does not work given that WRAPI does not always associate with an AP on the same floor as the client. In order for this approach to work, a set of criteria must be developed that would identify APs that are most likely to be on the same floor as a client. Such criteria would likely start with a check to find out which AP has the strongest signal strength value. Once the criteria have been defined, then a new method that would, for example, associate a floor to the AP would have to be added to WRAPI. This method would return a MAC address of the AP or APs that are on the same floor.

In short, to solve this problem, rather than mapping the relationship between a mobile device and a single access point, a different approach compares the strength of the device's Wi-Fi card signal to that of three base stations. With this information, the system can then use triangulation to calculate and map the device's location.

One of Place Lab's projects, indoor location estimation with Place Lab, investigated increasing Place Lab's accuracy in determining indoor location, using Wi-Fi access point measurements. (Note that the current version of Place Lab supports war driving as opposed to walking.) This allows building-scale applications to be built that could use this increased accuracy. This project extended Place Lab's particle filter to understand the difference between floors (before, the positions of all APs and estimates were in two dimensions). Other methods for floor determination were also evaluated. The steps that were taken include the following:

- Changing the beacon class to support extra metadata, in the case floor. (x, y) coordinates and other data for the indoor area APs were added and populated the database.
- Adding the notion of floor to each particle estimate.
- Changing the motion model to deal with walking indoors, instead of driving outdoors.
- Changing the sensor model to account for wireless signal attenuation due to floors.

The modified Place Lab code is available from the University of Washington (http://www.cs.washington.edu/education/courses/590gb/04wi/projects/hile-liu/myplacelab.tgz): the main classes that were changed were org.placelab.core.Beacon, MapExporter, and MapLoader, to enable the new database, and then classes from org.placelab.core.tracker were extended for TwoHalfDParticleFilterTracker, TwoHalfDMotionModel, TwoHalfDSensorModel, and a variety of support classes. The mapping and testing application is called MultiFloorMap, in the default package.

Because the system works in three dimensions it avoids placing users on the wrong floor. Such a system would compensate for the radio interference that inevitably occurs with Wi-Fi signal propagation.

ALGORITHMS

In terms of algorithms, there are two basic algorithms for computing the position of the mobile device: (1) choose the location corresponding to the fingerprint with the minimum distance to the measured fingerprint; and (2) choose the closest database entries (those with the smallest signal distance) and estimate the location based on the average of the coordinates of these points. These are both based on the signal distance between the measured fingerprint and fingerprint entries in the database. A positioning algorithm may return the location information in terms of a position relative to reference points that are meaningless outside a specific context, e.g., "15 m northeast of entrance to lecture hall B."

With dense Wi-Fi coverage, the specific algorithm used for positioning is not as important as other factors, including composition of the neighborhood (lots of tall buildings vs. low rises), age of training data, density of training data sets, and noise in the training data. In sparser neighborhoods, sophisticated algorithms that can model the environment more richly win out. All of this positioning accuracy (although lower than that provided by precise indoor positioning systems) can be achieved with substantially less calibration effort: half an hour to map out an entire city neighborhood compared to over 10 h for a single office building.

PERFORMANCE

Issues related to the performance of positioning are very important, and include the accuracy of the location information (the error in the calculated position of the mobile device). The architecture of a building could result in layers of accuracy — for instance, accurate to 5 ft in certain areas, accurate to 10 ft outside of these areas but within other limits, and accurate to 20 ft outside these limits.

Also pertaining to a performance measure is the delay in making the position calculation available for the application due to bandwidth reasons. Positioning systems invoke the transmission of overhead packets over the wireless LAN in order to implement positioning algorithms. This additional overhead can significantly lessen throughput available to users. As a general tip, choose a solution that minimizes the transmission of packets. A vendor's test results related to overhead and throughput can be a beneficial piece of information. If the positioning system requires most of the usable bandwidth of the WLAN, then there will not be much left for other applications.

The decision to deploy the live tracking data collection on the mobile device or on the network can be influenced by a number of practical considerations. In the case of the mobile device, if communications bandwidth is scarce, then it may be preferable to have the mobile device merely communicate and have all tracking-related data collection occur on the back-end network. In the case of the network, if the computational resources cannot scale to support computing the locations of all the mobile devices to be tracked, the mobile device can instead collect signal strength data from the APs, apply the statistical model locally to compute its location, and relay the result to the back-end network. Or, depending on the deployment scenario and the specific capabilities of the mobile device, a suitable mixture of both techniques can be applied.

Also, in spite of the additional load, location fingerprinting — a database-centric approach — has good applicability for indoor positioning systems, given the complexity of triangulation considering indoor wireless propagation patterns over time. Part of this is the capacity of the system (how many requests for position calculation can be processed in unit time), as well as the coverage (the area where the LBS is available). The infrastructure capabilities influence the delay, capacity, and coverage. In the case of RF location fingerprinting, the size of DB containing the fingerprint information can have an impact on the delay, accuracy, and capacity (in addition to cost) of the positioning system.

In general, the accuracy of position depends on the technology employed, the characteristics of the radio channel, the bandwidth, and the complexity of signal processing techniques employed.

As WLANs (especially public) become more widely deployed and used, it is crucial to consider techniques that effectively handle the dynamically varying user load without compromising network performance. According to Microsoft Research (Balachandran and Voelker, 2002; Balachandran et al., 2002a, 2002b), a workload of 195 users, their high-level results indicate that (1) web browsing and secure shell are the dominant applications (64% of all bytes and 58% of all flows); (2) individual user sessions are relatively short (<10 min); and (3) longer sessions tend to be idle for the majority of the time. Moreover, the same research found that both average and peak individual bandwidth requirements are relatively small. Further, the studies showed that the ability of the network to achieve effective load balancing is dependent on the individual user workloads and not merely on the number of users associated to an AP. Therefore, inter-AP load-balancing algorithms that are implemented by some AP hardware vendors based purely upon the number of associated users can perform poorly, and such algorithms can actually benefit from the additional complexity of incorporating individual user workloads.

In terms of data bandwidth, there is the need to address quality of service issues, such as bandwidth management and enhanced service provisioning due to WLANs serving more than just simple transfer of data, e.g., multimedia applications.

Bandwidth limits are hard restrictions when communicating with mobile devices, irrespective of the transmission medium. In order to save bandwidth, two-dimensional vector graphics can be generated instead of a bitmap. These vector graphics have two main advantages over bitmaps: they consume less bandwidth and memory, and they can be scaled and rotated without loss of quality. In addition, generating vector representations after using graphical abstraction techniques can further reduce the amount of data. Another major advantage of vector graphics is that they can be transmitted incrementally to a mobile device, thereby reducing the apparent transmission time (delay between the start of transmission and the first element displayed).

Figure 10.1 shows an example of incremental generation and transmission. First, metagraphics are generated (the current position being indicated by a black dot and an arrow depicting the path) and transmitted as one pack age. Then the rest of the floorplan is rendered and the corresponding vector graphics are derived. These

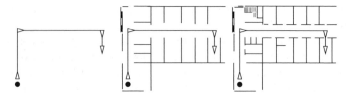

FIGURE 10.1 Incremental transmission of vector graphics.

vectors are next sorted by their size and transmitted in several packages, starting with the larger and possibly more significant line segments.

INTEGRATING DIFFERENT SYSTEMS

Outside buildings, the mobile telephone systems or GPS can give a rough position. When entering a building, however, the measurement of position is not detailed enough to be useful, in particular since it cannot give location information in three dimensions.

Communication (service) continuity and handoff issues, and roaming when combined with location, add another unique challenge: positioning handover. Seamless handover of location positioning is when an LBS is seamless or uninterrupted from one zone (that uses Global System for Mobile Communications (GSM)) to another zone (that uses Wi-Fi).

Seamless handover is when a handoff from one area (cell) to another takes place without perceptible interruption of the communication connection. Two worlds can coexist consisting of different communication networks. For discussion purposes, the two worlds are the indoor world and the outdoor world, having Wi-Fi and GSM networks, respectively.

The cell base stations within the GSM network send out RF signals. This enables the mobile station to monitor the signal quality of available cells. Based on the measurement of the strength of those RF signals, the station decides when to switch to a new cell. The switching process is called handoff, or handover. Moreover, the GSM network already has some basic location capability. The system knows the cell number that an active GSM user is currently located at, with an accuracy that varies from a few hundred meters in urban areas to a few kilometers in rural areas. Similarly, the Wi-Fi network has APs that are analogous to the GSM network's base stations, where the cell ID positioning method can be used.

The significance of enabling seamless communication and positioning handover can be seen when a company wants to expand its services into a new zone or world. This seamless world notion can be of two types. First, there is the outdoor-to-indoor world seamless handover. Imagine a mobile user (standing outdoors) is trying to find an ATM. The starting location would be either defined from a cellular network measurement (or GPS) or entered manually by the user (i.e., address that will be geocoded). Upon receiving the information, the user is directed to the ATM destination, which is within a shopping mall. In a cellular-to-Wi-Fi handoff-enabled world, the user would receive additional directions once inside the mall, based on Wi-Fi location capabilities, and would quickly find the ATM.

Second, there is the seamless handover within the subzones (i.e., floors) of the indoor world. Each zone or floor can be equipped with a different type of positioning infrastructure, meaning one LPS can provide absolute positioning ((x, y) coordinates), such as MIT Cricket or MS radar, while another LPS can provide relative positioning that is purely based on Wi-Fi cell ID (here the Wi-Fi AP has no fixed coordinates but a location that is represented symbolically, such as "in Room 101").

Absolute positioning is represented using geometry in a single coordinate frame (i.e., floor map of a building). It is important to realize that there can be several zones (i.e., floors) in a building, hence frames of reference. Potentially, these maps can be from different mapping infrastructure providers that would need to be in the same spatial reference system in order to enable data/map overlay.

There are two obvious problems with combining conventional cellular networks to Wi-Fi networks that make cellular-to-Wi-Fi service handoff an issue. First, not all mobile devices support both cellular bearers (GSM, code division multiple access (CDMA), iDEN, etc.) and IEEE 802.11b. Even if they do, the client software may not be able to support positioning over the Wi-Fi network. And second, when a user on a cellular network enters into a Wi-Fi coverage area, there is no mechanism in place to hand over the network connection from the cellular network to the Wi-Fi network (even if these two networks would be provided by the same provider, like T-mobile).

One possible solution is the integration of cellular and Wi-Fi networks. Taking from a real-world integration solution example, future versions of the T-Mobile software codeveloped with Boingo will allow users to manage their connections between T-Mobile's Wi-Fi and GPRS networks. Customers will be able to designate their network preferences or choose to get connected at the best available network speed, and the software will automatically connect them to the network of their choice.

Another possible solution is the use of location and geographic information system (GIS). Location capabilities (using cell ID, for example) in cellular networks can be used as a mechanism to manage Wi-Fi service handoff. Wi-Fi networks are typically confined to small geographic areas like a campus, and these areas can be mapped and represented by GIS databases as geographic zone features. Conventional cellular networks that already support location capabilities (explained earlier) are capable of tracking device locations. Combining this cellular location capability with a Wi-Fi coverage zone map gives the ability to intelligently trigger a cellular-to-Wi-Fi services handoff based on location when a user enters into a Wi-Fi zone (which can be indoors or outdoors).

This handoff trigger is based on the same premise of location-based presence for the purposes of zone-based alerting in cellular networks, but the zone happens to be a Wi-Fi area in this case.

Overall, the integration of the various components from each of the infrastructure types (and from different providers) highlights the complexities of effectively delivering indoor LBS and raises a number of questions:

- How do two providers integrate their services?
- Is it possible to integrate any kind of services?
- How do these services exchange information?
- How do you initiate/request a bundled service?

- How does a bundled service adhere to industry regulations/standards? (technical viewpoint)
- How do you achieve authentication and authorization between the services?

DEVELOPING THE LOCATION DATABASE

How to obtain and handle the content that the user will access in the application is probably the least researched but most important aspect of the system. Merely translating a floor map to a coordinate system will mean major work, and may not be particularly meaningful to humans accessing the system — but it will be tremendously useful to a cleaning robot, which has to navigate which rooms it has entered and cleaned, and which room it should go to next.

In either case, geometric or symbolic, the directions should probably be given by means of symbolic reasoning (i.e., "When you reach store A, make a right turn," instead of "After 10 ft, make a right turn"), as people associate with this type of reasoning better. In the case of the Segway or an automatic wheelchair, directions need to be given in terms of geometric reasoning since the machine is not capable of symbolic reasoning.

CREATING THE SOFTWARE INFRASTRUCTURE

In terms of the software (service) infrastructure, there are a number of possible models. The one that seems to be most favored so far is that the user will invoke a location-based concierge application that would determine her location, and she would subsequently receive navigation and routing details about how to get to the destination. Optionally, the service is able to specify some waypoints (or these points have be specified by the user), and the user may specify route determination criteria. The criteria might be fastest, shortest, etc., and can also specify the preferred mode of transport of the user (elevator vs. staircase).

The details of the returned information might include directions to the site and other relevant information (according to the user's profile, for instance). The route can also be optionally stored on the terminal or application server. The user may store it for as long as needed, thus requiring the means to also fetch a stored route.

Regardless of how the endpoints and waypoints are established, the system consists of different servers that provide the services. The route information is calculated by a separate route server. Applying OGC's OpenLS services concept, a subscriber would seek navigation advice from the service provider via a personal navigation service. This is an application service that utilizes OpenLS Core Services (Gateway, Directory, Geocoder, Presentation, and Route Determination).

Application servers sit between the mobile network, the databases, and the end user. The position is derived from the mobile network (or directly from the terminal) by using a standardized application programming interface (API).

Today, however, we do not have the standardized APIs. What we have are network-specific APIs. If you were to develop your own application server, you would have to provide specific interfaces to all of them, which is what the application server providers do today. The application server interfaces to the network of the mobile operator in positioning systems such as assisted GPS and network-based positioning. As a developer, you use the API to call the position information. You do not have to write applications that interface directly to the system of the operator (indeed, it is unlikely that you would ever be allowed to do so, because the operators are not too keen on letting people into their networks — and for most applications it is more practical to use an application server anyway). The application server probably does not interact directly with the core network infrastructure, nor does it include SS7 interfacing and the like. Querying the proper mobile positioning center (MPC) retrieves the user location.

Your application resides in the application server. Some application server providers have created proprietary or semiproprietary APIs, which your application can call. The most barebones application server imaginable is actually the web server, using Hypertext Transfer Protocol (HTTP) to call the MPC and retrieve the position document, but application servers incorporate web servers and provide database interfaces as well. How this process is managed is different between different application server providers, and architectures can vary a great deal. So can business models. The application server is housed in a central location, like an Internet Service Provider (ISP), and the service providers can house their applications in it, and it takes care of the interfaces to the network. Some of the network operators who have deployed location-dependent services see themselves as application service providers (ASPs); others plan to sell the data to companies that want to provide services. The services would then be provided in the same way as web services today — by companies that essentially are publishers but run their own infrastructure.

An example is the API developed by the French firm Webraska. The company intends to work as an ASP, making functions available to the programmer through an API on its own server. The functions that it makes available are fairly typical and comprise the following:

- Geocoding: Address validation and conversion into latitude–longitude coordinates
- Reverse geocoding: Conversion of precise latitude–longitude coordinates into an address
- Positioning: Conversion of approximate latitude–longitude coordinates (as provided by network positioning technologies) into a range of possible addresses
- Mapping: Customized dynamic generation of maps in gif, bmp, or wbmp format to developer-defined scale, zoom, and size
- Routing: Optimized route and journey time calculation for transport by car, on foot, or by public transport
- Enhanced spatial searching (ranking): Provision of ranked lists of potential destinations in terms of journey time from departure point

The web server typically acts as a front end to the application server, using the mechanisms of the web and WAP (HTTP and Hypertext Markup Language (HTML), Wireless Markup Language (WML), or XHTML) to deliver the presentation. It is often integrated into the application server. Because web servers are so prevalent, I will not go deeper into development using HTTP or other functions of web servers. If there are developers who are not familiar with the web today, there are many good books you can buy.

The web server senses what display capacities the requesting terminal has and delivers content accordingly. It can perform this task by using proprietary means (which is by far the most common today), or it can use a standard such as CC/PP.

The application server also shields application developers from changes in the back end, such as the type of location determination technology used, the selection of content providers or the mapping engine used, and the databases used. Most application servers are integrated with a database with spatial processing functions, which is used to handle the conditionality of the queries (in other words, drawing conclusions based on the coordinates and areas presented and determining how what data should be presented). Personalized data are managed in the same way.

Most application servers contain some kind of directory function that enables developers to add applications and data sources and manage them. Depending on the architecture, this situation can be handled very differently. In the Pacific Ocean server, for instance, the directory is managed by using Enterprise Java Beans; in other systems, it can be anything from a proprietary database to an LDAP directory. You call other modules — created by you or created by the application server developer — by calling functions through this directory service (which of course means that it is done very differently in different servers). Typically they are of a number of different types:

Locator: The component that gets the user's position

Billing client: Used to bill a user for the service access and the positioning that is done

Management client: Connected to the management system used by the operator

Logger: Used for logging different events for statistics, alarming, and tracing

Data sources: Mapping, information (such as restaurant guides), geolocation, and so on

Generic location-dependent functions: Routing, geolocation, and so on

Depending on the architecture, the services can be triggered in different ways. If the system includes an event handler, traps can be created that are triggered at different events. In principle, all location-dependent applications can be seen as based on events (including those that are purely request based), because an incoming request is an event that is trapped by the database. Tracking applications will create events when the periodic or triggered position reports come in. The trick is to direct them to the right places and perform the right operations on them.

Another question is whether it has to be located on one single machine or whether the functionality can be distributed. While there might be some slight security advantages of having everything in one place in a normal data center, there are several other systems with distributed functionality, and they will work in the same way. Performance will be better if functions can be offloaded to different CPUs (especially if this function can be automated), and tuning of the operations is vastly improved. If the processes can be distributed, it is also more likely that the system can provide failover (in other words, when one machine fails, another takes over). If there are load-balancing functions, it can also help the data center operator to maintain a smooth operation. But depending on philosophy, this function can reside in the application or in the operating system. It depends on how you have architected the data center, and the application server has to consider that.

QUESTIONS TO ASK WHEN SELECTING APPLICATION SERVERS

First, the application server has to have interfaces to the user information and positioning system used by the core network. Then, there are additional functionalities you can use to select one:

Is it integrated into a management framework, such as Tivoli or OMAP?
Does it implement standard APIs?
Does it have interfaces to the user information databases of portal engines?
What interfaces does it have to provide position information?
Does it have an integrated web server?
Does it have an interface to a map engine or a GIS?
Does it run on a single machine only, or can the functions be distributed?
    Does it provide load-balancing functions, or is that delegated to the operating system?
Does it provide for redundancy and failover?
Does it use a standardized data format?
Is there an interface to a billing system, and how difficult is it to customize?
Which version of the servlet's specification does it implement?

The application servers increasingly use a standardized data format, both for their internal communications and for the communication with MPCs. This feature is increasingly based on eXtensible Markup Language (XML), and XML libraries are available in most programming languages and are relatively easy to implement. Depending on how your data center is built, the application server should be possible to divide over serveral CPUs (probably using some clustering solution) to remove the risk of a single point of failure.

Different content providers (for instance, map providers) will have different specializations and abilities to provide content (for instance, being specialized in maps of one area, or only restaurant information). To connect to them, the emerging mechanism is XML channels; some might be so established that they have developed their own APIs, which the application server will implement. Content routing is a different problem altogether and can be done over the Internet.

How far distribution should be carried depends on how the data are created and organized. In principle, having all data in one data center will make access much faster, and you will not be at the mercy of the Internet to retrieve data. That might not be practical, however, if the data you are working with are updated frequently, are proprietary, or otherwise cannot be placed on your site. The simplest way is to reference the data by using a URI, but then you will have the same latency as you would have over the Internet, in addition to the latency that is native to the application. That might be detrimental in an environment where the user demands that all data be no more than three clicks and three seconds away.

THE APPLICATION DATA FLOW

How the data flows in the application depends on whether it is a triggered application, in which case the MPC will return periodic positioning messages and the application server has to forward them to the appropriate application (which will take the appropriate action). Or, if it is a request–response application, the user would generate a request that starts the data flow. The following example is adapted from the LocatioNet application server, where the different modules are called gateways. A typical request could be, "Give me John's location on an aerial photo. Show the nearest gas station and all the relevant advertisements and coupons within 1 mile according to John's personalized profile."

The request is sent to the application gateway. The gateway checks the user authentication and sends a location request to the location determination gateway.

The location determination gateway checks which network is the user's home network and sends a location request to the MPC of the network. Once the location information returns, the location determination gateway directs it to the application gateway.

The application gateway then sends simultaneous requests to the mapping gateway and the information gateway, which hold the location data that were just retrieved by the location gateway. If multiple information or mapping providers need to be accessed, the answers from all providers are aggregated and returned as a single response.

The application gateway then calls any personalization information included in the application, such as personal preferences (selecting a certain subset of the information held in the information gateway, for instance). Other personal information can be included or used for the structuring of the information as well.

Once all information has been returned from the information and mapping gateways, the application gateway formats the answer according to the original request (for example, drawing the content and locations on a map if the application requires this function to be performed). It then builds an XML response.

The XML response is then formatted according to the device capabilities that were reported with the user's request (included in an HTTP header field) or based on a database of device characteristics (if there is a different way to determine the characteristics of the device). If the response is to be sent as an SMS, the formatting is negligible.

As the data are forwarded to the web server of SMSC to be passed on to the user, the application server also triggers a billing event that is sent to the billing system. The originating application then returns the response to the user, using the appropriate channel for the format (it sends the data to the SMSC with the phone number in case the request came in over SMS, and to the web server in case it came in over HTTP).

Because any web request is serviced with a unique response, the adaption of each response does not represent any significant additional load on the system. The data have to be composed from the database anyway, so adding the personal adaption does not make much difference in terms of performance, but it enhances the user experience quite drastically.

PERSONALIZATION

When you adapt content to a set of user preferences, you are performing personalization. Location is just one example of personal profile data; the device characterization and the user's preferences for how it should be used are others. The service can collect a lot of information, such as how the user prefers to get what information in which places. Making the choices unobtrusive will mean that the user provides the information without thinking about it.

Personalization, especially combined with location data and logging by the service provider, has the potential to be extremely contentious. In many countries, it is illegal to use data for other purposes than those for which they were collected; in any case, you will seriously damage the user's trust in you (and your future business) if you try to use data for other purposes than those which the user has approved. The line is fine, and users are willing to give up quite a great deal of privacy in the interest of convenience, provided that they can trust you.

An example of how the personalization database is integrated with the location determination is the LocatioNet server. Each location determination vendor is identified and registered in the platform database with a different ID. The same location determination ID is kept at the subscriber's personalization database. Once a location request is generated for a subscriber, the location determination gateway directs that request to the appropriate location determination system. This mechanism enables multiple location determination technologies support per subscriber, such as cell ID vs. A-GPS (and so on).

The location determination channel is integrated by using the protocols of the various location providers. Because multiple location determination systems can be connected to the systems simultaneously, a unified protocol is implemented (essentially a precursor to the API of the Location Interoperability Forum, or LIF). This protocol enables location systems to input subscriber location to the platform upon request or synchronous to any request. The location determination channel needs to support the most usual coordinate and datum formats.

There are two major problems with the personalization mechanisms in most application servers. The first is that they build up a very extensive bank of personal data without any privacy guarantees. The second is that they use proprietary formats. If you have one profile on Yahoo! and another on the Netscape portal, you cannot merge them today. Actually, you cannot even look in them because they are inside the database systems of the services. But there is no interchange format, and the efforts that have been done are all hampered by the fact that the service providers regard the user's profile as something that they own (and can sell). No application server I have looked at even tries to alleviate this problem.

Of course, once you have characterized the data that you want to collect and defined interfaces and database fields, it is trivial to gather them. Taking data from the log files into a database is barely a starting point, because you can collect information about where the user is when he makes requests and if he has any preferences for what should happen with the data in different positions (for instance, does he usually look for beer on the way home on Fridays?). Combined with other personal data, such as credit card data, this situation can be a major threat to the user's privacy.

As database problems go, it is relatively simple, though. Any relational database can be used to handle it, and information can be collected by using filters. The problem is not collecting the information; rather, it is deciding how to adapt the information when it should be outputted. For instance, if there are conflicts, such as the user wanting advertisements for handbag shops when he or she passes shops that sell dresses, but never advertisements after 5 P.M., how do you resolve the conflict?

This example is simple, and when you start collecting more information, you will continue to run into conflicts that most application servers will try to solve *ad hoc*. There is no prioritization filter in the service, and even if there were, how would you know which rule the user wanted to be applied to the content in which order (essentially a rule in itself)? You would have to ask, because the more information you collect, the more complicated the process becomes until you would end up implementing a minor *artificial intelligence* (AI) system just to keep track of how different users want their personalization applied (because, of course, you can never assume that users will want the rule sets to be applied the same way or even be able or willing to provide the same information).

It is not enough that you have to adapt the data that are sent in the scope of the transaction. In mobile data networks, the network usually maintains a database containing the services that the user has selected. This information can be synchronized with a database on the handset, which makes it possible to have a personalized presentation. This process is called provisioning, and it can be done over the air — at least in GSM and IMT-2000 — by using an SMS (actually, this function is what SMS was invented for). Usually, the typical process is as follows.

The operator sets up the default provisioning profile for groups of users (young users, business users, and so on) in batch mode by associating those users with a profile number from a list of defined profiles in the database. Access is typically from the operator customer care center (and its specific databases) or a dedicated application in the application server, which interfaces to the computers of the customer care center.

Enterprise customers can potentially customize their users' profiles by accessing the database and modifying their users' provisioning attributes. This procedure can be done by using a web-based application, a WAP phone, or by calling the customer care center.

End users can access the system anytime over WAP, SMS, and web interfaces and personalize their profile and service behavior according to their preferences. This process requires the operator to allow this action, of course. Some operators will not let users into the configuration files for fear either of giving away a lot of data about their networks or that the user will require an impossible configuration. Given the amount of data that has to be configured in a normal mobile phone for it to work simply, this situation is probably justified (most people do not realize how many parameters can be set and set them wrong in a mobile phone). Still, a mobile phone is very simple compared to a more advanced client, such as a personal digital assistant (PDA), which is almost as complex as a PC (when I got my latest PC, it took our corporate IT department almost half a day to configure it, and that was without the Japanese settings).

Many application server providers have provided starting points for personalization, but none have understood the problem even if they claim to do so. This situation will be a major problem in the next generation of application servers, especially when considering the need to provide provisioning information.

USING FILTERS IN SERVLETS

As the vendors of application services provide APIs, you as a programmer can create your application by creating a program that runs in the application server and provides what is unique for your service.

This service typically means a Java program, because Java seems to be on the way to establishing itself as the primary programming language for position-dependent services. Most application servers implement the servlet's APIs in addition to the proprietary APIs, which call the different services. Java programs in a web server mean servlets, because servlets have established themselves more or less as the unique standard for Java web server programming since it was first launched a few years ago.

Servlets run in the server in much the same way as applets run in a browser, but of course they have access to the operating system and other APIs and system functions in the server and run on a real processor.

With version 2.3 of the servlet's specification, a new function was created that dynamically intercepts requests and responses to transform or use the information contained in the requests or responses. Called filters, they typically do not themselves create responses, but instead provide universal functions that can be attached to any type of servlets or Java Server Page (JSP).

The filter mechanism provides a way to encapsulate common functionality in a component that can be reused in many different contexts. Filters are easy to write and configure, as well as being portable and reusable. In summary, filters are an essential element in a web developer's tool kit. Filters can be mapped to one or more servlets and vice versa. Filters can be chained indirectly via filter mappings. The order of the filters in the chain is the same as the order that filter mappings appear in the web application deployment descriptor.

Filters are useful if you create location-based services, because they provide a simple way to encapsulate recurring tasks in reusable units. They also enable you to transform the response from a servlet or a JSP. A common task for the web application is to format data sent back to the client (for instance, Scalable Vector Graphics (SVG) or VRML instead of Geographic Markup Language (GML)). Filters can be written in

Extended Stylesheet Language — Transformations (XSLT) as well as Java. But filters can be used for anything you can do with a Java program in a web server.

Programming the filter is only half the job of using filters — you also need to configure how they are mapped to servlets when the application is deployed in a web container. This decoupling of programming and configuration is a prime benefit of the filter mechanism, because nothing has to be recompiled and you do not need to change the input or output of your web application. You just edit a text file or use a tool to change the configuration. For example, adding compression to a PDF download is just a matter of mapping a compression filter to the download servlet.

The filter API is defined by the Filter, FilterChain, and FilterConfig interfaces in the javax.servlet package. You define a filter by implementing the Filter interface.

There are many ways for a filter to modify a request or response. For example, a filter could add an attribute to the request or it could insert data in or otherwise transform the response. A filter that modifies a response must usually capture the response before it is returned to the client. This filter can be, but does not have to be, an XSLT filter. You can, for instance, set it to use the style sheets to transform the response depending on the value of a request parameter. The filter sets content type of the response according to the request parameter. The response is then wrapped in a CharResponseWrapper and passed to the doFilter method of the filter chain. The last element in the filter chain is a servlet that returns the inventory response. When doFilter returns, the filter retrieves the response data from the wrapper and transforms it by using the style sheet.

To map a filter to a servlet, you must declare the filter by using the <filter> element in the web application deployment descriptor. This element creates a name for the filter and declares the filter's implementation class and initialization parameters. You must also map the filter to a servlet by defining a <filter-mapping> element in the deployment descriptor. This element maps a filter name to a servlet by name or by universal resource locater (URL) pattern.

USING PREPOSITION FILTERS

Back in Chapter 5 we talked about location models and the role of prepositions:

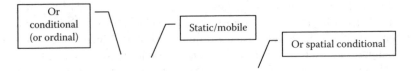

When/If <logical> <object> + is/are <relationship> <location> + when <temporal> then <action> +

where <location> could be one of the following:

- In (true symbolic relationship or static state); could be also within (a Wi-Fi cell)
- Alone in (true symbolic relationship or static state)
- Not in (true symbolic relationship or static state)

where <relationship> could be one of the following

- Entering
- Leaving

The following table shows how English prepositions can be broken down into logical, spatial, and temporal components of location (or context).

Preposition	<Logical>	<Object>	<Relationship>	<Location>	<Temporal>
About	About a dozen			About the room	About noon
Above	Numbers above 6			Above the table	(Not temporal)
Across	(Not ordinal)			Across the table	Across the years
After	After John alphabetically		After the table?	(Not spatial)	After July
Against	(Not ordinal)			Against the table	(Not temporal)
Ahead	Ahead of the others			(Not spatial)	Ahead of schedule
Alone			Alone in	(*In* is spatial)	
Along	(Not ordinal)			Along the river	(Not temporal)
Among	Among the attendees		(Not temporal)	?	
Amongst	(Not ordinal)			(Not spatial?)	(Not temporal)
...					
In (not in)			Entering/leaving	Inside the room	

CONTENT COMPARISON AND EQUIVALENCE

The following operators can be used to compare events of the same class, or compare specific event fields against specified values:

$$=, <, >, !=, <=, =>$$

For example, the following returns a set of events whose <field1> and <field2> both compare to the respective values x and y as specified:

$$((a)<field1> <operator> x \text{ and } <field2> <operator> y)$$

The following is a set of events (type a or a combination of type a and type b) whose <field1> or <field2> (as applicable) compares to the respective parameters as specified. These types do not have to denote the same event class, nor do they have the same set of attribute fields.

$$((a) <field1> <operator> x \text{ or } (b) <field2> <operator> y)$$

TEMPORAL CONTEXT

(<event>|<interval>) before (<event>|<interval>) [without <event>] [within <timevalue>])

The above returns the set of all a, which are followed at some point by type a b, where type a and type b can be either events or intervals. (Type a before type b within <timevalue>) returns the set of all a that are followed by type a b within the specified time period denoted by <timevalue>. (Type a before type without type c) returns the set of all a that are followed by type a b and no type c occurred in between.

The following is similar to the previous construct, but in this case the type a is preceded at some point by type a b.

((<event>|<interval>) after (<event>|<interval>) [without <event>] [within <timevalue>])

In the following, (type a equal type b) returns the set of all a that occurred at the same time as type a b:

(<event> equal <event>)

INTERVAL OPERATORS

An interval is defined as having a start time and an end time, and can encapsulate any number of events occurring within those inclusive boundaries. When an interval is defined within a query, a new object instance of the interval class type is constructed, and this provides the following methods that can be employed within the query itself. Some maintenance functions are provided, like start(), which returns the time value denoting the start of the interval; end(), the end boundary; events(), the number of events within it; and eventAt(x), the event at position x.

An interval is specified as follows:

(: (<timevalue> | <event>) (((until | to) (<timevalue> | <event>)) | for <timeamount>) :)

(: and :) denote that an interval is being specified and to differentiate between excluding or including the second time value itself. These operators default to a start-point consumption model. A start time is matched with the closest endpoint and then not used again in any further matches. If *for* is used, the second parameter must be a relative amount of time, which is then added to the starting boundary to compute the ending boundary of the interval.

There are a number of relationships between intervals, all of which can be specified in terms of the start() and end() operators of the interval object class. *Before* and *after* have already been discussed in the context of event operators. The other operators we have provided reflect the ways in which an ordered pair of intervals can be related. These are *equal, meets, overlaps, during, starts, finishes, contains,* and *their inverses. Intersect* and *join* are interval operators that create new intervals on which queries can be applied. At the end, see the OGC Filter Encoding specification's filters, which can be adopted for such an event language. This specification already has the *contains*, *intersects*, *joins*, and *overlaps* elements.

COMPOSITE/SEQUENCE PATTERN OPERATORS

A composite event is denoted as an event defined in terms of a sequence of other basic or composite events. It exists only within the context of the current query and is defined by specifying a path template, also equivalent in meaning to a sequence or pattern. Operations on paths are carried out by enclosing the template in [: :], as in

[: A followed by (B or C) without E :]

A composite path does not return single-event instances, but rather constructs a set of new composite event objects (if successful), which each contain pointers to all the basic events they were made up from within the session timeline. Functions are provided that can extract relevant information from the resulting construct.

SYNTAX EXAMPLES

Consider the following event: "When Cliff entered Room 101 and Kris was there."
    One way of expressing this is to find the period when Kris was in Room 101 and then see if Cliff was seen within that period in the same room:

fromwhere (badge.Name/Id='Cliff.K and badge.location='R101)

during (: (badge.Name/Id='Kris.K and badge.Location='R101)

until (badge.Name/Id='Kris.K and badge.Location!='R101) :)

Or "When Kris walked from the office room to his office (Room 101) and started editing a document." This will return all event sequences that match the specification provided. A simple way of specifying this is:

eventsOf([:(badge.Name='Kris.K and badge.Location='Office-Room')

followedby (badge.Name='Kris.K and badge.Location='R101)

followedby (workstation.documentedit.user='Kris.K):]

during (: 10:00:AM to 11:00:AM :))

To summarize, it is possible to find events relative to other events that have occurred before or after them, as well as with respect to specified bounded intervals. Intervals can be operated on to determine overlapping periods, and if desired, output can be filtered so that only specific event types, or events with specific parameters, are returned.

Preposition	Temporal Example	Spatial Example	Ordinal Example
1. About	About noon	About the room	About a dozen
2. Above	(Not temporal)	Above the table	Numbers above 6
3. Across	Across the years	Across the table	(Not ordinal)
4. After	After July	(Not spatial)	After John alphabetically
5. Against	(Not temporal)	Against the table	(Not ordinal)
6. Ahead (of)	Ahead of schedule	(Not spatial)	Ahead of the others
7. Along	(Not temporal)	Along the river	(Not ordinal)
8. Among	(Not temporal)	Among the daisies	Among the leaders
9. Amongst	(Not temporal)	(Not spatial?)	(Not ordinal)
10. Around	Around 1776	Around the table	Around 100
11. As			
12. At	At noon	At the table	At the halfway point
13. Before	Before 6 P.M.	Before the fireplace	I was served before you
14. Behind	Behind schedule	Behind the table	Behind John in line
15. Below	(Not temporal)	Below the table	Below average
16. Beneath	(Not temporal)	Beneath the table	(Not ordinal)
17. Besides			
18. Beside	(Not temporal)	Beside the table	(Not ordinal)
19. Between	Between 6 and 7 o'clock	Between table and lamp	Between 40 and 49
20. By	By Saturday	By the table	(Not ordinal)
21. Down	(Not temporal)	Down the hill	(Not ordinal)
22. During	During June and July	(Not spatial)	(Not ordinal)
23. For	For an hour	(Not spatial)	(Not ordinal)
24. From	From 6 to 7 o'clock	From the table	From 95 to 97 octane
25. In	In the Pennsylvanian	In the room	In the top 50
26. Inside	Inside of an hour	Inside the cupboard	(Not ordinal)
27. Like			
28. Near	Near Christmas	Near the table	Near par
29. Nearby	(Not temporal)	Nearby the table	(Not ordinal)
30. Next, next to	(Not temporal)	Next to the table	Next in line
31. Of	(Possessive only)		
32. Off	(Not temporal)	Off the table	Off the scale
33. On	On Thursdays	On the table	(Not ordinal)
34. Onto	(Not temporal)	Onto the table	(Not ordinal)
35. Out	(Not temporal)	Out of the box	Out of order
36. Outside	(Not temporal)	Outside the lines	(Not ordinal)
37. Over	Over an hour	Over the table	Over 60 mph
38. Till, until	Until midnight	(Not spatial)	(Not ordinal)
39. Through	Through the week	Through the field	(Not ordinal)
40. To	Monday to Friday	To the table; to the left	Cooled to 50°
41. Toward	Toward midnight	Toward the table	(Not ordinal)
42. Under	Under an hour	Under the table	Under 100 lb
Preposition	Temporal Example	Spatial Example	Ordinal Example
43. Until	Until Saturday	(Not spatial)	Until age 7

44. Up	Time is up	Up the wall	Up the scale
45. Upon	Once upon a time	Upon the table	(Not ordinal)
46. While	While the TV is on	(Not spatial)	(Not ordinal)
47. Whilst	(Not temporal)	(Not spatial)	(Not ordinal)
48. With	(Not temporal)	(Not spatial)	(Not ordinal)
49. Within	Within an hour	Within the room	Within 3°
50. Without	(Not temporal)	(Not spatial)	(Not ordinal)

OGC Filter Encoding Specification's Filter Expressions

Filter: Defines Filter Expressions

Element Name	Description
	– Spatial Operators <spatialOps> –
Equals	Defined in Section 3.2.19.2 of the Open GIS Simple Feature specification for Structure Query Language (SQL)
Disjoin	Same as above
Touches	Same as above
Within	Same as above
Overlaps	Evaluates whether the value of the specified geometric property and the specified literal geometric value (expressed using GML) spatially overlap; defined in Section 3.2.19.2 of the Open GIS Simple Feature specification for SQL
Crosses	Defined in Section 3.2.19.2 of the Open GIS Simple Feature specification for SQL
Intersects	Same as above
Contains	Same as above
DWithin	Tests whether the value of a geometric property is within or beyond a specified distance of the specified literal geometric value; distance values are expressed using the <Distance> element; the content of the <Distance> element represents the magnitude of the distance, and the units attribute is used to specify the units of measure; the units attribute is of type anyURI, so that it may be used to reference a units dictionary
Beyond	Same as above
BBOX	Encodes the bounding box constrained based on the gml:Box geometry; it is equivalent to the spatial operation <Not><Disjoint> ... </Disjoint></Not>, meaning that the <BBOX> operator should identify all geometries that spatially interact with the box in some manner
	– Comparison Operators <comparisonOps> –
	Note: For our purposes, *Property* can mean *Object* or *User*
PropertyIsEqualTo	Defined in the OGC Common Catalog Query Language
PropertyIsNotEqualTo	Defined in the OGC Common Catalog Query Language
PropertyIsLessThan	Defined in the OGC Common Catalog Query Language

OGC Filter Encoding Specification's Filter Expressions (continued)

Filter: Defines Filter Expressions

Element Name	Description
PropertyIsGreaterThan	Defined in the OGC Common Catalog Query Language
PropertyIsLessThanOrEqualTo	Defined in the OGC Common Catalog Query Language
PropertyIsGreaterThanOrEqualTo	Defined in the OGC Common Catalog Query Language
PropertyIsLike	Encodes a character string comparison operator with pattern matching
PropertyIsNull	Encodes an operator that checks to see if the value of its content is NULL; a NULL is equivalent to no value present; the value 0 is a valid value and is not considered NULL
PropertyIsBetween	Defined as a compact way of encoding a range check; the lower and upper boundary values are inclusive

– Logical Operators <logicalOps> –

AND	Used to combine scalar, spatial, and other logical expressions to form more complex compound expressions
OR	Same as above
NOT	Same as above

– Arithmetic Operators <arithmeticOps> –

Add	Encodes the operation of addition and contains the arguments that can be any expression that validates according to the OGC Filter Encoding specification
Sub	Encodes the operation of subtraction where the second argument is subtracted from the first; the <Sub> element contains the argument that can be any expression that validates according to the OGC Filter Encoding specification
Mul	Encodes the operation multiplication; the <Mul> element contains the two arguments to be multiplied, which can be any expression that validates according to the OGC Filter Encoding specification
Div	Encodes the operation of division, where the first argument is divided by the second argument; the <Div> element contains the arguments that can be any valid expression that validates according to the OGC Filter Encoding specification; the second argument or expression cannot evaluate to zero

INTERFACES TO EXTERNAL SERVICES

We already discussed the location determination systems. You will use an API to get the location information and define how your application will use it. The application server uses the API to send location requests (for single or multiple subscribers) from the platform to a location, sends a subscriber's position from the location system to the platform, and sends a subscriber's status from the location system to the platform, as well as sending commands from the platform to the location system.

One significant problem occurs if you want to use location information in the client and the client is not generating that information. This situation is, for instance, true if you have a periodically triggered application that generates position information and sends it to the application server. If the client does not have a web server, there is no way to get the information to the client — except possibly WAP Push or another protocol like SMTP, which is used for e-mail.

The application server can take care of that for you, essentially working as an interface between the MPC and the terminal. That is the way it is intended to work in the LIF API. In theory, this situation means that the clients can connect to the application server by using any protocol (for instance, Domain Name System (DNS), SNMP, SMTP, or File Transfer Protocol (FTP)). It could function as a gateway toward the positioning system.

The application server can also direct the request for positioning from the user to the right network. If your site is on the Internet, there is no direct connection between the user's identity and the mobile network, which can provide the position. You can receive requests from many users in many networks. That is not necessarily true in the "walled garden" model that many network operators prefer, where the user can access only information provided by the network he or she is accessing — a strategy that has been proven wrong time and time again.

The application server can also serve as the interface to other services, for instance, the billing system and other services in the network, such as portals. There is no standardized interface for user information, however, nor for billing information. How you connect the portal to the application server depends on the interfaces of each. This area could deal with standardization, but because the portals are based on their proprietary interfaces, that is unlikely to happen in the near future. It will be true a lot sooner for application servers, where there are now a number of standardization efforts going on (too many, actually). Some application servers provide for personalization beyond the use of position information, as well as the transcoding of information. But neither of these are standards, and I will not discuss them in any depth — although it is clear that personalized services are the key to the future, and location-based services are just a branch of situated services where the context of the user in terms of preferences, previous experiences, and so on, is used as a base for the personalization.

If the application server sits in a data center, it is an advantage if it is integrated into a systems management framework, such as Tivoli or a similar system from an operator. Here, as well, the standard interfaces are lacking. But if your application server is integrated into the management framework, it will be able to send alerts, give performance measures, and generally act as any of the other application servers that are situated in the data center. If you are thinking about setting up as an ASP, having this type of functionality is a must.

MANAGEMENT SYSTEM INTERFACES

If the application server is used in a central location, such as in an ASP, then the management of the server becomes a crucial issue. This statement is especially true if the system sits in the data center of a network operator who typically has higher requirements on up time (99.99% is not unusual) than the average PC user. This requirement might seem excessive on the surface, but considering that people actually trust their lives to the telephone system, it is not inappropriate.

On the Internet, there is a standard protocol for network adminstration: the *Standard Network Management Protocol* (SNMP). There is a *management information base* (MIB) for web servers, but it is not used in many systems. There is no MIB for application servers in location-based systems, and the requirement is higher to integrate with the management system of the network operator, in case the application server sits in its data center.

There are no standards for management systems in the telecommunications world, however. Each manufacturer has its own, and delivering it is a way of achieving a lock on the customer — because once you have a system running thousands of systems, you do not change easily.

One way is to create an API that maps to the internal interfaces of the management system. The application server connects to the application server, and any event that causes a management trigger is sent to the management system through the API.

The operators of mobile networks see the location-based services as a potential revenue spinner. They would like to offer their information to as many different providers as possible to maximize their revenues. But at the same time, they do not want to expose and share their most sensitive systems and information. For this reason, it is probable that you have to use an application server or even an application service provider that has an exclusive agreement with the mobile network operator.

This situation is precisely the use case for application servers functioning as a middleware between the different systems providing the information and integrating them into a presentation that can be sent to the user. Some operators are likely to be even more careful (like NTT DoCoMo of Japan) and only allow application service providers (ASPs) access to their network (that DoCoMo started its own ASP does not, of course, make its business case worse). For the application developer, however, it is a fact of life that the position data in their raw form are inaccessible, and if you want to get them, you have to use an application server (which, of course, also brings a number of advantages, such as access to a number of other services).

The application server can sit in the service owner's data center, or it can be outsourced. It can be used to provide one single service, or it can be shared by several users (if provided by an ASP). The application providing the service addresses the functions provided by the application server through an API.

Application servers will also have to connect to content distribution networks, which are now emerging. Syndicated information is typically only news and stock prices, but the same mechanisms used to personalize that information could be used to adapt it to the location. Then the data flow would come from an external source, and the application server would only contain the template and profile management. The platform would enable each content database to support multiple information categories.

REFERENCES

Balachandran, A. and Voelker, G.M., *Real-Time Monitoring and Control of a Wireless LAN Using NDIS Wireless Extensions*, The Wireless Research API, Technical Report, University of California, San Diego, December 2002.

Balachandran, A., Voelker, G.M., and Bahl, P., Hot-spot congestion relief in public-area wireless networks, in *Proceedings of the Workshop on Mobile Computing Systems and Applications, WMCSA'02*, June 2002a.

Balachandran, A., Voelker, G.M., Bahl, P., and Rangan, P.V., Characterizing user behavior and network performance in a public wireless LAN, in *Proceedings of the ACM SIGMETRICS'02*, June 2002b.

11 Standards

Standards are concerned with ensuring that technology (including data) from one supplier will work and operate as it was meant to with technology from another supplier. Standards are expensive with respect to both financial and human capital. When developing standards, the leading experts in a field must be donated by their employers to the process, and in addition, those employers must put up the cash to transport those people to meetings, organize conferences, etc. As standards in general have become more important, the costs and, more importantly, the return on investment of the standards process have come under more intense scrutiny. In addition, understanding standards brings a cost. A developer has to understand what the standard implies — even when the specification is crystal clear, it requires interpretation, which may not be documented in the public documentation of the standards body. This usually means that even if you are only a user, not a developer, you have to send someone to the meeting of the standards body and follow the mailing lists. Another frequent misconception is that using a standard is a solution to a technical problem. Regardless of how it has been developed, standards are the result of compromise in committees. Standards are a solution to a market problem — enabling companies to agree on a common solution, which widens the market and simplifies the sales and purchasing of products.

Enterprise applications were initially developed on closed, homogeneous mainframe architectures. In the explosively growing heterogeneous landscape of IT systems in the 1980s and 1990s, integration of intra- and intercompany business processes became one of the most important and most cost intensive tasks of the IT economy. Due to missing or nontransparent standards, many enterprises pursued integration by extremely expensive *ad hoc* solutions. With the spreading of the Internet and the increasing importance of electronic business, open Internet-oriented solutions have emerged. Enterprise-internal monolithic software was broken into smaller, autonomous, and flexible components. This enabled the access to services not only enterprise-internal, but along the whole value chain to suppliers, distributors, and customers.

Currently, most of the indoor location-based service (LBS) applications are based on self-contained systems (vertical implementation of all the necessary components, such as location positioning servers, middleware, applications; all are in one package from one vendor) that provide a one-size-fits-all solution. This usually means suboptimizations in the different parts of the system, even if the whole of it works. The optimum strategy is to use an open architecture and platform that is capable of uniting and integrating different features and functions from various infrastructure providers in a highly distributed way to create a rich variety of different value propositions for a diverse customer audience.

The purpose of this chapter is to introduce the standardization approaches of the Location Interoperability Forum and Open Mobile Alliance (LIF and OMA) and the Open Geospatial Consortium (OGC).

INTEROPERABILITY AND THE ROLE OF STANDARDIZATION

The process of standardization has been important in creating and growing global markets for computing and communications systems. For example, communication network standards like Global System for Mobile Communications (GSM), Ethernet, and IEEE 802.11 enable interoperability between equipment from different manufacturers, lower costs, and reassure users that their investment in technology will be viable beyond the short term. However, since the market for indoor location-based systems is in its infancy, and the requirements for indoor location-based systems applications are just beginning to be understood, caution must be taken to avoid making early decisions that will impede market adoption. Much of the standardization that has taken place so far, even though it is applicable to indoor location-based services, has been driven by the mobile system vendors and their requirements for interoperability, which in turn comes from their customers.

Standardization can be a serious success factor for any new technology, and location-based systems are not an exception to this rule. Standardization can also be a requirement from strong customers, which will drive the market; and here, the indoor location-based systems market is not mature enough. As yet, customers have not taken to indoor location-based systems, and the market is still looking for applications beyond the smallest niches.

Standardization activities for location-based technologies should be rooted in the market because of the huge value to the overall market in the long term. When standards are adopted, the ultimate technical benefit will be interoperability between systems and software from different vendors, allowing for the reuse and exchange of data, with seamless integration of the location information, into any existing network infrastructure. From the business point of view, standards lower costs and reassure users that their investment in technology will be viable beyond the short term. Interoperability translates to the ability of heterogeneous, typically distributed systems to exchange data in real time to provide services. Interoperability also has a temporal aspect, which refers to evolution of systems over time with backward and forward compatibility (McKee, 2003).

For example, the process of standardization has been important in creating and growing global markets for computing and communications systems. Standards like GSM, Ethernet, and IEEE 802.11 enable communication network interoperability between equipment from different manufacturers, lower costs, and reassure users that their investment in technology will be viable beyond the short term.

In terms of the software (services) infrastructure, interfaces and protocols should be published (open) and standardized as a general business and technical requirement. This would reduce the engineering complexity of communication among location-based services content and service providers, as well as developers. Here, the best example is probably Linux and other open-source systems, which would not have come as far as they have without the open standards of the World Wide Web. Open interoperability standards will support the commercial viability of a location-based service provided by a community of cooperating yet competing providers. Moreover, open standards define open standards as being, of course, nonproprietary and created in an open, international, participatory industry process in which anyone can participate. Also, openness also means an open license, which does not require any royalty or fee. And all interface specifications of an open standard must be freely and publicly available and also freely distributable. Additionally, an open standard has to be nondiscriminative against persons or groups and technology neutral (McKee, 2003).

Without interoperability standards, application domains will continue to remain stand-alone implementations (niche markets), when there is the potential for a broader market with the bundling of services. For example, a standardized application programming interface (API) for location data access from any positioning technology would enable interoperability among distributed indoor LBS applications and services. This was not the case for the first generation of LBS, which were closed, stand-alone applications. They were not primarily designed to communicate with other applications and systems. Spinney (2003) refers to this first generation as the first phase of LBS, taking place from 1997 to 2001. It began with the implementation of first proprietary network location nodes within SS7 networks. These location nodes were later standardized and are now referred to as either mobile positioning center (MPC) or gateway mobile location center (GMLC). Third-party application providers were able to access operators MPC/GMLC only via proprietary interfaces. There were no standardized interfaces to MPC/GMLC or geographic information systems (GIS) servers.

There are a variety of location models that were developed by different vendors and organizations, expressing location information in an individual format resulting in interoperability problems and the need for standards. Organizations such as the 3rd Generation Partnership Project (3GPP), LIF/OMA, OGC, and the World Wide Web Consortium (W3C) have addressed the interoperability problem by developing and proposing standards aimed at ensuring interoperability between location-aware services and applications.

In the context of geospatial information and services, there are standards dealing with the manipulation of geospatial data (spatial standards, providing interfaces for standard methods), but also standards for access to geospatial data — metadata standards — that are used for catalogs (or search engines) that describe the content of information resources that are needed to search and retrieve data, services (i.e., Open GIS Web Service), and resources on the World Wide Web. As a result, they need to be in standard formats. This has meant that collaboration or service chaining among the different applications is possible.

Nevertheless, considering the fact that standardization can also be a failure factor, care must be taken to avoid making early decisions that will impede market adoption. For example, the Magic Services API is the fourth method to get position information from a network. Unfortunately, in contrast to the other methods, Parlay, LIF Mobile Location Protocol (MLP), and the WAP Forum APIs, the Magic API had no providers of mobile positioning centers committed to implementing it. Magic, which stands for Mobile Automotive Geoinformation Services Core, was created by a loose industry group, led by Microsoft, and biased toward the automotive industry.

Another example is the failure of WAP in Europe, which is mainly because the technology was oversold (due to bad marketing) to the detriment of services. The relationships between operators and service providers/content providers was another reason: mobile operators did not yet enter into agreement for resource sharing. When they did, for instance, with Vodafone Live, the market took off. In Japan, the success of mobile services is mainly due to the cooperative business model between NTT DoCoMo and its service providers for i-mode. The Japanese network operator deducts a commission of 9% on services offered on its portal. Although the business model for location-based services is different, it is still an open business model, which enables the providers of services to calculate their costs in advance of deploying the service.

Even though the indoor location-based service market is in its infancy — application requirements are still not well understood — the indoor location-based services field must settle on a few consensus-derived and well-proven standards and practices. This is especially true considering the general fact that location-based services are dependent not only on a number of direct enabling technologies, but also on a number of indirect facilitating technologies of added-value services. The need for a focused effort concerning location interoperability is also evident considering the multilayered nature of the location-based services industry responsible for developing, operating, and supporting the location services value chain. Hence, interoperability among indoor location-based systems and applications should be attained as a result of a coordinated effort of these diverse players. For this reason, OGC's OpenLS initiative (discussed below) brings together key industry players to build and consolidate the standards infrastructure for these interoperating location-based services.

Moreover, while standards are important, it should also be recognized that these are often more accurately discovered than imposed, evolved and adopted as a result of real-world pressures, rather than through a top-down process. Standardization processes are political processes, vulnerable to pressure from large vendors with their own particular interests to support, as well as large customers (for instance, the mobile operators) with their own axes to grind. For communication infrastructure providers, and especially service developers, these factors can limit the freedom to innovate and respond rapidly to threats and opportunities.

STANDARDS FRAMEWORKS

There are many standards organizations with activities that touch on the indoor location-based services. These include the Internet Engineering Task Force (IETF), the World Wide Web Consortium (WC3), and the Organization for the Advancement of Structured Information Standards (OASIS), which have initiatives that touch on the location and geospatial area, yet location is not their core focus. Even the proposed ZigBee (IEEE 802.15.4) standard for ubiquitous communication represents one current effort toward the goal of location awareness that is already being discussed in the context of that proposed standard.

The drivers of location interoperability are two other organizations, which were founded especially with this in mind: the Location Interoperability Forum (LIF), which has since become a part of the Open Mobile Alliance (OMA), and the Open Geospatial Consortium (OGC). Both OGC and OMA are the standard holders regarding wireless location interoperability. Their standards ensure the smooth flow of information between content repositories, the Internet, and end-user devices through open protocols. These two organizations are accomplishing this task through a cooperative strategy of working with other standards bodies to promote a cohesive set of wireless location interoperability standards. One basic requirement for LBS solutions is to integrate multiple vendors and vendors of multiple kinds of products (positioning technology, geographic information systems, application development frameworks, etc.) with the telecommunications infrastructure of a wireless service provider. This process requires open standards for data exchange and application interfaces, or otherwise it goes uphill very quickly. The objective of such a common platform is to provide a foundation that makes the development of location-based service applications much easier and allows for a better integration of and interaction between different applications. The technical environment in which location platforms operate has grown considerably more complex, with new standards bodies and new application areas continuing to appear. For the industry to fully exploit the revenue potential of location-based services, as well as provide all the related management and support functions, a wide range of open and standardized APIs and features must be supported for as seamless a way as possible.

As a result, standards will provide application and content providers with a ubiquitous contextual meaning to location, regardless of the location positioning technologies utilized. As standards bodies continue to define network nodes, functionality, and interfaces, communication infrastructure providers (i.e., mobile operators) face the reality of integrating their location servers with network elements that have varying levels of standards compliance and back-end systems with unique interfaces. In short, no one-size-fits-all solution will work. The most important standard issuing bodies in the geospatial and location services fields are Open Geospatial Consortium, Inc., and Open Mobile Alliance's Location Working Group (LOC). There are several other organizations working in the LBS sector, such as ISO/TC 211 IETF and W3C. ISO/TC 211, "Geographic Information/Geomatics," consists of more than 20 separate specifications that address the acquisition, management, analysis, access, presentation, and transfer of geographic information among different systems, users, and locations. Among these specifications, both a model for location and standards related to the acquisition of location information can be found. According to the model developed by the ISO, location is described by means of geometric or topological objects. Location information may refer to different coordinate systems and includes data concerning size, shape, and orientation. In addition, the standard defines a taxonomy of spatial operators that serve to use, query, create, delete, and manipulate location information. Several services for coordinate transformation and other types of conversions, as well as positioning services, are defined and described. The positioning services standard specifies an interface that provides both general and technology-specific data structures used to access various location sensors. IETF has proposed a common data set and an extensible framework for expressing location information on the Internet. The data set consists of mandatory and optional elements. It stipulates that location is expressed as latitude and longitude with reference to the World Geodetic System 1984 (WGS84) system and a time stamp is attached. Optional elements include orientation, speed, accuracy, course, direction, and unspecified attributes. In addition, the Spatial Location BOF proposes an eXtensible Markup Language (XML) encoding of the common data set. The group has also addressed the representation of location that is not specified by geographic coordinates by proposing a common syntax and coding for descriptive location. This approach subdivides locations into three types: civil objects such as countries, towns, rooms, etc.; geographic objects like rivers or tunnels; and opaque objects, which serve to represent any other locations not belonging to the former two types. For civil and geographic objects, an enumeration of possible values for each of these types is given.

W3C's Composite Capabilities/Preference Profiles (CC/PP) framework is another standard, but is not related to location. CC/PP serves to specify device capabilities and user preferences by means of profiles. Profiles are transmitted from client devices to content servers, which accordingly adapt the delivery of content to the capabilities and preferences described in the profiles. CC/PP is based on the Resource Description Framework (RDF) designed by the W3C. A CC/PP profile contains a structured set of attributes and values for these. The framework defines a standard set of CC/PP attributes, their permissible values, and associated meanings, which constitute a CC/PP vocabulary. Since CC/PP is designed with an emphasis on flexibility and extensibility, different vocabularies may be defined.

The framework for OGC, OMA LOC (OMA/LIF), and related standard bodies is presented in Figure 11.1.

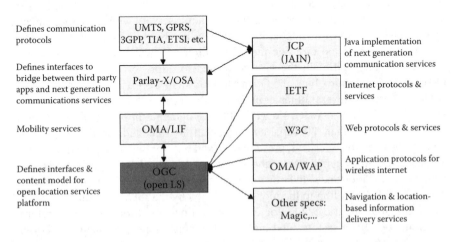

FIGURE 11.1 The standards framework. (From Open Geospatial Consortium, Inc., *OpenGIS Location Services (OpenLS): Core Services*, Parts 1–5, OpenGIS Implementation Specification 03-006r3, January 2004.)

There are two widely accepted standards that promise to expedite integration within the communication and positioning infrastructure and stimulate application development within the software (service) infrastructure:

1. The LIF's MLP API (for the communication and positioning infrastructure)
2. The Open GIS Consortium's (OGC) OpenLS API for spatial processing (for the software (services) infrastructure)

3GPP and 3GPP2 are making progress on the overall incorporation of LBS in the network architecture. OMA covers e2e applications and content interfaces. It also addresses privacy, security, charging, billing, and roaming issues. User's privacy, security, and safety are dealt with at the EICTA, EC, CTIA, FCC, and UMTS Forum.

Overall, most of these standardization efforts focus on geometric locations; however, indoor LBS location frequently is represented in a manner that abstracts from geometry. In many indoor domains the relevant locations are usually rooms, corridors, buildings, users, etc., where symbolic representation makes more sense. The only approach that explicitly addresses symbolic locations is the common syntax proposed by the IETF's Spatial Location BOF. However, its classification of locations into three groups containing a finite enumeration of location names or arbitrary text is far too restricted and lacks formality. Expressing locations in symbolic format eliminates the need to transform each sensor-determined or user-defined location into geographic coordinates, which often results in a significant and unnecessary processing overhead, in particular when location sensors that do not provide geometric location information are employed.

OPEN GEOSPATIAL CONSORTIUM

OGC's focus is on the application, data, and presentation layers of the Internet stack. The types of services that are fundamental to any spatial data infrastructure (SDI) include data catalogs, online/mobile mapping, and access. Other services include coordinate transformation, classification, data authentication and validation, data analysis, data fusion, custom symbolization, multiperson collaboration, gazetteers, processing algorithms, and service catalogs allowing discovery of required services.

In the location-based services market, OGC accomplishes this through the OpenLS initiative, which engineers location services application interfaces designed to support interoperable solutions that "geo-enable" the Internet and wireless location-based services, and mainstream IT. The OpenLS initiative focuses on developing interface standards "needed by industry to support implementation of the location services invoked by mobile or wireless Internet devices in end to end settings". Open GIS Location Services (OpenLS) (OGC, 2004) is an OGC specification for a platform of open location services, known as GeoMobility Server (GMS). OpenLS comprises an open platform for position access and location-based applications targeting mobile terminals. The OpenLS feature set is defined by core services and abstract data types (ADTs) that comprise this platform. OpenLS defines access to GMS, which consists of core services and abstract data types. The five core services that GMS offers, and upon which the location-based applications are built, are directory, gateway, location utility, presentation, and route services. Additionally, there is a navigation service, an enhanced route service, which is an independent specification (OGC, 2003). The introduction of new network elements, namely, the location-enabling middleware and the GMS-provided third-party application providers (APIs) to interface the core network and embed functionality from the location-enabling middleware and the GMS, dramatically reduces implementation costs. In this architecture, privacy, presence, and personalization are taken care of by the middleware. Standards play a very important role in this integration model, especially the Mobile Location Protocol (MLP) by Open Mobile Alliance's Location Working Group and OpenLS by the Open Geospatial Consortium (OGC). GMS uses open interfaces to access a core network's positioning entity (GMLC/MPC) and provides a set of interfaces allowing applications to access the OpenLS Core Services. For application developers, these standard APIs provide a one-stop shop for all their location-based application development needs. On the other hand, operators can control what applications they introduce and how those applications are integrated with their systems (Figure 11.2).

The GMS also provides content, such as maps, routes, addresses, points of interest (POIs), and traffic information. The request–response messages and associated abstract data types for the GMS are encoded in XML for Location Services (XLS). However, despite the fact that there has been a lot of progress in standardization of location services, GIS and mobile wireless services in indoor location systems lack standardization, making integration into open LBS architecture difficult. At the time of writing, the OpenLS services and information model is limited to outdoor navigation, but the OWS-3 OpenLS thread started within OGC is aimed at adding a tracking service that supplies a position management and access capability, and making first steps toward path planning and navigation in buildings and other environments beyond the limits of road networks. An enhancement to the OpenLS services and information model is to support seamless indoor–outdoor

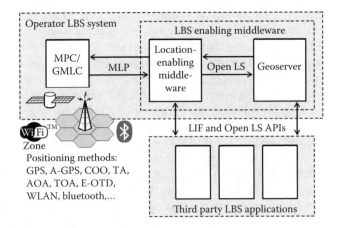

FIGURE 11.2 OpenLS platform.

navigation. The OpenLS services and information model will have to be modified to accommodate indoor location and navigation constructs. Also, OGC's Gazetteer Service transforms symbolic locations into geometric locations. In order to make use of external services, the model for indoor positioning should be interoperable with other models for location, and in particular with proposed standards. Thus, the location model should support a conversion of location information into a different model's representation and vice versa.

In terms of positioning and geoservers, consider the following. The Ekahau Positioning Engine (EPE) is a software-based positioning server. A wireless local-area network (WLAN) requires a separate location server that provides an open solution interface, which in this case is MLP. The EPE location server is suitable to construct such a server. There already exist software packages that provide MLP interfaces to servers, but more work needs to be done to achieve the compatibility of the existing packages and EPE, or to assess the possibility of constructing such an interface.

The Open GIS Implementation Specification's focus is the functionality a software component (such as a map viewer) should support, and the interfaces required to connect to and extract data from such a component. The specifications do not specify implementation details, meaning that a component should be coded in C++ or Java, be specific to an operating system, etc. Similar to the way that Hypertext Transfer Protocol (HTTP) enabled the growth of activity in the World Wide Web, OpenLS standards, resulting from OGC's cooperative test bed process, have the potential to enable significant growth of location services markets in the wireless web.

In addition, OGC Sensor Web is an open platform for exploiting web-connected sensors. This work includes SensorML, which is an information model, and XML encodings for discovering, querying, and controlling web-resident sensors. SensorML defines an XML schema that serves to represent the general, geometric, and observational characteristics of sensors. Adherence to a common schema makes it possible to search for sensors and sensor data with more precision than is available with text searches using a search engine. For example, searching for particular kinds of sensors and data in a particular geographic region, with data collected within a particular time window, will be easy. Sensor Web is a neutral interoperability framework for web-based discovery, access, control, integration, analysis, and visualization of online sensors, sensor-derived data repositories, and sensor-related processing capabilities. It addresses the problem of isolated, custom-designed, single-application sensor networks, incompatible sensor standards, lack of real-time availability of data, and lack of common and consistent schemas for sensor description, control, and data. Sensor Web applies to web-accessible sensors with discoverable sensors and sensor data; sensors will be self-describing to humans and software (using a standard encoding), and most sensor observations will be easily accessible in a timely fashion over the web. The Sensor Web framework involves several OGC encoding and service specifications designed for general geospatial uses, as well as schema and service specifications that are specifically sensor related. Some studies define services to parse and mark free text messages or to transform symbolic locations into a geometric representation.

A fundamental unit in Sensor Web data messages is the feature, a descriptive container for almost all data and sensor entities. Each data entity submitted, stored, and operated upon, and each physical entity in the infrastructure, has a feature type associated with it. The features are organized hierarchically and stored in an application-accessible directory. A dictionary contains the mapping between terms and features so that applications can correlate them in meaning and operate across data sets. Features are encoded in OGC's Geographic Markup Language (GML). GML regards locations as geographic features, the state of which is defined by a set of properties. Properties may be either simple data types or geometric types such as points, polygons, or collections of other geometries. Geometric types are in turn represented by either a tuple of geographic coordinates, which may refer to different coordinate systems, or coordinates contained in a single character string. By allowing for the definition of application-specific schemas and geometric types, GML is an extensible language. Although it is possible to make use of properties that do not refer to geographic values, the focus of GML lies upon geometric locations. In addition, several discussion papers exist that address issues of acquiring and processing location information. And to facilitate passage of feature data between clients and servers, OGC developed the Web Feature Server (WFS) web interface.

OGC SensorML and IEEE 1451

IEEE 1451 defines communication protocols for different hardware media (1451.2 point-to-point asynchronous, 1451.3 bus based, 1451.4 analog, and 1451.5 wireless) and has proposed an object-based protocol (1451.1) to make sensors accessible to clients over a network. Sensor Web is linked with IEEE 1451 in developing plug-and-play standards for smart transducers. A transducer is smart when it includes sufficient descriptive information to allow control software to automatically determine the transducer's operating parameters, decode its electronic data sheet, and issue commands to read and actuate the transducer. Information engineering under Sensor Web will develop, integrate, and harmonize information models and schema for SensorML, TransducerML, and IEEE 1451. Results of this activity are anticipated to be made part of the SensorML specification. In short, Sensor Web will enable methods for sensor descriptions in the hardware-oriented IEEE 1451 to be available in web-oriented SensorML. IEEE1451 sensors will be made available as web services using an OGC Sensor Observation Service.

The Open Mobile Alliance Location Interoperability Forum (OMA LIF)

In contrast to OGC, which deals with the application side (geoprocessing), LIF, now part of the Open Mobile Alliance, focuses on interoperability from the wireless side of the LBS market. LIF is dealing with specifications pertaining to the query and response for the actual location or position of the mobile device. The vision of LIF is that LBS are seamlessly integrated and available to all mobile users wherever they are.

LIF has developed the Mobile Location Protocol (MLP) standard API for utilization by carriers, wireless infrastructure providers, and mobile application developers. The role of MLP is to seamlessly integrate location from the location carriers/operator communication network. Moreover, MLP is an application-level protocol for the positioning of mobile terminals and is independent of the underlying network technology and, in turn, the position method.

To quote from the LIF MLP specification abstract:

> The purpose of this specification is to define a simple and secure access method that allows Internet applications to query location information from a wireless network, irrespective of its underlying air interface technologies and positioning methods.
>
> This specification covers the core of a Mobile Location Protocol that can be used by a location-based application to request MS location information from a location server (GMLC/MPC or other entity in the wireless network).
>
> … The API is based on existing and well-known Internet technologies as HTTP, SSL/TLS and XML, in order to facilitate the development of location-based applications.

The LIF was set up in 2000 to make sure that the location industry did not fragment into a number of incomplete technology islands, specifically to address the growing concern of location-centric barriers to wireless interoperability. This was especially urgent when considering the question of how application servers could address different mobile position gateways (or work directly with the Home Location Registers (HLRs) of the mobile operators). Otherwise, an application server might need a different way of connecting to a particular manufacturer's network, despite having all of them use the GSM network. OMA has agreements with other organizations, such as 3rd Generation Partnership Project (3GPP). The 3GPP has standardized the representation of location that is to be used within GSM and UMTS networks and by subscriber applications. In the corresponding standard called Geographical Area Description, location is expressed as latitude and longitude according to the World Geodetic System 1984 (WGS84) reference system. The standard also covers geometric shapes, namely, ellipsoid points, ellipsoid arcs, and polygons. Ellipsoid points may be associated with information concerning their uncertainty and their altitude. 3GPP's location model additionally allows the representation of entities' horizontal and vertical velocities. The standard furthermore defines the coding of location information and the message format by means of which this information is exchanged.

Interoperability between LIF and OGC standards is critical when it comes to the convergence of location technologies and the widespread adoption of end-to-end solutions for location-based services. OMA and OGC work closely to ensure that the Internet and wireless standards driven by either organization are mutually supported. Specifically, OGC works closely within the OMA LIF group to ensure that the LIF document type definition (DTD) and OpenLS XML schemes work in harmony, allowing content transmission between LIF and OpenLS-based software without the loss of information and minimal or no translation. This is critical to supporting the wireless-to-Internet interchange of location-specific content and service support. Even though there is considerable overlap between LIF and OGC concerning wireless location interoperability, LIF provides for expanded awareness and adoption of the combined standards.

OMA LOCATION WORKING GROUP

Open Mobile Alliance's Location Working Group (LOC), which was formerly known as Location Interoperability Forum, continues the work of the former Location Interoperability Forum (LIF) and Location Drafting Committee of the former WAP Forum. OMA LOC develops specifications to ensure interoperability of mobile location services on an end-to-end basis. The most well known result of OMA LOC's work is the Mobile Location Protocol (MLP). OMA is targeted at providing an end-to-end architectural framework for mobile location-based services. It addresses issues such as interoperability, privacy and security, billing, and roaming.

MOBILE LOCATION PROTOCOL

The Mobile Location Protocol is an application-level protocol for querying the location of mobile terminals. The latest version of the MLP specification is 3.0.0, released in June 2002. The purpose of MLP is to deliver a simple and secure access method that allows Internet applications to query location information from a wireless network, irrespective of its underlying air interface technologies and positioning methods (ISO/TC 211 website, http://www.isotc211.org). The MLP defines how applications can access location information from a wireless network over the Internet. Like 3GPP's location model, this specification allows location to be expressed as various shapes and, additionally, with reference to different geographic coordinate systems. The Mobile Location Protocol stipulates that any implementation must at least support the WGS84 system. Furthermore, information concerning the horizontal speed of entities, their direction of movement, and the quality of location information is taken into consideration.

MLP works as an API between location server and location client (usually an application). The location server is a logical unit, which means that the actual implementations vary from the gateway mobile location center (GMLC) in GSM/UMTS to the mobile positioning center (MPC) standardized by the American National Standards Institute (ANSI). Other implementations are also possible. In practice, it is common that the LCS client initiates the dialog with an MLP request to the location server. The server responds with an MLP response message. MLP supports different transport mechanisms, since the transport protocol is separated from the XML content. Possible MLP transport protocols include HTTP, WSP, and Simple Object Access Protocol (SOAP). Figure 10.3 shows a layered view of MLP. The MLP specification provides several basic services for different usages. These are SLIS, ELIS, SLRS, ELRS, and TLRS. Basic MLP services are based on location services defined by 3GPP.

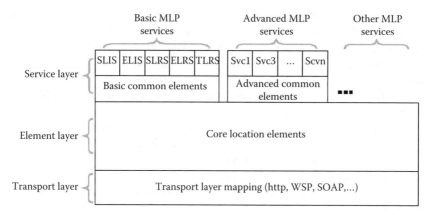

FIGURE 11.3 Layered view of Mobile Location Protocol.

MLP is based on well-known Internet technologies, such as HTTP/1.1, Secure Sockets Layer (SSL)/Transport Layer Security (TLS), and XML. MLP is an XML interface that enables communication between location-based services and location infrastructure. Because MLP includes the user ID and password as clear text in the MLP header, it is standard practice to use SSL or virtual private networks (VPNs) for security. Supporting MLP is important, as it ensures the widest possible compatibility in the industry.

THE SUCCESS OF STANDARDS: ADOPTION

Standards adoption is a means to an important end — building critical mass in the development of interoperable data and services. Note that critical mass here means that the creation of many diverse interoperability solutions will not necessarily improve the situation with stand alone (stovepipe) application integration.

It is easy to see the cumulative nature of standards — it is impossible to conceive of a Transmission Control Protocol (TCP)/Internet Protocol (IP) standard without a multitude of other standards that make such communication even possible in the first place. What we are primarily concerned with is those fields where a standard is nonexistent, immature, less than satisfactory, or where competing standards exist. We are interested in questions like:

- Does the existing set of practices or evolving standards of OMA and OGC meet the developers' needs?
- Do the developers have the clout or resources to successfully create or contribute to a new standard?
- Would these efforts serve the developers' business ends?

Perhaps the answers to these questions are ultimately personal and need to be found through discussion and exploration at a personal or company level. In short, the question of "Why have standards?" is not difficult, and can be answered with minimal reflection. We know that in the long run standards are beneficial.

The question of concern is at what level is standardization required, and at what level is the technology implementation left up to the vendors. There needs to be room for those "killer applications." Combined with the disparate wireless systems, local positioning systems, Internet, and GIS, the ubiquitous wireless location interoperability standard will be a long way off from industry-wide adoption. In any case, without standards, there can be no end-to-end location implementation utilizing the best-in-class technologies.

The users of these applications are not concerned with the underlying architecture or technologies powering their user experience. Location information by itself has no value to them. Instead, customers will welcome the use of location technologies in their everyday lives, provided that these technologies are as mobile, seamless, and ubiquitous as their lifestyles and provide them with services that add value to their lives.

The only way for the indoor location-based industry to meet this challenge is if it works together through the development and adoption of location wireless interoperability standards. The overall scope in terms of adoption should be to raise awareness of the need for, and the potential of, implementing these standards as an avenue toward interoperability.

However, obligation or not, a great deal of financial and human capital must be spent in standard creation. Few companies or groups can afford to do this merely for unselfish reasons; companies need a clear, compelling rationale why their contribution to or adoption of a given standard is a wise use of their valuable resources. IT managers need to be able to develop this rationale as part of the business case in a development project. The examples of how simple it is to develop indoor location-based services that we have presented in this book should provide them with that rationale.

The advantages of adopting a standard location interface would enable various location providers to provide information to the location service so that the location service can use the information to determine its location. Location providers might provide location information in different forms. For example, a global positioning systems (GPS) location provider might provide location information that is GPS specific. Similarly, an IP/subnet location provider might provide information that is specific to an Internet protocol. A mobile phone location provider might provide location information in the form of a cell ID or user input of an address. The location information can be provided with confidence and accuracy estimates to enable the location service to evaluate the relative quality of the information before it is used. The various providers also have the ability to monitor themselves, which assists in the providers' ability to intelligently convey information to the location service. By having a common interface, the collection of location providers is dynamically extensible; that is, location providers can be added or removed from the collection of location providers without any interference of the functionality performed by the location service or device. By having a common interface, the location providers are also extensible (to support future providers) in that they can be dynamically added or removed from the collection of location providers. All that is required of a particular location provider is that it be written to support the common interface. To determine the current device location, the location service may have to consult with an active directory, a web service, or a location database. The active directory might, for example, maintain a federation of the location model and other networking metadata, such as subnet and site information that can help determine location based on networking connectivity. Web service can hold the mail location model, the attributes of which can be used to find the location. For example, if a cell phone knows its base station ID, then the location provider can query the secondary federation to ascertain the nodes that match that base station ID. The location database is basically a version of the web service that is hosted or cached locally.

REFERENCES

McKee, L., The Importance of Going "Open," white paper presented at the Open GIS Consortium, September 2003.

Open Geospatial Consortium, Inc., *OpenGIS Location Services (OpenLS)*, Part 6, *Navigation Service*, Version 0.5.0, OpenGIS Implementation Specification 03-007r1, April 2003.

Open Geospatial Consortium, Inc., *OpenGIS Location Services (OpenLS): Core Services*, Parts 1–5, OpenGIS Implementation Specification 03-006r3, January 2004.

Spinney, J., A Brief History of LBS and How OpenLS Fits into the New Value Chain, *Directions Magazine*, July 2003.

Index

G

H

M

Q

R

T